# Water Quality Monitoring

# Water Quality Monitoring

A practical guide to the design and implementation of freshwater quality studies and monitoring programmes

*Edited by*

Jamie Bartram and Richard Ballance

Published on behalf of

United Nations Environment Programme

World Health Organization

**First Published by E&FN Spon**
**11 New Fetter Lane, London EC4P 4EE**

First edition 1996

Reprinted by Spon Press 2001

*Spon Press is an imprint of the Taylor & Francis Group*



Printed in Great Britain by TJ International Ltd, Padstow, Cornwall

ISBN 0 419 22320 7 (Hbk) 0 419 21730 4 (Pbk)



A catalogue record for this book is available from the British Library

**Publisher's note**
This book has been produced from camera ready copy supplied on behalf of the co-sponsoring organizations.

∞ Printed on acid-free text paper, manufactured in accordance with ANSI/NISO Z39.48-1992 and ANSI/NISO Z39.48-1984 (Permanence of Paper).

# TABLE OF CONTENTS

# FOREWORD

According to Agenda 21 of the United Nations Programme of Action following the Earth Summit in Rio de Janeiro:

> *"The complex interconnectedness of freshwater systems demands that freshwater management be holistic (taking a catchment management approach) and based on a balanced consideration of the needs of people and the environment. The Mar del Plata Action Plan has already recognised the intrinsic linkage between water resource development projects and their significant physical, chemical, biological, health and socio-economic repercussions".*

The approaches and methods for water quality monitoring described in this handbook are based upon the experience gained, over two decades, with the design and establishment of the global freshwater quality monitoring network, GEMS/WATER. The GEMS/WATER programme is co-sponsored by the United Nations Environment Programme (UNEP) and the World Health Organization (WHO), together with the United Nations Educational, Scientific and Cultural Organization (UNESCO) and the World Meteorological Organization (WMO). One of the goals of GEMS/WATER is:

> "*to strengthen national water quality monitoring networks in developing countries, including the improvement of analytical capabilities and data quality assurance*".

This handbook supports this goal by providing a practical tool for use in water quality management by national and local agencies and departments dealing with water quality issues.

*Water Quality Monitoring* and its companion guidebook *Water Quality Assessments. A Guide to the Use of Biota, Sediments and Water in Environmental Monitoring*, Second edition (edited by Deborah Chapman and published on behalf of UNESCO, WHO and UNEP by Chapman & Hall, London, 1996) constitute principal methodology guidebooks developed and used in the monitoring and assessment activities of GEMS/WATER. Together they make a direct contribution to capacity building in the area of water quality monitoring and assessment.

This book brings together the information necessary to design and implement a water quality monitoring programme and provides a basis for water quality assessments and studies of the impact of pollution on the natural environment. Freshwater quality is addressed in a holistic way, considering both surface waters and groundwaters. Emphasis is given to monitoring the natural environment and to detecting and monitoring trends in water quality.

The book outlines general considerations related to water quality monitoring, provides a general protocol for a monitoring programme and includes such elements as staff requirements, staff training and the equipping of analytical laboratories. It also includes consideration of the problems that may be encountered when implementing programmes in remote areas and developing countries and when establishing water quality monitoring programmes from scratch.

It is hoped that the approaches and methods described will be useful for anyone concerned with water quality monitoring whether they have a scientific, managerial or engineering background and including, particularly, field staff and those who may not be water quality experts. Potential users may be from local, regional or national government agencies, research groups, consulting firms or non-governmental organisations. This book will be especially useful to those who plan and manage the various aspects of water quality monitoring outside the framework of large established programmes, as may be the case in many developing countries and for specific projects and studies world-wide.

Material on sampling and analytical methods for the most important physical, chemical and bacteriological variables has been brought together here in a convenient form, with an emphasis on field techniques at a level suitable for those implementing water quality programmes. An overview of the principles underlying, and the importance of hydrological, biological and sediment measurements and their relevance to water quality monitoring is also included. Information on some of the commercially available analytical kits for use in the field is included in an appendix.

## ACKNOWLEDGEMENTS

The co-sponsoring organisations, UNEP and WHO, wish to express their appreciation to all of those whose efforts made the preparation of this book possible. Special thanks are due to the Department of International Development and Co-operation, Ministry of Foreign Affairs of Finland which provided generous and continued financial support to the preparation of the book as well as for the organisation of several regional review workshops. The Tampere University of Technology, Finland and the National Board of Waters and the Environment, Finland both supported the initiation of this handbook. Thanks are also due to Dr Veerle Vandeweerd of UNEP/GEMS for her enthusiasm and energy during the initiation of the project which led to the preparation of this document and for her tireless support throughout the subsequent development and refinement of this book.

An international group of authors provided material and, in most cases, several authors and their collaborators contributed to each chapter. It is difficult to identify precisely the contributions made by individuals and we apologise for any oversights in the following list of principal contributors.

Dick Ballance, formerly of World Health Organization, Geneva, Switzerland (editor, and contributor)

Jamie Bartram, European Centre for Environment and Health, World Health Organization, Rome, Italy, formerly of the Robens Institute, University of Surrey, Guildford, UK (editor and contributor)

Ray Briggs, Robens Institute, University of Surrey, Guildford, UK (contributions to Chapter 9)

Deborah Chapman, Environment Consultant, Kinsale, Ireland (contributions to Chapter 11)

Malcolm Clarke, University of Victoria, BC, Canada (contributions to Chapter 14)

Richard Helmer, Urban Environmental Health, World Health Organization, Geneva, Switzerland (contributions to Chapter 1)

John Jackson, formerly of UNEP GEMS Monitoring and Assessment Research Centre, Kings College London, UK (contributions to Chapters 11 and 14)

Merete Johannessen, Norwegian Institute for Water Research (NIVA), Oslo, Norway (contributions to Chapters 3 and 9)

Falk Krebs, Federal Institute of Hydrology, Koblenz, Germany (contributions to Chapter 11)

Esko Kuusisto, National Board of Waters and the Environment, Helsinki, Finland (contributions to Chapters 1, 2 and 12)
John Lewis, Data Processing, Aberdare, UK (contributions to Chapter 14).
Ari Mäkelä, National Board of Waters and the Environment, Helsinki, Finland (contributions to Chapters 1, 2, 3 and 5)
Esko Mälkki, National Board of Waters and the Environment, Helsinki, Finland (contributions to Chapters 1, 2 and 5)
Michel Meybeck, Université de Pierre et Marie Curie, Paris, France (contributions to Chapters 2 and 3)
Harriet Nash, Wardel Armstrong, Newcastle-under-Lyme, UK (contributions to Chapters 5 and 14)
Ed Ongley, Canada Centre for Inland Waters, Burlington, Ontario, Canada contributions to Chapters 12, 13 and 14)
Steve Pedley, Robens Institute, University of Surrey, Guildford, Surrey, UK (contributions to Chapter 10)
Alan Steel, Consultant, Kinsale, Ireland (contributions to Chapter 14)
Paul Whitfield, University of Victoria, BC, Canada (contributions to Chapter 14)

Dr Richard Helmer, WHO, Geneva, Dr Veerle Vandeweerd, UNEP, Nairobi and Dr Jeffrey Thornton, International Environmental Management Services, Wisconsin critically reviewed early drafts. Later drafts were reviewed by John Chilton, British Geological Survey and Ms Harriet Nash, Wardel Armstrong, who made useful suggestions concerning groundwater and basic data checks respectively.

Drafts of the text were reviewed at a series of regional workshops convened in the framework of the GEMS/WATER programme in Tanzania (1992), Zimbabwe (1993), Uganda (1994) and Jordan (1994). More than 60 people contributed to the review process in this way and the orientation of the final document towards real, practical problems is largely due to their efforts.

Thanks are due to Verity Snook, Mary Stenhouse and John Cashmore for secretarial services to the review meetings and for secretarial and administrative assistance at various stages of the project. The editorial assistance of Sarah Ballance is also much appreciated during the preparation of the final manuscript. Thanks are also due to Helen MacMahon and Alan Steel (preparation of illustrations) and to Deborah Chapman (editorial assistance, layout and production management).

United Nations Environment Programme
World Health Organization

# Chapter 1

# INTRODUCTION

Freshwater is a finite resource, essential for agriculture, industry and even human existence. Without freshwater of adequate quantity and quality sustainable development will not be possible. Water pollution and wasteful use of freshwater threaten development projects and make water treatment essential in order to produce safe drinking water. Discharge of toxic chemicals, over-pumping of aquifers, long-range atmospheric transport of pollutants and contamination of water bodies with substances that promote algal growth (possibly leading to eutrophication) are some of today's major causes of water quality degradation.

It has been unequivocally demonstrated that water of good quality is crucial to sustainable socio-economic development. Aquatic ecosystems are threatened on a world-wide scale by a variety of pollutants as well as destructive land-use or water-management practices. Some problems have been present for a long time but have only recently reached a critical level, while others are newly emerging.

Gross organic pollution leads to disturbance of the oxygen balance and is often accompanied by severe pathogenic contamination. Accelerated eutrophication results from enrichment with nutrients from various origins, particularly domestic sewage, agricultural run-off and agro-industrial effluents. Lakes and impounded rivers are especially affected.

Agricultural land use without environmental safeguards to prevent over-application of agrochemicals is causing widespread deterioration of the soil/water ecosystem as well as the underlying aquifers. The main problems associated with agriculture are salinisation, nitrate and pesticide contamination, and erosion leading to elevated concentrations of suspended solids in rivers and streams and the siltation of impoundments. Irrigation has enlarged the land area available for crop production but the resulting salinisation which has occurred in some areas has caused the deterioration of previously fertile soils.

Direct contamination of surface waters with metals in discharges from mining, smelting and industrial manufacturing is a long-standing phenomenon. However, the emission of airborne metallic pollutants has now reached

*This chapter was prepared by J. Bartram and R. Helmer*

such proportions that long-range atmospheric transport causes contamination, not only in the vicinity of industrialised regions, but also in more remote areas. Similarly, moisture in the atmosphere combines with some of the gases produced when fossil fuels are burned and, falling as acid rain, causes acidification of surface waters, especially lakes. Contamination of water by synthetic organic micropollutants results either from direct discharge into surface waters or after transport through the atmosphere. Today, there is trace contamination not only of surface waters but also of groundwater bodies, which are susceptible to leaching from waste dumps, mine tailings and industrial production sites.

The extent of the human activities that influence the environment has increased dramatically during the past few decades; terrestrial ecosystems, freshwater and marine environments and the atmosphere are all affected. Large-scale mining and fossil fuel burning have started to interfere measurably with natural hydrogeochemical cycles, resulting in a new generation of environmental problems. The scale of socio-economic activities, urbanisation, industrial operations and agricultural production, has reached the point where, in addition to interfering with natural processes within the same watershed, they also have a world-wide impact on water resources. As a result, very complex inter-relationships between socio-economic factors and natural hydrological and ecological conditions have developed. A pressing need has emerged for comprehensive and accurate assessments of trends in water quality, in order to raise awareness of the urgent need to address the consequences of present and future threats of contamination and to provide a basis for action at all levels. Reliable monitoring data are the indispensable basis for such assessments.

Monitoring is defined by the International Organization for Standardization (ISO) as: "the programmed process of sampling, measurement and subsequent recording or signalling, or both, of various water characteristics, often with the aim of assessing conformity to specified objectives". This general definition can be differentiated into three types of monitoring activities that distinguish between long-term, short-term and continuous monitoring programmes as follows:

- *Monitoring* is the long-term, standardised measurement and observation of the aquatic environment in order to define status and trends.
- *Surveys* are finite duration, intensive programmes to measure and observe the quality of the aquatic environment for a specific purpose.
- *Surveillance* is continuous, specific measurement and observation for the purpose of water quality management and operational activities.

The distinction between these specific aspects of monitoring and their principal use in the water quality assessment process is described in the companion guidebook *Water Quality Assessments. A Guide to the Use of Biota, Sediments and Water in Environmental Monitoring*, 2nd edition (edited by D. Chapman and published on behalf of UNESCO, WHO and UNEP by Chapman & Hall, London, 1996).

It is important to note the emphasis given to collection of data for a purpose in the definitions of water quality monitoring above. This purpose is most commonly related to water quality management, which aims to control the physical, chemical and biological characteristics of water. Elements of management may include control of pollution, use and abstraction of water, and land use. Specific management activities are determined by natural water quantity and quality, the uses of water in natural and socio-economic systems, and prospects for the future.

Water quality requirements or objectives can be usefully determined only in terms of suitability for a purpose or purposes, or in relation to the control of defined impacts on water quality. For example, water that is to be used for drinking should not contain any chemicals or micro-organisms that could be hazardous to health. Similarly, water for agricultural irrigation should have a low sodium content, while that used for steam generation and related industrial uses should be low in certain other inorganic chemicals. Preservation of biodiversity and other conservation measures are being recognised increasingly as valid aspects of water use and have their own requirements for water quality management. Water quality data are also required for pollution control, and the assessment of long-term trends and environmental impacts.

## 1.1 Elements of a water quality monitoring programme

Before the planning of water sampling and analysis can be started, it is necessary to define clearly what information is needed and what is already available and to identify, as a major objective of the monitoring programme, the gaps that need to be filled. It is useful to prepare a "monitoring programme document" or "study plan", which describes in detail the objectives and possible limitations of the monitoring programme. Figure 1.1 outlines the contents of this book in relation to the process of developing such a plan, its implementation and the interpretation of the findings. If the programme's objectives and limitations are too vague, and the information needs inadequately analysed, the information gaps will be poorly identified and there will be a danger of the programme failing to produce useful data.

Many reasons can be listed for carrying out water quality monitoring. In many instances they will overlap and the information obtained for one

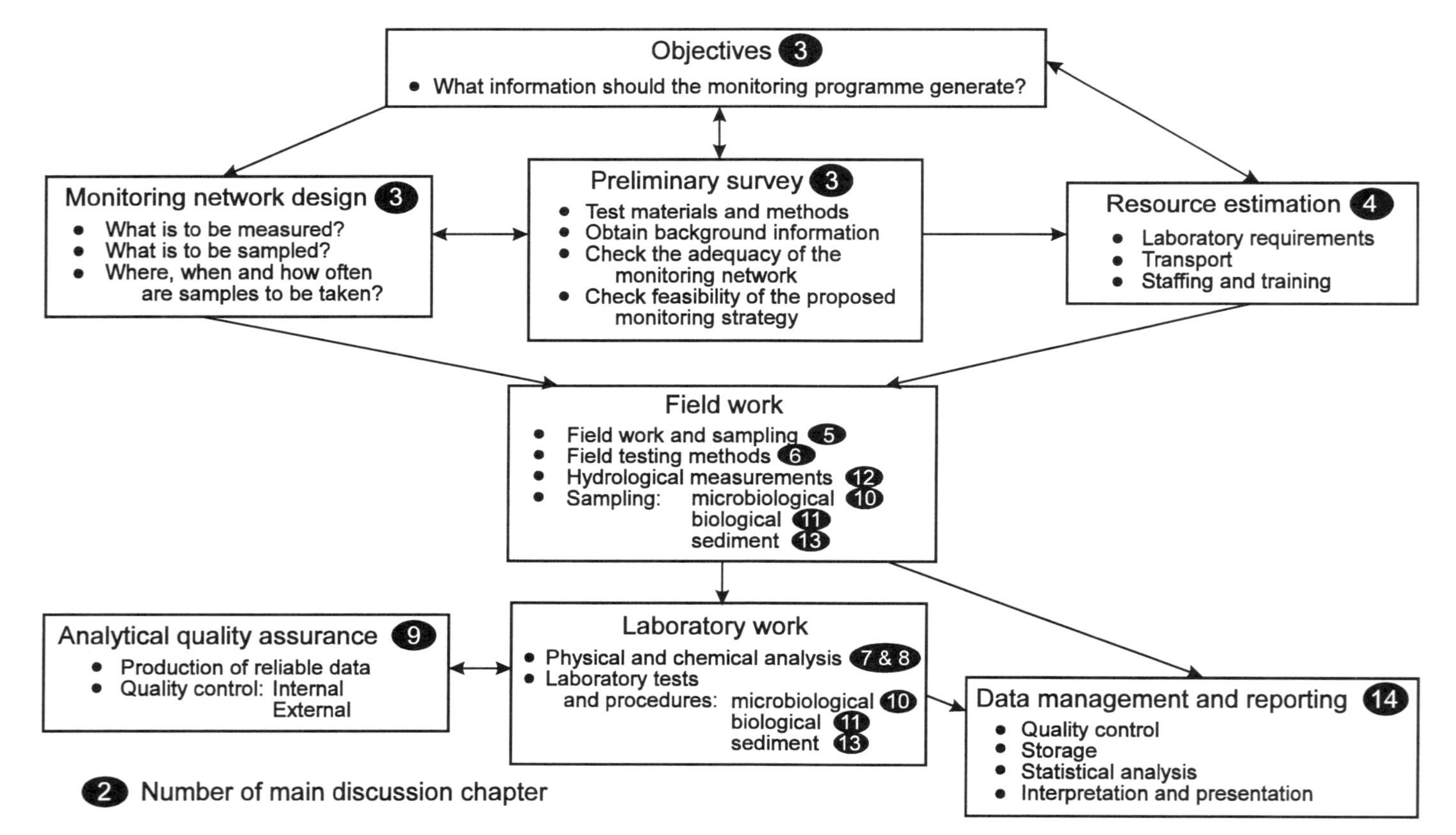


**Figure 1.1** Links between the critical elements of water quality monitoring discussed in the various chapters of this book

purpose may be useful for another. Water quality monitoring data may be of use in the management of water resources at local, national or international level. Where water bodies are shared by more than one country, a water quality monitoring programme can yield information that may serve as a basis for international agreements regarding the use of these waters, as well as for evaluation of compliance with any such agreements.

Water quality monitoring is the foundation on which water quality management is based. Monitoring provides the information that permits rational decisions to be made on the following:

- Describing water resources and identifying actual and emerging problems of water pollution.
- Formulating plans and setting priorities for water quality management.
- Developing and implementing water quality management programmes.
- Evaluating the effectiveness of management actions.

To fulfil these functions some preliminary survey work must first be done to provide basic, background knowledge of existing water quality conditions (Figure 1.1). Subsequent monitoring efforts will identify problems and problem areas, short- and long-term trends and the probable cause of the problems. Once sufficient data have been gathered, it is possible to describe the average conditions, the variations from average and the extremes of water quality, expressed in terms of measurable physical, chemical and biological variables. In the meantime, priorities may have been set, plans may have been made and management programmes may have been implemented. Ultimately, information from the monitoring programme is fed back into the management system so that any necessary changes to priorities and plans can also be made.

Specifications for the collection of data should be uniform so as to ensure compatibility and make it possible to apply to any particular location the experience gained in another. Networks for water quality monitoring should be developed in close co-operation with other agencies actively collecting water data. This not only minimises the cost of establishing and operating the network but also facilitates the interpretation of water quality data. The particular hydrological measurements and water characteristics required will differ from one water body to another. For example, in rivers and streams it is necessary to measure the velocity of flow and intensity of longitudinal mixing, while thermal regimes are important considerations when monitoring lakes. Measurement of wastewater discharges containing nutrients may not be necessary for many rivers but may be important in lakes (which act as traps for nutrients) because the additional input of nutrients may accelerate the eutrophication process.

Networks for water quality monitoring must conform to programme objectives (Figure 1.1). A clear statement of objectives is necessary to ensure collection of all necessary data and to avoid needless and wasteful expenditure of time, effort and money. Furthermore, evaluation of the data collected will provide a basis for judging the extent to which programme objectives were achieved and thus justify the undertaking. Before observations begin, it is also essential to specify the location of sampling stations, the frequency of sampling and the water quality variables to be determined.

Monitoring programmes should be periodically reviewed to ensure that information needs are being met. As greater knowledge of conditions in the aquatic system is gained, a need for additional information may become apparent. Alternatively, it may be concluded that some of the information being collected is unnecessary. In either case, an updated monitoring programme document must be prepared and distributed to the information users. If such users are not kept fully informed of the exact scope of the programme they may expect more than it can deliver and may not support its continuation.

## 1.2 Monitoring for management

The elements of water quality monitoring and assessment described above and discussed in detail in the following chapters are only part of a wider picture of water quality and quantity management, environmental protection and policy formulation and of development concerns. The critical element stressed is the development of objectives. It may be that long-term objectives (such as integrated monitoring for environmental and health protection) may be set for the monitoring programme but that the programme operates, and is practically structured, to meet specified short-term objectives (such as monitoring for immediate health priorities). This illustrates that water quality monitoring operates within a larger structure of policy decisions and management priorities. A programme may need to be flexible to meet short-term objectives but still be capable of developing over longer periods to meet new concerns and priorities.

The elements outlined above and described in more detail in the following chapters should be implemented flexibly according to the different priorities of the monitoring programmes. The initial phases of surveys and design may extend over periods of months or even years before a clear idea of needs and priorities is achieved. The time required for other elements of the monitoring and assessment process varies. Sampling missions may take several days, as may sample transport to the laboratory. Complete chemical analysis of a sample may take a week. Data treatment can take weeks, depending on the amount of data that is to be dealt with, while interpretation of results and the writing and publication of reports can take from a few months to a few years. If operational surveillance is one of the aspects of a multi-purpose monitoring programme, the field work, data assessment and reporting should be accomplished within a time limit appropriate to the operational requirements. For example, reports on the surveillance of drinking water quality should be made very quickly so that corrective actions can be taken when a contaminated supply threatens public health.

The importance of the use of information should be stressed. There is little point in generating monitoring data unless they are to be used. It is essential that the design, structure, implementation and interpretation of monitoring systems and data are conducted with reference to the final use of the information for specific purposes.

In some countries, water quality standards may be laid down by national legislation. A government authority is then charged with monitoring the extent to which the standards are fulfilled. This is particularly common for water intended for drinking and is carried out as a public health protection measure. The monitoring objectives in this case will be concerned with detecting any deterioration in raw water quality so that appropriate source protection or treatment can be applied. In other instances, it may be necessary to develop a new water source in order to meet increasing demands; the objective may then become that of monitoring the quality and quantity of sources that might fulfil this need.

Where water quality legislation is rudimentary or non-existent, the water authority's mandate may be to develop legislation and regulations appropriate to the country's economic development plans. In this case the monitoring objectives will probably focus, in the first instance, on acquiring background information on water quality. The objectives will change as information on water quality is accumulated, as problems emerge and solutions are developed, and as new demands are made on the water resources.

## 1.3 Monitoring and assessment

This handbook concentrates on providing the practical information which is necessary to design, to implement and to carry out monitoring programmes in freshwaters. Emphasis is placed on the collection and analysis of water samples because, at present, this activity forms the principal component of most monitoring programmes. The fundamental techniques associated with the use of sediment and biota in monitoring programmes are also presented in order to illustrate how they may be incorporated into existing programmes based on water sample analysis. Further details of the growing application of these approaches is available in the companion guidebook *Water Quality Assessments*.

Monitoring, as a practical activity, provides the essential information which is required for an assessment of water quality. However, assessments require additional information, such as an understanding of the hydrodynamics of a water body, information on geochemical, atmospheric and anthropogenic influences and the correct approaches for analysis and interpretation of the data generated during monitoring. The companion guidebook gives further detail on the supporting information which is required for the full assessment of water quality in rivers, lakes, reservoirs and groundwaters, illustrated by examples of assessments from different world regions. A detailed description of different approaches available for the interpretation of monitoring data is also given.

# Chapter 2

# WATER QUALITY

"Water quality" is a term used here to express the suitability of water to sustain various uses or processes. Any particular use will have certain requirements for the physical, chemical or biological characteristics of water; for example limits on the concentrations of toxic substances for drinking water use, or restrictions on temperature and pH ranges for water supporting invertebrate communities. Consequently, water quality can be defined by a range of variables which limit water use. Although many uses have some common requirements for certain variables, each use will have its own demands and influences on water quality. Quantity and quality demands of different users will not always be compatible, and the activities of one user may restrict the activities of another, either by demanding water of a quality outside the range required by the other user or by lowering quality during use of the water. Efforts to improve or maintain a certain water quality often compromise between the quality and quantity demands of different users. There is increasing recognition that natural ecosystems have a legitimate place in the consideration of options for water quality management. This is both for their intrinsic value and because they are sensitive indicators of changes or deterioration in overall water quality, providing a useful addition to physical, chemical and other information.

The composition of surface and underground waters is dependent on natural factors (geological, topographical, meteorological, hydrological and biological) in the drainage basin and varies with seasonal differences in runoff volumes, weather conditions and water levels. Large natural variations in water quality may, therefore, be observed even where only a single water-course is involved. Human intervention also has significant effects on water quality. Some of these effects are the result of hydrological changes, such as the building of dams, draining of wetlands and diversion of flow. More obvious are the polluting activities, such as the discharge of domestic, industrial, urban and other wastewaters into the water-course (whether intentional or accidental) and the spreading of chemicals on agricultural land in the drainage basin.

*This chapter was prepared by M. Meybeck, E. Kuusisto, A. Mäkelä and E. Mälkki*

Water quality is affected by a wide range of natural and human influences. The most important of the natural influences are geological, hydrological and climatic, since these affect the quantity and the quality of water available. Their influence is generally greatest when available water quantities are low and maximum use must be made of the limited resource; for example, high salinity is a frequent problem in arid and coastal areas. If the financial and technical resources are available, seawater or saline groundwater can be desalinated but in many circumstances this is not feasible. Thus, although water may be available in adequate quantities, its unsuitable quality limits the uses that can be made of it. Although the natural ecosystem is in harmony with natural water quality, any significant changes to water quality will usually be disruptive to the ecosystem.

The effects of human activities on water quality are both widespread and varied in the degree to which they disrupt the ecosystem and/or restrict water use. Pollution of water by human faeces, for example, is attributable to only one source, but the reasons for this type of pollution, its impacts on water quality and the necessary remedial or preventive measures are varied. Faecal pollution may occur because there are no community facilities for waste disposal, because collection and treatment facilities are inadequate or improperly operated, or because on-site sanitation facilities (such as latrines) drain directly into aquifers. The effects of faecal pollution vary. In developing countries intestinal disease is the main problem, while organic load and eutrophication may be of greater concern in developed countries (in the rivers into which the sewage or effluent is discharged and in the sea into which the rivers flow or sewage sludge is dumped). A single influence may, therefore, give rise to a number of water quality problems, just as a problem may have a number of contributing influences. Eutrophication results not only from point sources, such as wastewater discharges with high nutrient loads (principally nitrogen and phosphorus), but also from diffuse sources such as run-off from livestock feedlots or agricultural land fertilised with organic and inorganic fertilisers. Pollution from diffuse sources, such as agricultural run-off, or from numerous small inputs over a wide area, such as faecal pollution from unsewered settlements, is particularly difficult to control.

The quality of water may be described in terms of the concentration and state (dissolved or particulate) of some or all of the organic and inorganic material present in the water, together with certain physical characteristics of the water. It is determined by *in situ* measurements and by examination of water samples on site or in the laboratory. The main elements of water quality monitoring are, therefore, on-site measurements, the collection and analysis of water samples, the study and evaluation of the analytical results, and the reporting of the findings. The results of analyses performed on a single water sample are only valid for the particular location and time at which that sample was taken. One purpose of a monitoring programme is, therefore, to gather sufficient data (by means of regular or intensive sampling and analysis) to assess spatial and/or temporal variations in water quality.

The quality of the aquatic environment is a broader issue which can be described in terms of:

- water quality,
- the composition and state of the biological life present in the water body,
- the nature of the particulate matter present, and
- the physical description of the water body (hydrology, dimensions, nature of lake bottom or river bed, etc.).

Complete assessment of the quality of the aquatic environment, therefore, requires that water quality, biological life, particulate matter and the physical characteristics of the water body be investigated and evaluated. This can be achieved through:

- chemical analyses of water, particulate matter and aquatic organisms (such as planktonic algae and selected parts of organisms such as fish muscle),
- biological tests, such as toxicity tests and measurements of enzyme activities,
- descriptions of aquatic organisms, including their occurrence, density, biomass, physiology and diversity (from which, for example, a biotic index may be developed or microbiological characteristics determined), and
- physical measurements of water temperature, pH, conductivity, light penetration, particle size of suspended and deposited material, dimensions of the water body, flow velocity, hydrological balance, etc.

Pollution of the aquatic environment, as defined by GESAMP (1988), occurs when humans introduce, either by direct discharge to water or indirectly (for example through atmospheric pollution or water management practices), substances or energy that result in deleterious effects such as:

- hazards to human health,
- harm to living resources,
- hindrance to aquatic activities such as fishing,
- impairment of water quality with respect to its use in agriculture, industry or other economic activities, or
- reduction of amenity value.

The importance attached to quality will depend on the actual and planned use or uses of the water (e.g. water that is to be used for drinking should not contain any chemicals or micro-organisms that could be hazardous to health).

Since there is a wide range of natural water qualities, there is no universal standard against which a set of analyses can be compared. If the natural, pre-polluted quality of a water body is unknown, it may be possible to establish some reference values by surveys and monitoring of unpolluted water in which natural conditions are similar to those of the water body being studied.

## 2.1 Characteristics of surface waters

### 2.1.1 Hydrological characteristics

Continental water bodies are of various types including flowing water, lakes, reservoirs and groundwaters. All are inter-connected by the hydrological cycle with many intermediate water bodies, both natural and artificial. Wetlands, such as floodplains, marshes and alluvial aquifers, have characteristics that are hydrologically intermediate between those of rivers, lakes and groundwaters. Wetlands and marshes are of special biological importance.

It is essential that all available hydrological data are included in a water quality assessment because water quality is profoundly affected by the hydrology of a water body. The minimum information required is the seasonal variation in river discharge, the thermal and mixing regimes of lakes, and the recharge regime and underground flow pattern of groundwaters.

The common ranges of water residence time for various types of water body are shown in Figure 2.1. The theoretical residence time for a lake is the total volume of the lake divided by the total outflow rate ($V/\Sigma Q$). Residence time is an important concept for water pollution studies because it is associated with the time taken for recovery from a pollution incident. For example, a short residence time (as in a river) aids recovery of the aquatic system from a pollution input by rapid dispersion and transport of waterborne pollutants. Long residence times, such as occur in deep lakes and aquifers, often result in very slow recovery from a pollution input because transport of waterborne pollutants away from the source can take years or even decades. Pollutants stored in sediments take a long time to be removed from the aquatic system, even when the water residence time of the water body is short.

River flow is unidirectional, often with good lateral and vertical mixing, but may vary widely with meteorological and climatic conditions and drainage pattern. Still surface waters, such as deep lakes and reservoirs, are characterised by alternating periods of stratification and vertical mixing. In addition, water currents may be multi-directional and are much slower than in rivers. Moreover, wind has an important effect on the movement of the upper layers of lake and reservoir water. The residence time of water in lakes is often more than six months and may be as much as several hundred years. By contrast, residence times in reservoirs are usually less than one year.

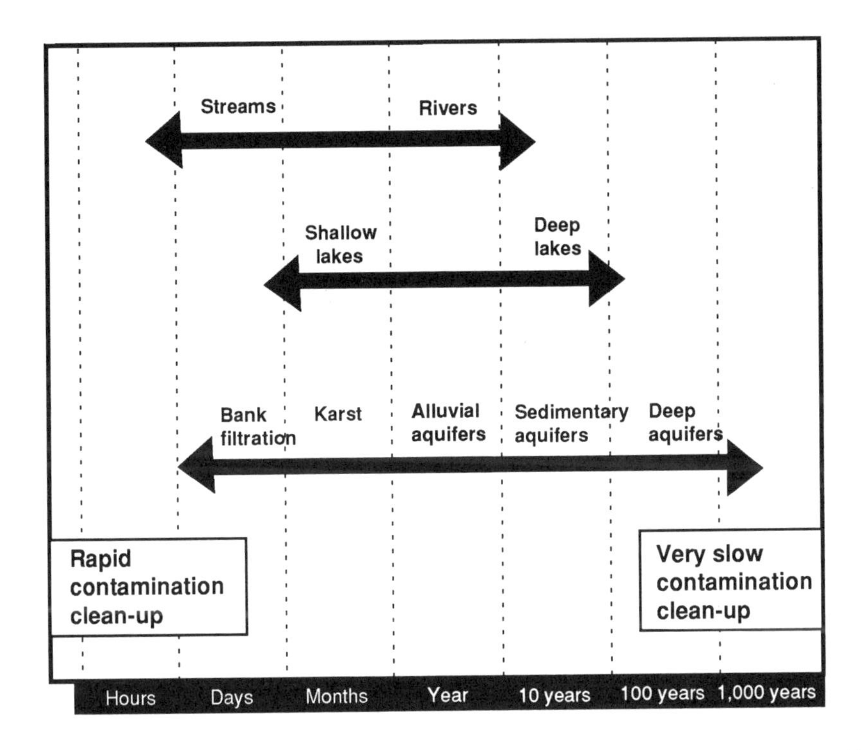


**Figure 2.1** Typical water residence times in inland water bodies. Note: Actual residence times may vary. Residence times in karstic aquifers may vary from days to thousands of years, depending on extent and recharge. Some karstic aquifers of the Arabian peninsula have water more than 10,000 years old

### 2.1.2 Lakes and reservoirs

An important factor influencing water quality in relatively still, deep waters, such as lakes and reservoirs, is stratification. Stratification occurs when the water in a lake or reservoir acts as two different bodies with different densities, one floating on the other. It is most commonly caused by temperature differences, leading to differences in density (water has maximum density at 4 °C), but occasionally by differences in solute concentrations. Water quality in the two bodies of water is also subject to different influences. Thus, for example, the surface layer receives more sunlight while the lower layer is physically separated from the atmosphere (which is a source of gases such as oxygen) and may be in contact with decomposing sediments which exert an oxygen demand. As a result of these influences it is common for the lower layer to have a significantly decreased oxygen concentration compared with the upper layer. When anoxic conditions occur in bottom sediments, various compounds may increase in interstitial waters (through dissolution or reduction) and diffuse from the sediments into the lower water layer. Substances produced in this way include ammonia, nitrate, phosphate, sulphide, silicate, iron and manganese compounds.

*Temperate lakes*

Thermal stratification has been studied for many years in temperate regions where, during spring and summer, the surface layers of the water become warmer and their density decreases. They float upon the colder and denser layer below and there is a resistance to vertical mixing. The warm surface layer is known as the epilimnion and the colder water trapped beneath is the hypolimnion. The epilimnion can be mixed by wind and surface currents and its temperature varies little with depth. Between the two layers is a shallow zone, called the metalimnion or the thermocline, where the temperature changes from that of the epilimnion to that of the hypolimnion. As the weather becomes cooler, the temperature of the surface layer falls and the density difference between the two layers is reduced sufficiently for the wind to induce vertical circulation and mixing in the lake water, resulting in an "overturn". This can occur quite quickly. The frequency of overturn and mixing depends principally on climate (temperature, insolation and wind) and the characteristics of the lake and its surroundings (depth and exposure to wind).

Lakes may be classified according to the frequency of overturn as follows (Figure 2.2):

- Monomictic: once a year — temperate lakes that do not freeze.
- Dimictic: twice a year — temperate lakes that do freeze.
- Polymictic: several times a year — shallow, temperate or tropical lakes.
- Amictic: no mixing — arctic or high altitude lakes with permanent ice cover, and underground lakes.
- Oligomictic: poor mixing — deep tropical lakes.
- Meromictic: incomplete mixing — mainly oligomictic lakes but sometimes deep monomictic and dimictic lakes.

Thermal stratification does not usually occur in lakes less than about 10 m deep because wind across the lake surface and water flow through the lake tend to encourage mixing. Shallow tropical lakes may be mixed completely several times a year. In very deep lakes, however, stratification may persist all year, even in tropical and equatorial regions. This permanent stratification results in "meromixis", which is a natural and continuous anoxia of bottom waters.

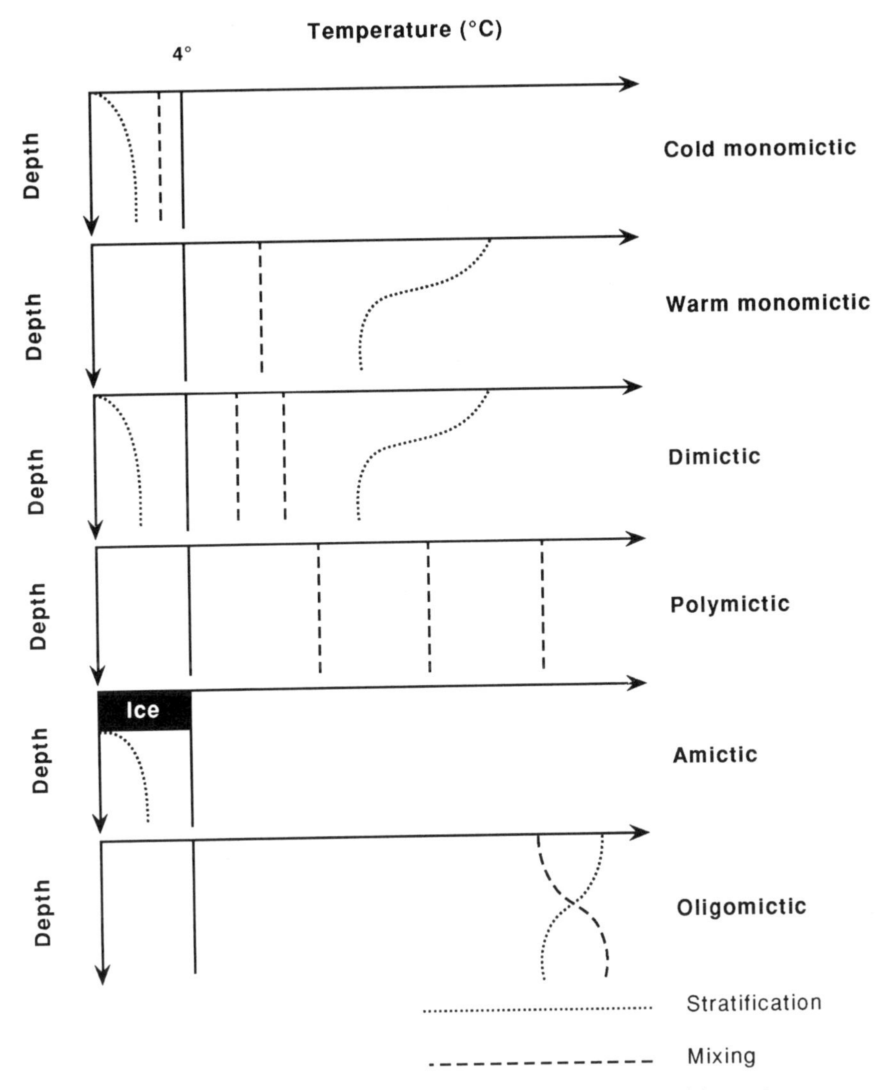


**Figure 2.2** The classification of lakes according to the occurrence of thermal stratification and mixing in the water column

*Tropical lakes*

A common physical characteristic of tropical lakes is that seasonal variations in water temperature are small, as a result of relatively constant solar radiation. Water temperatures are generally high but decrease with increasing altitude. The annual water temperature range is only 2–3 °C at the surface and even less at depths greater than 30 m. Density differences are minimal because water temperature is almost constant. Winds and precipitation, both of which tend to be seasonal, play an important role in mixing. The very limited seasonal temperature variation also results in a correspondingly low annual heat budget in tropical lakes. However, the relative variation in the heat budget in consecutive years may be considerable, because the peak value of heat storage may result from a single meteorological event.

In some tropical lakes, variations in water level of several metres may result from the large differences in rainfall between wet and dry seasons. Such variations have pronounced effects on dilution and nutrient supply which, in turn, affect algal blooms, zooplankton reproduction and fish spawning.

During the dry season, wind velocities are generally higher than at other times of the year and evaporation rates are at their maximum. The resulting heat losses, together with turbulence caused by wind action, promote mixing.

The classification of lakes based on seasonal temperature variations at different depths is not generally applicable to tropical lakes. A classification which considers size, depth and other physical characteristics, such as the following, is more relevant.

- Large, deep lakes all have a seasonal thermocline in addition to a deep permanent thermocline over an anoxic water mass. Recirculation of the deep water may occur but the responsible mechanism is not clear.
- Large, shallow lakes have a distinct diurnal temperature variation. Temperature is uniform in the morning, stratification develops in the afternoon and is destroyed during the night. The fluctuation in water level may be considerable relative to lake volume and the large flood plain that results will have profound effects on the productivity of biological life in the water.
- Crater lakes generally have a small surface area relative to their great depth and are often stratified. Despite such lakes being in sheltered positions, special weather conditions can cause complete mixing of lake contents.
- High-altitude lakes in climates where there is only a small diurnal temperature difference are unstable and experience frequent overturns. Where temperature differences are larger, a more distinct pattern of stratification can be identified. There may also be substantial losses of water by evaporation during the night.
- River lakes are created when areas of land are flooded by rivers in spate. When the water level in the river goes down, the lake water flows back towards the river. This annual or semi-annual water exchange affects the biological and chemical quality of the water.
- Solar lakes. In saline, dark-bottomed lakes an anomalous stratification can develop. A lower, strongly saline water layer may be intensely heated by

solar radiation, especially if it is well isolated from the atmosphere by the upper layer of lighter brine. Temperatures as high as 50 °C have been recorded in the lower levels of solar lakes.

- Temporary lakes occur in locations where the fluctuations of water level cause a shallow lake basin to dry up completely. In regions where there are pronounced wet and dry seasons this can occur annually, while in other regions the frequency of occurrence may be medium to long term. Temporary lakes often have an accumulation of salts on the lake bottom.

### 2.1.3 Rivers

An understanding of the discharge regime of a river is extremely important to the interpretation of water quality measurements, especially those including suspended sediment or intended to determine the flux of sediment or contaminants. The discharge of a river is related to the nature of its catchment, particularly the geological, geographical and climatological influences (see also Chapter 12).

*Tropical rivers*

The regime of a tropical river is largely determined by the annual cycle of wet and dry seasons. Some regimes, and some of the climatic and geographical conditions that affect regimes, are as follows:

- Equatorial rivers with one flow peak, resulting from heavy annual precipitation (1,750–2,500 mm) in areas with no marked dry season.
- Equatorial rivers with two flow peaks, produced by precipitation totalling more than 200 mm monthly and well over 1,750 mm annually. Equatorial forest predominates in the catchment area.
- Rivers in the moist savannah of tropical wet and dry lowlands exhibit pronounced seasonal effects of rainfall patterns. In these areas, the dry season persists for at least three months.
- In some areas of the tropical wet and dry highlands, the length of the dry season varies significantly. River basins in such areas are covered by woodland and relatively moist savannah.
- In the relatively drier regions of the tropical wet and dry highlands, river basins are located in the marginal parts of the dry climate zones. Precipitation rarely exceeds 500–700 mm annually, which is typical of semi-desert regions, and vegetation in the river basins is predominantly dry savannah.
- In areas where the dry season is prolonged, ecologists may divide the associated vegetation into wooded steppe and grass steppe. River regimes, however, do not differ between the two types of region, and flow is likely to be intermittent.

- In desert regions, where annual rainfall is less than 200 mm, river basins are covered with sand, desert grass or shrubs, and rivers are of the wadi type. The drainage network is poor and where it is traceable, it is likely to have developed before the area reached its present stage of aridity.
- Many of the rivers of tropical mountain regions have drainage basins of very limited size.

The great rivers of the tropics do not fall exclusively into any of these categories, because their drainage basins extend over many regions of differing climate and vegetation. The regime of the Congo (Zaire) River, for example, is largely a combination of regimes of the equatorial wet region and the tropical highland climate. Mean monthly flow is highest in April (76,000 $m^3$ $s^{-1}$) and lowest in July (32,000 $m^3$ $s^{-1}$). The Niger has its headwaters in a wet zone near the ocean, but then flows into a semi-arid region where it is subject to evaporation losses. Lower still in its course, the river flows into wet and dry tropical lowland. The Zambesi river basin is located largely in wet and dry tropical highland and semi-arid regions, while the Nile exhibits the most complex regime of all African rivers, extending over many widely different climatic zones.

The Amazon has a complex flow regime because of the different precipitation patterns of its main tributaries. Flow begins to increase in November, reaches a peak in June and then falls to a minimum at the end of October. The Orinoco, although draining a similar area to the Amazon, reaches peak flow a month later than the Amazon but also has minimum flow in October.

Most of the large rivers of Asia that flow generally southward have their sources in the mountains and flow through varied climatic conditions before discharging to the sea. Peak flows generally occur when run-off from melting snow is supplemented by monsoon rains. The Ganges and Irrawady receive snow-melt from the Himalayas and southern Tibet respectively from April to June, and the flow rate is just beginning to decline when the July monsoon begins. Flooding can occur from July to October. The Mekong is somewhat similar. It has its beginnings at an altitude of about 4,900 m in China's Tanglha Range. Snow-melt is later here, so peak flows do not occur until August/September in the upper reaches of the river and October in the lower reaches. Minimum flow in the Mekong occurs from November to May.

In western Asia, the Indus has some of its source tributaries in the Hindu Kush mountains, while the Tigris and Euphrates rise in the mountains of Turkey and Armenia respectively. Monsoon rains cause the Indus to flood between July and September. The Tigris and Euphrates are affected by seasonal rains that overlap with snow-melt run-off and cause flooding from March to June. The floods on the Euphrates inundate low-lying areas to form

**Table 2.1** Mean annual sediment loads of some major rivers

| River | Basin area ($10^3$ km$^2$) | Mean annual sediment load ($10^6$ t a$^{-1}$) |
|---|---|---|
| Amazon | 6,300 | 850 |
| Brahmaputra | 580 | 730 |
| Congo (Zaire) | 4,000 | 72 |
| Danube | 816 | 65 |
| Ganges | 975 | 1,450 |
| Indus | 950 | 435 |
| Irrawady | 430 | 300 |
| Orinoco | 950 | 150 [1] |
| Mekong | 795 | 160 [1] |
| Ob | 2,430 | 15 |
| Rhine | 160 | 2.8 |

[1] Average discharge taken from Meybeck *et al.* (1989) taking into account existing dams.

Source: World Resources Institute, 1988

permanent lakes that have no outlets. Water loss from these lakes is mainly by evaporation, although some of the water is withdrawn for irrigation.

Data on erosion in all of the world's river basins are far from complete. In general, however, erosion can be said to vary according to the following influences:

- amount and pattern of rainfall and resultant river regime,
- slope of the land,
- extent of destruction of vegetation,
- regeneration of vegetation, and
- soil type and resistance to the effects of temperature changes.

Erosion rates are thus extremely variable, with highest rates occurring usually in mountain streams where human intervention has resulted in extensive damage to vegetation. Erosion is primarily responsible for the amount of sediment transported to the sea. The mean annual sediment loads transported by several major rivers are given in Table 2.1.

## 2.2 Characteristics of groundwater

Groundwater is held in the pore space of sediments such as sands or gravels or in the fissures of fractured rock such as crystalline rock and limestone. The body of rock or sediments containing the water is termed an aquifer and the upper water level in the saturated body is termed the water table. Typically, groundwaters have a steady flow pattern. Velocity is governed mainly by the porosity and permeability of the material through which the water flows, and is often up to several orders of magnitude less than that of surface waters. As a result mixing is poor.

The media (rock or sediment) in an aquifer are characterised by porosity and permeability. Porosity is the ratio of pore and fissure volume to the total volume of the media. It is measured as percentage voids and denotes the storage or water volume of the media. Permeability is a measure of the ease with which fluids in general may pass through the media under a potential gradient and indicates the relative rate of travel of water or fluids through media under given conditions. For water it is termed hydraulic conductivity.

*Types of aquifer*

Underground formations are of three basic types: hard crystalline rocks, consolidated sedimentary formations and unconsolidated sediments. The hard crystalline rocks include granites, gneisses, schists and quartzites and certain types of volcanic rocks such as basalts and dolerites. These formations generally have little or no original porosity, and the existence of aquifers depends on fractures and fissures in the rock mass providing porosity and pathways for groundwater movement. Although these are often further enhanced by weathering, aquifers in hard rocks are usually small and localised and not very productive. Groundwater in volcanic formations in regions of "recent" volcanic activity frequently contain fluoride and boron in concentrations that are unacceptably high for certain uses.

Consolidated sedimentary formations are often thick and extensive, and sometimes artesian. Limestone and sandstone formations may be highly porous and permeable and form some of the largest, most important and highest-yielding aquifers in the world. The permeability of these formations is largely due to fissures (fractures, faults, bedding planes). Porosity is also significant for the movement and storage of some pollutants. Dissolution of the rock can increase the permeability. The dissolution of carbonates, in particular, is responsible for the formation of karst aquifers, which can have large underground caverns and channels yielding substantial quantities of water.

Unconsolidated sediments occur as thin, superficial deposits over other rock types or as thick sequences in the major river or lake basins. Porosity and permeability are related to grain size. Sand and gravel deposits can provide important and high-yielding aquifers, whereas silts and clays are less productive. In the largest river basins, thick sedimentary deposits may contain many layers of different materials built up over long periods of time, producing important multi-aquifer sequences.

Aquifers may be confined or unconfined (Figure 2.3). A confined aquifer is overlain by an impermeable layer that prevents recharge (and contamination) by rainfall or surface water. Recharge of confined aquifers occurs where the permeable rock outcrops at or near the surface, which may be some

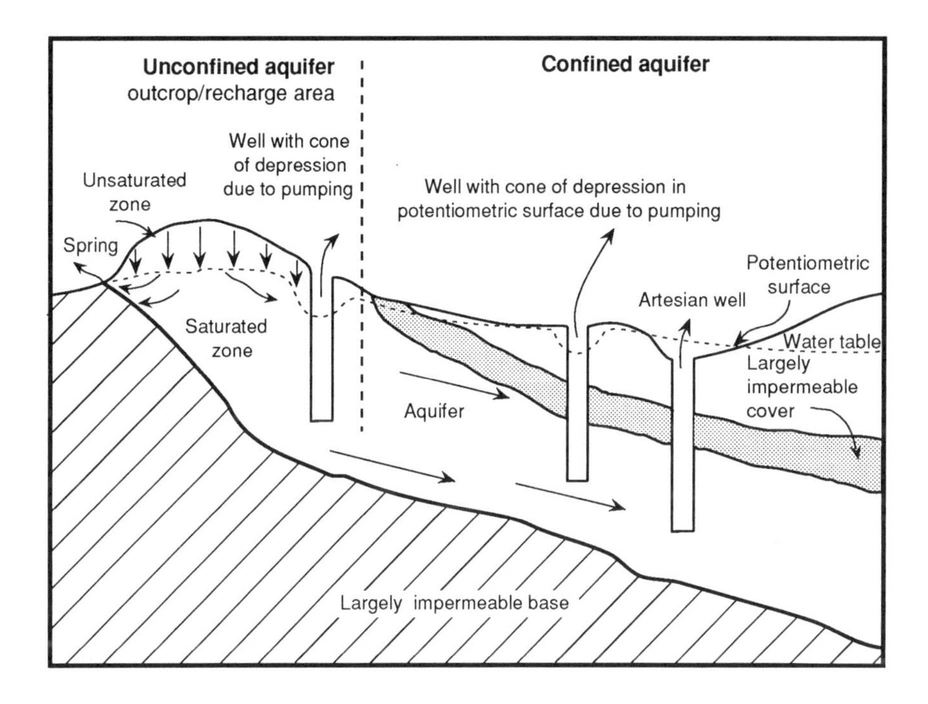


**Figure 2.3** Confined and unconfined aquifers (After Chilton, 1996)

distance from the area of exploitation. This feature may make control of quality and of pollution more difficult. Some aquifers are not perfectly confined and are termed semi-confined or leaky.

Unconfined aquifers are overlain by a permeable, unsaturated zone that allows surface water to percolate down to the water table. Consequently, they are generally recharged over a wide area and are often shallow with a tendency for interaction with surface water.

Confined aquifers are less vulnerable than unconfined aquifers to pollution outside their recharge zone because surface water and contaminants cannot percolate to the water table. If contamination does occur, however, it is often difficult to remedy because confined aquifers are usually deep and the number of points where contaminated water may be pumped out is limited. Given the limited outflow, contaminants may also be increasingly concentrated in confined aquifers and this may restrict abstraction of water. The greater vulnerability of unconfined aquifers to contamination is a result of the wider area over which they are recharged and in which contamination may enter, and the greater interaction with polluted surface water bodies which may lead to contaminant movement into groundwater. The risk of contamination will depend on the depth of the overlying unsaturated layer, the rate of infiltration to the water table and the land use in areas surrounding groundwater sources.

*Water quality*

The quality of groundwater depends on the composition of the recharge water, the interactions between the water and the soil, soil-gas and rocks with which it comes into contact in the unsaturated zone, and the residence time and reactions that take place within the aquifer. Therefore, considerable variation can be found, even in the same general area, especially where rocks of different compositions and solubility occur. The principal processes influencing water quality in aquifers are physical (dispersion/dilution, filtration and gas movement), geochemical (complexation, acid–base reactions, oxidation–reduction, precipitation–solution, and adsorption–desorption) and biochemical (microbial respiration and decay, cell synthesis).

Groundwater quality is influenced by the effects of human activities which cause pollution at the land surface because most groundwater originates by recharge of rainwater infiltrating from the surface. The rainwater itself may also have an increased acidity due to human activity. The unsaturated zone can help reduce the concentrations of some pollutants entering groundwater (especially micro-organisms), but it can also act as a store for significant quantities of pollutants such as nitrates, which may eventually be released. Some contaminants enter groundwaters directly from abandoned wells, mines, quarries and buried sewerage pipes which by-pass the unsaturated zone (and, therefore, the possibility of some natural decontamination processes).

*Contamination*

Artificial pollution of groundwater may arise from either point or diffuse sources. Some of the more common sources include domestic sewage and latrines, municipal solid waste, agricultural wastes and manure, and industrial wastes (including tipping, direct injection, spillage and leakage). The contamination of groundwaters can be a complex process. Contaminants, such as agricultural chemicals, spread over large sections of the aquifer recharge area may take decades to appear in the groundwater and perhaps longer to disappear after their use has ceased. Major accidental spills and other point sources of pollutants may initially cause rapid local contamination, which then spreads through the aquifer. Pollutants that are fully soluble in water and of about the same density (such as chloride-contaminated water from sewage) will spread through the aquifer at a rate related to the groundwater flow velocity. Pollutants that are less dense than water will tend to accumulate at the water table and flow along the surface. Dense compounds such as chlorinated solvents will move vertically downwards and accumulate at the bottom of an aquifer.

There is usually a delay between a pollution incident and detection of the contaminant at the point of water abstraction because movement in the unsaturated zone and flow in the aquifer are often slow. For similar reasons the time needed to "flush out" a pollutant is long and in some cases the degradation of groundwater quality may be considered irreversible.

Land use in areas surrounding boreholes and where aquifers are recharged should be carefully monitored as part of a pollution control programme. The vulnerability of the aquifer to pollution will depend, in part, on the human activity and land use in areas where rainfall or surface water may percolate into the aquifer. In these areas, contamination of surface water or of the unsaturated layer above an aquifer is likely to cause groundwater pollution.

Further details of the natural features of groundwaters, the quality issues particularly relevant to groundwaters and examples of monitoring and assessment programmes are available in the specialised literature (see section 2.6) and the companion guidebook *Water Quality Assessments*.

## 2.3 Natural processes affecting water quality

Although degradation of water quality is almost invariably the result of human activities, certain natural phenomena can result in water quality falling below that required for particular purposes. Natural events such as torrential rainfall and hurricanes lead to excessive erosion and landslides, which in turn increase the content of suspended material in affected rivers and lakes. Seasonal overturn of the water in some lakes can bring water with little or no dissolved oxygen to the surface. Such natural events may be frequent or occasional. Permanent natural conditions in some areas may make water unfit for drinking or for specific uses, such as irrigation. Common examples of this are the salinisation of surface waters through evaporation in arid and semi-arid regions and the high salt content of some groundwaters under certain geological conditions. Many groundwaters are naturally high in carbonates (hardness), thus necessitating their treatment before use for certain industrial applications. Groundwaters in some regions contain specific ions (such as fluoride) and toxic elements (such as arsenic and selenium) in quantities that are harmful to health, while others contain elements or compounds that cause other types of problems (such as the staining of sanitary fixtures by iron and manganese).

The nature and concentration of chemical elements and compounds in a freshwater system are subject to change by various types of natural process,

**Table 2.2** Important processes affecting water quality

| Process type | Major process within water body | Water body |
|---|---|---|
| Hydrological | Dilution | All water bodies |
| | Evaporation | Surface waters |
| | Percolation and leaching | Groundwaters |
| | Suspension and settling | Surface waters |
| Physical | Gas exchange with atmosphere | Mostly rivers and lakes |
| | Volatilisation | Mostly rivers and lakes |
| | Adsorption/desorption | All water bodies |
| | Heating and cooling | Mostly rivers and lakes |
| | Diffusion | Lakes and groundwaters |
| Chemical | Photodegradation | Lakes and rivers |
| | Acid base reactions | All water bodies |
| | Redox reactions | All water bodies |
| | Dissolution of particles | All water bodies |
| | Precipitation of minerals | All water bodies |
| | Ionic exchange[1] | Groundwaters |
| Biological | Primary production | Surface waters |
| | Microbial die-off and growth | All water bodies |
| | Decomposition of organic matter | Mostly rivers and lakes |
| | Bioaccumulation[2] | Mostly rivers and lakes |
| | Biomagnification[3] | Mostly rivers and lakes |

1 Ionic exchange is the substitution of cations, for example in clay; the commonest is the replacement of calcium by sodium.
2 Bioaccumulation results from various physiological processes by which an organism accumulates a given substance (mercury and lead are common examples) from water or suspended particles.
3 Biomagnification is the increase in concentration of a given substance within a food chain such as phytoplankton → zooplankton → microphage fish → carnivorous fish → carnivorous mammals.

i.e. physical, chemical, hydrological and biological. The most important of these processes and the water bodies they affect are listed in Table 2.2.

The effects on water quality of the processes listed in Table 2.2 will depend to a large extent on environmental factors brought about by climatic, geographical and geological conditions. The major environmental factors are:

- Distance from the ocean: extent of sea spray rich in $Na^+$, $Cl^-$, $Mg^{2+}$, $SO_4^{2-}$ and other ions.
- Climate and vegetation: regulation of erosion and mineral weathering; concentration of dissolved material through evaporation and evapotranspiration.
- Rock composition (lithology): the susceptibility of rocks to weathering ranges from 1 for granite to 12 for limestone; it is much greater for more highly soluble rocks (for example, 80 for rock salt).

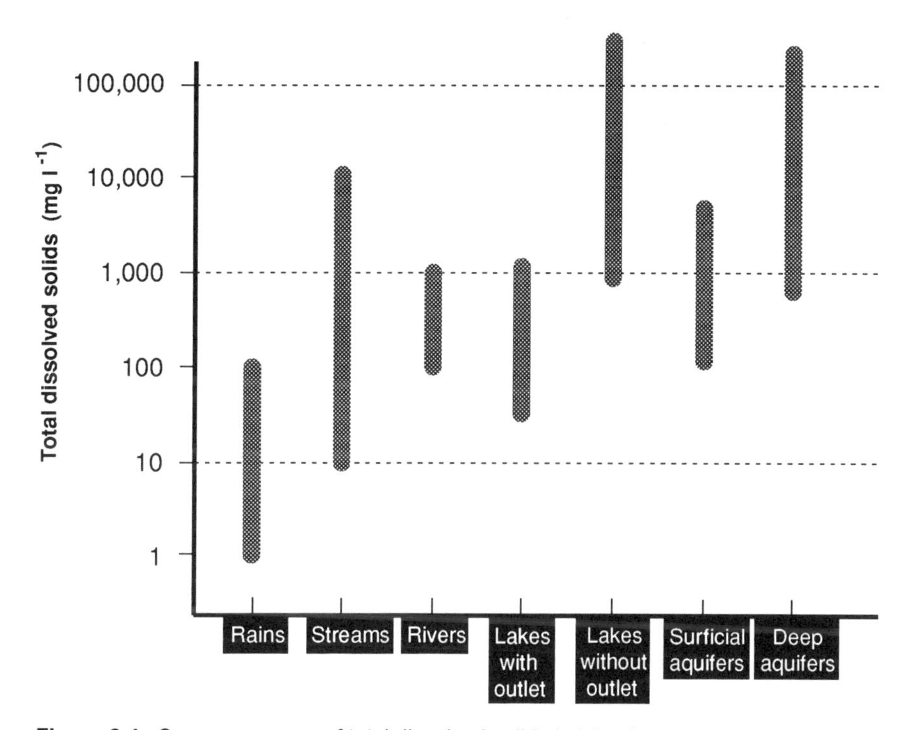


**Figure 2.4** Common ranges of total dissolved solids in inland waters

- Terrestrial vegetation: the production of terrestrial plants and the way in which plant tissue is decomposed in soil affect the amount of organic carbon and nitrogenous compounds found in water.
- Aquatic vegetation: growth, death and decomposition of aquatic plants and algae will affect the concentration of nitrogenous and phosphorous nutrients, pH, carbonates, dissolved oxygen and other chemicals sensitive to oxidation/reduction conditions. Aquatic vegetation has a profound effect on the chemistry of lake water and a less pronounced, but possibly significant effect, on river water.

Under the influence of these major environmental factors, the concentrations of many chemicals in river water are liable to change from season to season. In small watersheds (< 100 km$^2$) the influence of a single factor can cause a variation of several orders of magnitude. Water quality is generally more constant in watersheds greater than 100,000 km$^2$, and the variation is usually within one order of magnitude for most of the measured variables.

These natural variations are reflected in the range of total dissolved solids in various water bodies, as illustrated in Figure 2.4. Total dissolved solids is

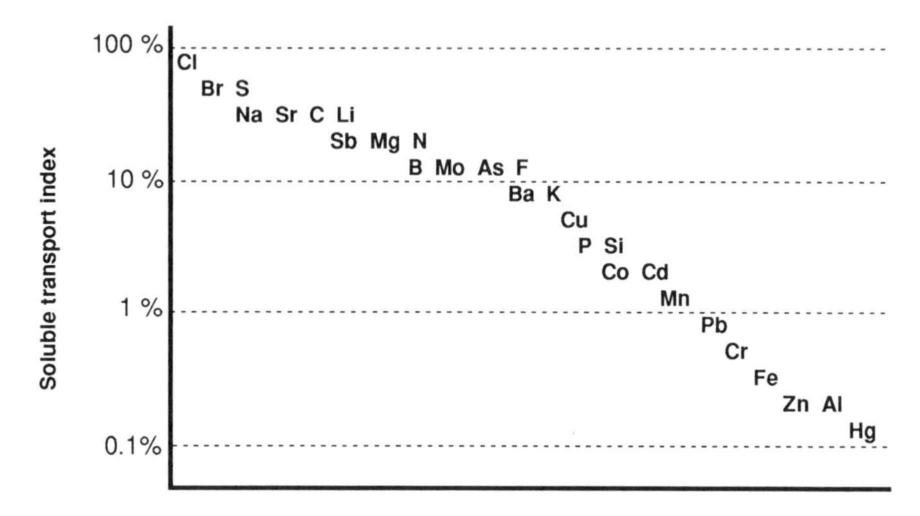


**Figure 2.5** Relative scale of Soluble Transport Index (the tendency for a chemical to be present in the soluble form)

the sum of the silica plus the major ions in the water. A close approximation of its value may be obtained by the measurement of electrical conductivity.

Some chemical elements have a strong affinity for particulate matter (see also Chapter 13) and, as a result of precipitation/dissolution and adsorption/desorption reactions, they may be found in only trace amounts in solution. Other elements, however, are highly soluble and rarely, if ever, present in water in particulate form. The tendency for a chemical to be present in the soluble form rather than associated with particulates is expressed as the Soluble Transport Index (Figure 2.5). Particulate matter is separated from material in solution by filtering a water sample through a filter with pore size 0.45 µm or 0.50 µm. It is recognised that some fine colloids such as the hydroxides of iron, aluminium and manganese may pass through the filter, but the use of finer filters is very costly and requires long filtration times.

The elemental composition of river particulates is less variable than the dissolved concentrations for the same elements. This is true for the major elements aluminium, iron, silicon, calcium, magnesium, sodium, potassium, manganese, titanium and phosphorus but less so for the trace elements arsenic, cadmium, cobalt, chromium, copper, mercury, nickel, lead, selenium, tin, zinc, etc. Except for aluminium, iron, calcium and sodium, the composition of major elements in particulates is very similar in tropical rivers and in temperate and cold rivers. In small watersheds, local geological conditions can lead to wide variations in the concentration of trace elements in particulates.

Within any one water body water quality can differ with time and with place. Differences due to time are of five types:

- Minute-to-minute and day-to-day differences resulting from water mixing and fluctuations in inputs, usually as a result of meteorological conditions. These differences are most evident in small water bodies.
- Diurnal (24-hour) variations resulting from biological cycles and day-light/darkness cycles which cause changes in, for example, dissolved oxygen and pH. Diurnal patterns also result from the cyclic nature of waste discharges from domestic and industrial sources.
- Irregular patterns. Irregular sources of pollution include fertilisers, pesticides and herbicides, present in the run-off from agricultural land, and wastes discharged from food processing plants. The resultant variations in water quality may be apparent over a matter of days or months.
- Seasonal biological and hydrological cycles.
- Year-to-year trends, usually as a result of increased human activities in the watershed.

Water quality differences may result from either internal or external processes. Internal processes are usually cyclic, with either daily or seasonal recurrence, and are not directly related to the size of the water body. External processes, such as the addition of pollutants, may be buffered by large water bodies (depending on flow regimes) and long water residence times. As a result, the average composition of a very large lake probably changes little from one year to the next. Similarly, the differences in water quality at different times of the year will be much greater for a stream than for a large river. This means that the sampling frequency necessary to allow average water quality to be described correctly is normally much greater for a stream than for a river; for lakes it is normally much lower than for rivers.

Water quality differences from place to place depend more on the homogeneity of the water body than on its size. The water in a round lake, for example, may be adequately described by one sample taken from near the centre of the lake. Long, thin lakes and lakes with many bays and inlets will require more samples; the minimum is 3 while the optimum number could be 10 or more. In lakes deeper than about 30 m, or that are stratified, it is especially important to obtain samples from different depths, so that a vertical profile can be obtained. Depending on the depth of the lake, up to five samples should be taken at depths determined by conditions measured in the field (see section 3.7.2).

The flow pattern, or regime, of any particular river will be the product of very specific conditions. Although similar to that of rivers in the surrounding geographical region, it will be extensively influenced by altitude, exposure of the slope to wind and variations in rainfall.

Further details of the natural factors and influences on water quality in rivers, lakes, reservoirs and groundwaters are given in the special chapters on each of these water bodies in the companion guidebook *Water Quality Assessments*.

## 2.4 Water use and water quality deterioration

Historically, the development of civilisations has led to a shift in the pattern of water use from rural/agricultural to urban/industrial, generally according to the following sequence: drinking and personal hygiene, fisheries, navigation and transport, livestock watering and agricultural irrigation, hydroelectric power, industrial production (e.g. pulp and paper, food processing), industrial cooling water (e.g. fossil fuel and nuclear power plants), recreational activities and wildlife conservation. Fortunately, the water uses with the highest demands for quantity often have the lowest demands for quality. Drinking water, by contrast, requires the highest quality water but in relatively small quantities.

Increasing industrialisation and the growth of large urban centres have been accompanied by increases in the pollution stress on the aquatic environment. Since ancient times, water in rivers, lakes and oceans has also been considered as a convenient receiver of wastes. This use (or abuse) conflicts with almost all other uses of water and most seriously with the use of freshwater for drinking, personal hygiene and food processing.

All water uses have impacts on the quality of the aquatic environment (Table 2.3), including hydrological changes such as storing water in reservoirs or transferring water from one drainage area to another. Human use of water for almost all purposes results in the deterioration of water quality and generally limits the further potential use of the water. The major types and the extent of deterioration in freshwater quality are summarised in Table 2.4 (for further details see the companion guidebook *Water Quality Assessments*).

A three-point strategy has been developed to resolve the conflicts between quality deterioration and water use as follows:

- The quality of water and of the aquatic environment is determined and water-use procedures that prevent deterioration are adopted.
- Wastes are treated before discharge to a water body in order to control pollution.
- Unsatisfactory water is treated before use in order to meet specific water quality requirements.

**Table 2.3** Common water uses

| Water uses | Consuming | Contaminating |
|---|---|---|
| Domestic use | Yes | Yes |
| Livestock watering | Yes | Yes |
| Irrigation | Yes | Yes |
| Aquaculture | Yes | Yes |
| Commercial fisheries | Yes | Yes |
| Forestry and logging | No[1] | Yes |
| Food processing | Yes | Yes |
| Textile industry | Yes | Yes |
| Pulp and paper industry | Yes | Yes |
| Mining | Yes | Yes |
| Water transportation | No | Yes |
| Hydroelectric power generation | No | No[2] |
| Nuclear power generation | Yes | Yes |
| Recreation | No | Yes |

1 Water availability may be altered due to changes caused in run-off regimes

2 Thermal characteristics of the water body may be altered

**Table 2.4** Freshwater quality deterioration at the global level[1]

| | Rivers | Lakes | Reservoirs | Groundwaters |
|---|---|---|---|---|
| Pathogens | xxx | x[2] | x[2] | x |
| Suspended solids | xx | oo | x | oo |
| Decomposable organic matter[3] | xxx | x | xx | x |
| Eutrophication[4] | x | xx | xxx | oo |
| Nitrate as a pollutant | x | o | o | xxx |
| Salinisation | x | o | o | xxx |
| Heavy metals | xx | xx | xx | xx[5] |
| Organic micro-pollutants | xx | x | xx | xxx[5] |
| Acidification | x | xx | x | o |
| Changes to hydrological regimes[6] | xx | xx | xx | x |

Radioactive and thermal wastes are not considered here.

xxx Globally occurring, or locally severe deterioration

xx Important deterioration

x Occasional or regional deterioration

o Rare deterioration

oo Not relevant

1 This is an estimate. At the regional level, these ranks may vary greatly according to the degree of economic development and the types of land use

2 Mostly in small and shallow water bodies

3 Other than that resulting from aquatic primary production

4 Algae and macrophytes

5 From landfills and mine tailings

6 Water diversion, damming, over-pumping, etc.

Source: Modified from Meybeck and Helmer, 1996

**Table 2.5** Sources and significance of pollutants resulting from human activities

| Sources | Bacteria | Nutrients | Trace metals | Pesticides and herbicides | Industrial organic micro-pollutants | Oils and greases |
|---|---|---|---|---|---|---|
| *Atmos. transport* | | x | xxxG | xxG | xxG | |
| *Point sources* | | | | | | |
| Urban sewage | xxx | xxx | xxx | x | xxx | |
| Industrial effluent | | x | xxxG | x | xxxG | xx |
| *Diffuse sources* | | | | | | |
| Agriculture | xx | xxx | x | xxxG | | |
| Urban waste and run-off | xx | xx | xxx | xx | xx | x |
| Industrial waste disposal | | x | xxx | x | xxx | x |
| Dredging | | x | xxx | x | xxx | x |
| Navigation and harbours | x | x | xx | | x | xxx |
| *Internal recycling* | | xxx | xx | x | x | |

x Low local significance
xx Moderate local or regional significance
xxx High local or regional significance
G Global significance

Source: Modified from Meybeck and Helmer, 1996

Human activities are the source of particulate, dissolved and volatile materials which may eventually reach water. Dissolved materials and many particulates are discharged directly to water bodies, while the particulate and volatile materials that pollute the atmosphere are picked up by rain and then deposited on land or in water. Some sources and the polluting material released are listed in Table 2.5.

Specific locations where pollution resulting from human populations and human activities occurs, such as discharges from sewage treatment works, industrial wastewater outlets, solid waste disposal sites, animal feedlots and quarries, can be described as point sources. The effect of a point source on the receiving water body is dependent on:

- the population, or size and type of activity, discharging the waste,
- the capacity of the water body to dilute the discharge,
- the ecological sensitivity of the receiving water body, and
- the uses to which the water may be put.

Pollutants may also be derived from diffuse and multi-point sources. Diffuse sources are often of agricultural origin and enter surface waters with run-off or infiltrate into groundwaters (particularly pesticides and fertilisers). Multi-point sources, such as latrines and septic tanks in rural and urban areas may be treated as diffuse sources for the purposes of monitoring and assessment because it is not possible to monitor each source individually.

Point sources of pollution can usually be identified and the polluting material eventually collected and treated. This cannot be done, however, with diffuse terrestrial sources, atmospheric depositions and the internal recycling of nutrients, metals and some organics. Pollution from these sources can be controlled only by prevention. Internal recycling is a particularly difficult problem because it occurs mostly under the anoxic conditions present in the interstitial water of some lake sediments and in some groundwaters. Pollution from accidental spills is unpredictable and its prevention requires the strict observance of safety procedures.

## 2.5 Water and human health

Water, although an absolute necessity for life, can be a carrier of many diseases. Paradoxically, the ready availability of water makes possible the personal hygiene measures that are essential to prevent the transmission of enteric diseases. Infectious water-related diseases can be categorised as waterborne, water-hygiene, water-contact and water-habitat vector diseases. Some water-related diseases, however, may fall into more than one category.

Waterborne infectious diseases are those in which the pathogen, or causative organism, is present in water and ingested when the water is consumed. Most of the pathogens involved are derived from human faeces, and the diseases transmitted by consumption of faecally contaminated water are called "faecal–oral" diseases. All of the faecal–oral diseases can also be transmitted through media other than water, for example faecally contaminated food, fingers or utensils. The principal faecal–oral diseases are cholera, typhoid, shigellosis, amoebic dysentery, hepatitis A and various types of diarrhoea.

One disease that is exclusively waterborne is dracunculiasis, or guinea worm disease, which is caused by *Dracunculus medinensis*. An individual can become infected with *Dracunculus* only by consuming water contaminated with the microscopic crustaceans (Cyclops) that contain the larvae of the pathogen. Dracunculiasis is not a faecal–oral disease.

The incidence, prevalence and severity of water hygiene diseases can be reduced by the observance of high levels of personal, domestic and community hygiene. Almost all waterborne diseases (excluding dracunculiasis) are, therefore, also water hygiene diseases. Other water hygiene diseases include tinea, scabies, pediculosis and skin and eye infections. Tinea, a skin disease, trachoma, an eye disease, and insect infestations such as scabies and pediculosis (lice) occur less frequently when personal hygiene and cleanliness are of a high standard. Water must be available in adequate quantities to permit hand washing, bathing, laundering, house cleaning, and the cleaning of cooking and eating utensils. The quantity required for these purposes is substantially greater than that needed for drinking.

Water contact diseases are transmitted when an individual's skin is in contact with pathogen infested water. The most important example is schistosomiasis (bilharziasis), in which the eggs of the pathogen (*Schistosoma* spp.) are present in the faeces and/or urine of an infected individual. The eggs hatch when they reach water and the larvae invade a suitable snail host where they multiply and develop, finally escaping from the snail as free swimming cercariae that infect humans by penetrating immersed or wetted skin.

Water-habitat vector diseases are transmitted by insect vectors that spend all or part of their lives in or near water. The pathogenic organisms that cause these diseases spend a portion of their life cycles in a specific vector. The best known examples are malaria and filariasis (mosquito vector) and onchocerciasis (aquatic fly vector). One method of controlling these diseases is to control the vector which, in some cases, involves some physical or chemical treatment of the water habitat.

Health effects from chemicals in water occur when an individual consumes water containing a harmful amount of a toxic substance. Infant methaemoglobinaemia, caused by the consumption of water with a high nitrate concentration by infants (usually those which are bottle fed), is an example. The occurrence of methaemoglobinaemia is usually related to nitrate (often in groundwaters) which has been derived from extensive use of nitrate fertilisers. Fluorosis, damage to the teeth and bones, results from long-term consumption of water containing excess fluorides (usually from natural sources).

## 2.6 Source literature and further reading

Balek, J. 1977 *Hydrology and Water Resources in Tropical Africa*. Elsevier, Amsterdam.

Beadle, L.C. 1974 *The Inland Waters of Tropical Africa*. Longman, London.

Chapman, D. [Ed.] 1996 *Water Quality Assessments. A Guide to the Use of Biota, Sediments and Water in Environmental Monitoring*. 2nd edition. Chapman & Hall, London.

Chilton, J. 1996 Groundwater. In: D. Chapman [Ed.] *Water Quality Assessments. A Guide to the Use of Biota, Sediments and Water in Environmental Monitoring*. 2nd edition. Chapman & Hall, London.

Foster, S.S.D. and Gomes, D.C. 1989 *Groundwater Quality Monitoring: An Appraisal of Practices and Costs*. Pan American Centre for Sanitary Engineering and Environmental Science (CEPIS), Lima.

Foster, S.S.D. and Hirata. R. 1988 *Groundwater Pollution Risk Assessment: A Method Using Available Data*. Pan American Centre for Sanitary Engineering and Environmental Science (CEPIS), Lima.

GESAMP 1988 *Report of the Eighteenth Session, Paris 11–15 April 1988.* GESAMP Reports and Studies No. 33, United Nations Educational, Scientific and Cultural Organization, Paris.

Hem, J.D. 1984 *Study and Interpretation of the Chemical Characteristics of Natural Water., 3rd edition. Water Supply Paper 2254, United States Geological Survey, Washington, DC.*

McJunkin, F.E. 1982 *Water and Human Health*. United States Agency for International Development, Washington, DC.

Meybeck, M. and Helmer, R. 1996 Introduction. In: D. Chapman [Ed.] *Water Quality Assessments. A Guide to the Use of Biota, Sediments and Water in Environmental Monitoring*. 2nd edition. Chapman & Hall, London.

Meybeck, M., Chapman, D. and Helmer, R. 1989 *Global Freshwater Quality: A First Assessment*. Blackwell Reference, Oxford.

Nash, H. and McCall, J.G.H. 1994 *Groundwater Quality*. Chapman & Hall, London.

Serruya, C. and Pollingher, U. 1983 *Lakes of the Warm Belt*. Cambridge University Press, Cambridge.

WHO 1993 *Guidelines for Drinking-water Quality. Volume 1 Recommendations*. 2nd edition. World Health Organization, Geneva.

WHO (In prep.) *Guidelines for Recreational Water Use and Beach Quality.* World Health Organization, Geneva.

World Resources Institute 1988 *World Resources 1988-89*. Basic Books Inc., New York.

# Chapter 3

# DESIGNING A MONITORING PROGRAMME

The design of a monitoring programme should be based on clear and well thought out aims and objectives and should ensure, as far as possible, that the planned monitoring activities are practicable and that the objectives of the programme will be met.

It is useful to prepare a programme document or study plan, which should begin with a clear statement of the objectives of the programme and a complete description of the area in which the monitoring is to take place. The geographical limits of the area, the present and planned water uses and the present and expected pollution sources should be identified. Background information of this type is of great help in preparing a precise description of the programme objectives and in deciding on some of the elements of the study plan. Subsequent sections of the study plan should cover the locations and frequency of sampling and the variables for analysis. The plan should also specify whether the analyses will be done in the field or in the laboratory. This decision must take into consideration the resources available for all the necessary field and laboratory work, data handling, analysis and interpretation, and the preparation and distribution of reports.

The principal elements of a study plan are:

- a clear statement of aims and objectives,
- information expectations and intended uses,
- a description of the study area concerned,
- a description of the sampling sites,
- a listing of the water quality variables that will be measured,
- proposed frequency and timing of sampling,
- an estimate of the resources required to implement the design, and
- a plan for quality control and quality assurance.

## 3.1 Purpose of monitoring

The principal reason for monitoring water quality has been, traditionally, the need to verify whether the observed water quality is suitable for intended uses. However, monitoring has also evolved to determine trends in the quality of the aquatic environment and how the environment is affected by the release of contaminants, by other human activities, and/or by waste

*This chapter was prepared by A. Mäkelä and M. Meybeck*

treatment operations. This type of monitoring is often known as impact monitoring. More recently, monitoring has been carried out to estimate nutrient or pollutant fluxes discharged by rivers or groundwaters to lakes and oceans, or across international boundaries. Monitoring for background quality of the aquatic environment is also widely carried out, as it provides a means of comparing and assessing the results of impact monitoring.

No monitoring programme should be started without critically scrutinising the real needs for water quality information. Since water resources are usually put to several competing beneficial uses, monitoring should reflect the data needs of the various water users involved.

The implementation of a monitoring and assessment programme may focus on the spatial distribution of quality, on trends, or on aquatic life. Full coverage of all three elements is virtually impossible, or very costly. Consequently, preliminary surveys are necessary in order to determine the focus of the operational programme.

The monitoring and assessment of water quality is based, ultimately, upon the fundamental physical, chemical and biological properties of water. However, water quality monitoring and assessment is a process of analysis, interpretation and communication of those properties within the wider context of human activity and use, and the conservation of the natural environment. It is not a fixed process, therefore, but is adapted in the light of local, national or international needs. The final aim is to provide information useful for management. Styles and strategies of management vary greatly, depending on institutions, resources and priorities.

Water quality monitoring and assessment can be conducted from a number of different perspectives which may combine the following goals in different ways:

- Uses of water. Does water meet user requirements for quantity and quality? (For example, with respect to meeting use-defined standards. In this context conservation of biodiversity may be considered a water use.)
- Influences on water quality from direct use or from other human activities or natural processes. What are these influences?
- Impacts on water quality (e.g. water as a medium for pollutant transport and exposure).
- Control and regulation of water quality. What is the capacity of water to assimilate pollutants? Are standards met? Are control strategies and management action appropriate and effective?
- How does water quality differ geographically in relation to uses and quality influences?

- How have past trends in water quality, influences and policies led to the present status?
- What factors in present water quality and in the past, present and planned activities, give an insight into future trends? What will these be?
- How does water quality influence other parts of the environment, such as marine coastal waters, soils, biota, wetlands?

These are examples of the types of goals, answers or information that are sought in undertaking water quality monitoring. They approach water quality from different perspectives in terms of basic variables and present status, time trends and spatial differences, uses, pollution impacts and management needs for information for decisions and action. These differences will result in different approaches to the design and implementation of monitoring programmes, to the selection of variables to be measured, to the frequency and location of measurements, to the additional information needed for interpretation and to the way in which information is generated and presented to meet particular information requirements.

## 3.2 The need for information for management

When a water quality monitoring programme is being planned, water-use managers or similar authorities can reasonably expect that the programme will yield data and information that will be of value for management decision-making. The following are examples of the type of information that may be generated by a monitoring programme:

- How the quality and quantity of water in a water body relate to the requirements of users.
- How the quality and quantity of water in a water body relate to established water quality standards.
- How the quality of water in a water body is affected by natural processes in the catchment.
- The capacity of the water body to assimilate an increase in waste discharges without causing unacceptable levels of pollution.
- Whether or not existing waste discharges conform to existing standards and regulations.
- The appropriateness and effectiveness of control strategies and management actions for pollution control.
- The trends of changes in water quality with respect to time as a result of changing human activities in the catchment area. Quality could be declining as a result of waste discharges or improving as a result of pollution control measures.

- Control measures that should be implemented to improve or prevent further deterioration of water quality.
- The chemical or biological variables in the water that render it unsuitable for beneficial uses.
- The hazards to human health that result, or may result, from poor water quality in the water body.
- How developments in the catchment area have affected or will affect water quality.
- The effects that deteriorating water quality have on plant and animal life in, or near, the water body.

The list above is not exhaustive, it merely provides examples. The information required from a monitoring programme does, however, provide an indication of the type of programme that should be implemented. Some monitoring programmes will be long-term and intended to provide a cumulative body of information; others will have a single objective and will usually be of short duration.

## 3.3 Objectives of water quality monitoring

It is particularly important that the objectives of a water quality monitoring programme be clearly stated and recorded. The very act of writing them down generally results in careful consideration being given to the possible options. Written objectives help to avoid misunderstandings by project participants, are an effective way of communicating with sponsors, and provide assurance that the monitoring programme has been systematically planned. They are also important when the programme is evaluated to determine whether or not the objectives are being met.

The objectives may be very general, such as when monitoring is for several purposes or when it would be premature to prepare highly detailed objectives, whereas statistical descriptions of objectives are usually reserved for the more advanced types of monitoring programme. To help with the establishment of objectives, the following questions might be addressed:

- Why is monitoring going to be conducted? Is it for basic information, planning and policy information, management and operational information, regulation and compliance, resource assessment, or other purposes?
- What information is required on water quality for various uses? Which variables should be measured, at what frequency and in response to which natural or man-made events?
- What is practical in terms of the human and financial resources available for monitoring? There is little point in setting unrealistic objectives.

- Who is responsible for the different elements of monitoring?
- Who is going to use the monitoring data and what are they intending to do with the information? Will it support management decisions, ensure compliance with standards, identify priorities for action, provide early warning of future problems or detect gaps in current knowledge?

The following is a list of typical monitoring objectives that might be used as the basis for design of sampling networks. The list is not intended to be exhaustive, merely to provide some examples.

- Identification of baseline conditions in the water-course system.
- Detection of any signs of deterioration in water quality.
- Identification of any water bodies in the water-course system that do not meet the desired water quality standards.
- Identification of any contaminated areas.
- Determination of the extent and effects of specific waste discharges.
- Estimation of the pollution load carried by a water-course system or subsystem.
- Evaluation of the effectiveness of a water quality management intervention.
- Development of water quality guidelines and/or standards for specific water uses.
- Development of regulations covering the quantity and quality of waste discharges.
- Development of a water pollution control programme.

There are two different types of monitoring programmes: those with a single objective which are set up to address one problem only, and multi-objective programmes which may cover various water uses, such as drinking water supply, industrial manufacturing, fisheries, irrigation or aquatic life. It is rare that a monitoring programme has a single objective. In practice, programmes and projects generally combine multiple objectives and data are used for multiple purposes. Monitoring with several objectives is commonly the first major, national programme to be established in a country. The design of such programmes requires preliminary survey work, so that the selection of sampling sites takes into account such considerations as actual and potential water uses, actual and potential sources of pollution, pollution control operations, local geochemical conditions and type(s) of water body.

Data may also be shared between agencies with similar or distinct objectives. In the case of the river Rhine, early warning monitoring for contamination is carried out in the upstream part of the river by the authorities in Switzerland. The information is passed on to downstream countries (Germany and France), who contribute to the cost of the exercise.

## 3.4 Preliminary surveys

When a new programme is being started, or a lapsed programme is being reinstated, it is useful to begin with a small-scale pilot project. This provides an opportunity for newly trained staff to gain hands-on experience and to confirm whether components of the programme can be implemented as planned. It may also provide an opportunity to assess the sampling network and provide indications of whether more (or possibly fewer) samples are needed in order to gain knowledge of the water quality at various points throughout a water body.

During the pilot project or preliminary survey it is important to test assumptions about the mixing of lakes, reservoirs and rivers at the selected sampling sites and times. It might be appropriate, therefore, to consider variations in water quality through the width and depth of a river at selected sampling sites throughout an annual cycle in order to confirm the number of samples required to produce representative data (see section 3.7.1).

In a lake or reservoir, it may be necessary to sample at different points to determine whether water quality can be estimated at a single point or whether the lake or reservoir behaves as a number of separate water bodies with different water quality characteristics. It is also essential to investigate variation in water quality with depth and especially during stratification. Lakes and reservoirs are generally well-mixed at overturn (i.e. when stratification breaks down) and sampling from a single depth or the preparation of a composite sample from two depths may adequately represent the overall water quality (see section 3.7.2).

For groundwaters, it is important during preliminary surveys to confirm whether or not the well casing is perforated, allowing access to more than one aquifer. If this is the case then an alternative site should be sought or measures taken to sample from a single aquifer only. The latter is generally problematic.

Preliminary surveys also help to refine the logistical aspects of monitoring. For example, access to sampling stations is tested and can indicate whether refinements are necessary to the site selection. Sampling sites could also be found to be impractical for a variety of reasons, such as transport difficulties. Similarly, operational approaches may be tested during the pilot project and aspects such as the means of transport, on-site testing techniques or sample preservation and transport methods, can be evaluated. Sample volume requirements and preservation methods can then also be refined.

Preliminary surveys also provide opportunities for training staff and for ensuring that staff are involved in the planning process. Such involvement, together with the undertaking of preliminary surveys, may often avoid major problems and inefficiencies which might otherwise arise.

## 3.5 Description of the monitoring area

The description of a monitoring area should consider as a minimum:

- definition of the extent of the area,
- a summary of the environmental conditions and processes (including human activities) that may affect water quality,
- meteorological and hydrological information,
- a description of the water bodies, and
- a summary of actual and potential uses of water.

A monitoring programme commonly covers the water-course system of a catchment area (i.e. a main river and all its tributaries, streams, brooks, ditches, canals, etc., as well as any lakes or ponds that discharge into the river or tributaries). The catchment area is defined as the area from which all water flows to the water-course. The land surface that slopes in such a way that precipitation falling on it flows towards the water-course is called the topographic catchment area. In some cases groundwater enters the water-course system from a groundwater catchment area, all or part of which may lie outside the topographic catchment. Topographic and groundwater catchment areas are rarely coincident.

Since a water-course system may be very large, it is often convenient to divide the catchment into several small sub-catchments. A catchment area, and its associated water-course, is hydrologically and ecologically discrete and, therefore, constitutes a logical unit for the planning and management of water use and for the monitoring of water quality. The dynamics of upstream water quality and sources of pollution can be related to downstream effects. A description of the catchment area includes its size (in $km^2$), its geographical location and the identification of each water body in the water-course system.

### 3.5.1 Conditions and processes

Environmental conditions in the catchment area should be described as fully as possible, because these have an effect on water quality and it may be useful to refer to this material when data are being evaluated. The natural processes that affect water quality have been described in Chapter 2. Reference should be made to these processes when the description of environmental conditions is prepared. In particular, rock composition, vegetation (both terrestrial and aquatic), wildlife, land form, climate, distance from salt-water bodies, and human activities in the catchment area should be reported. Human activities should be described mainly in terms of population and land use. The presence of cities and towns and the use of land for industry, agriculture, forestry or recreation are of interest.

### 3.5.2 Meteorological and hydrological information

Rainfall and ensuing run-off are of vital importance, especially when the programme includes the monitoring of fluxes or suspended loads of eroded materials. Some data interpretation techniques also require reliable hydrological information. If there is a gauging station or hydroelectric power plant near a sampling location, reliable data on river flow should be available. If they are not, estimates of flow can be based on data from the closest stream gauging station. As a last resort, it may be possible to estimate run-off from a calculation involving precipitation on the catchment area, the surface area of the catchment, a run-off factor and a time-of-flow factor. Measured values, however, are always much better than estimates. Where estimates have to be made, it is appropriate to request the assistance of experts in the hydrological services. Further information on hydrological measurements and calculation of flux is available in Chapter 12.

### 3.5.3 Water bodies

Reservoirs and lakes in the catchment should be described by their area, depth and, if possible, volume (calculated from bathymetric maps). The theoretical retention or residence time of a lake (see section 2.1.1) can vary from months for some shallow lakes to several decades (or more) for the largest lakes. In some cases rivers discharge into continental water bodies without outlets, such as the Dead Sea and the Aral Sea. These water bodies act as terminal recipients for surface waters. The characteristics of lakes frequently result in incomplete mixing (see section 2.1.2) and actual residence times vary widely because lake waters are, as a consequence, rarely homogeneous.

Impounded rivers and lakes are subject to a wide range of influences. The composition of the water in a lake or reservoir is influenced by the water budget, i.e. by the balances between inputs and outputs, by interchanges between the water column and sediments, and by the build-up of organic matter as a result of biological activity. However, the major inputs are usually from tributary rivers and streams that may carry a range of materials of both natural and artificial origin. There may also be point discharges directly into the lake, e.g. sewage and industrial wastes, and diffuse discharges from agricultural drainage, underground sources and rainfall. The major outputs

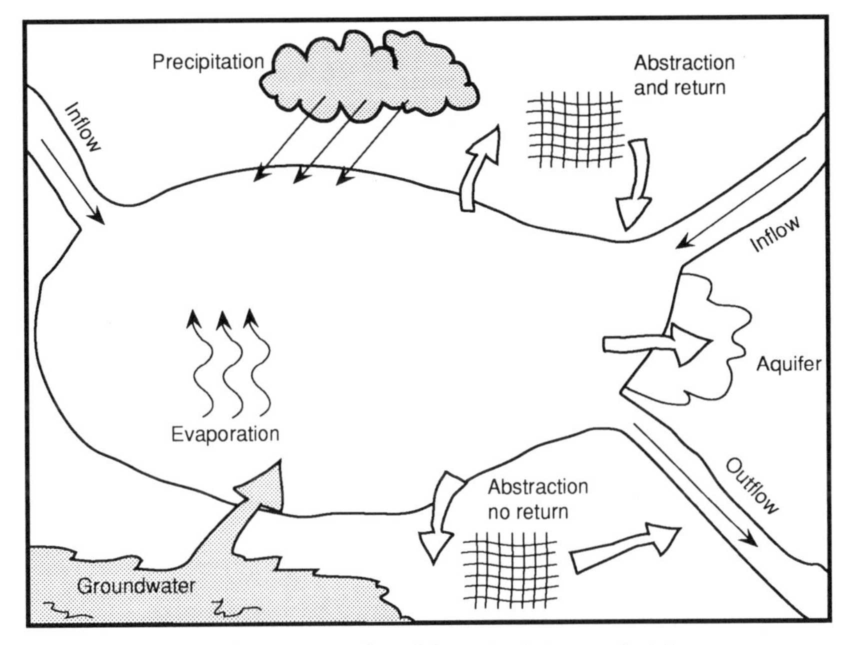


**Figure 3.1** Schematic representation of the water balance of a lake

from lakes are rivers and direct abstraction for public, agricultural and industrial use. Other outputs include evaporation and loss to groundwater. The water balance of a lake is shown schematically in Figure 3.1.

Rivers and streams should be described by their dimensions and their flows. A hydrograph, if available, provides particularly useful information. A map showing the catchment area of the river and the course that it follows is also useful. Features such as waterfalls, rapids, narrow and broad sections should be shown on the map and any unusual features described in the report. The location of man-made structures, e.g. bridges, dams, wharves, levees, gauging stations and roads leading to the river, should be shown on the map and described in the text of the report. The location of roads and trails through the monitoring area should also be shown, together with information on whether they are usable in all seasons of the year.

The description of aquifers will include extent, rock types, whether confined or unconfined, charts of water levels or piezometric head, main recharge and discharge areas and surface water-courses in hydraulic continuity with the aquifer (Figure 3.2).

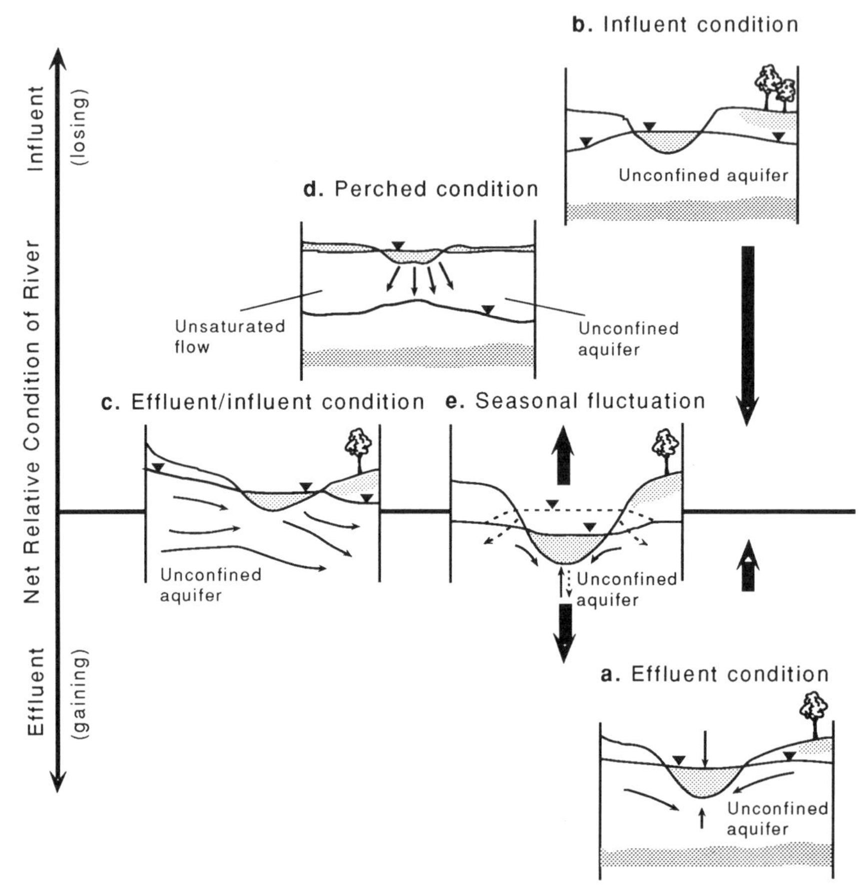


**Figure 3.2** Interactions between surface water and groundwater (After Foster and Hirata, 1988)

### 3.5.4 Water uses

Present uses of water should also be identified in the description of the monitoring area and the quantities withdrawn should be determined and listed. Many users contaminate the water during its use and return it to the stream or lake in a worse condition than when it was withdrawn. A possible exception to this is hydroelectric power generation where some aspects of water quality may be completely changed, for example by reduction of suspended solids and the increase of dissolved oxygen. Some of the common

**Table 3.1** Links between types of monitoring site and programme objectives

| Type of site | Location | Objectives |
|---|---|---|
| Baseline site | Headwater lakes or undisturbed upstream river stretches | To establish natural water quality conditions<br>To provide a basis for comparison with stations having significant direct human impact (as represented by trend and global flux stations)<br>To test for the influence of long-range transport of contaminants and the effects of climatic change |
| Trend site | Major river basins, large lakes or major aquifers | To test for long-term changes in water quality<br>To provide a basis for statistical identification of the possible causes of measured conditions or identified trends |
| Global river flux site | Mouth of a major river | To determine fluxes of critical pollutants from river basin to ocean or regional sea<br>Some trend stations on rivers also serve as global flux stations |

Source: Based on the GEMS/WATER monitoring programme

uses of water were listed in Table 2.3, together with their classification as consuming and/or contaminating. Any water uses that are planned but not implemented should also be included in the report, and the list may be supplemented by an identification of point and non-point sources of pollution.

## 3.6 Selecting sampling sites

Processes affecting water quality and their influence should be taken into account when sampling sites are selected. A sampling site is the general area of a water body from which samples are to be taken and is sometimes called a "macrolocation". The exact place at which the sample is taken is commonly referred to as a sampling station or, sometimes, a "microlocation". Selection of sampling sites requires consideration of the monitoring objectives and some knowledge of the geography of the water-course system, as well as of the uses of the water and of any discharges of wastes into it (Table 3.1). Sampling sites can be marked on a map or an aerial photograph, but a final decision on the precise location of a sampling station can be made only after a field investigation.

The following examples illustrate how sampling sites may be chosen. The choices are made with respect to the hypothetical catchment area in Figure 3.3 and are based on some of the monitoring objectives listed in section 3.3.

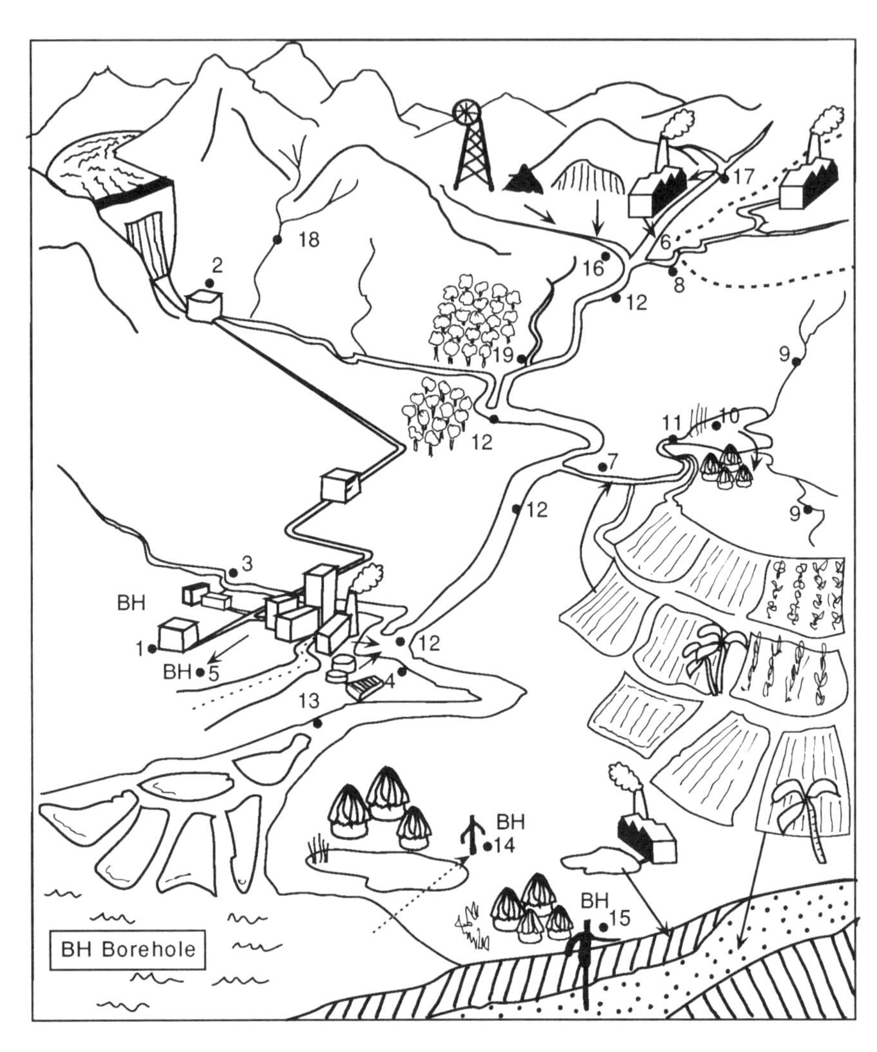


**Figure 3.3** Typical sampling sites for multi-purpose monitoring. The numbers are the sampling sites quoted in Examples 1–6 (see section 3.6)

**Example 1**

| | |
|---|---|
| Objective | Identification of baseline conditions in the water-course system. |
| Key question | What are the background levels of variables in the water? |
| Monitoring | Analyse for dissolved oxygen, major ions and nutrients.<br>Characterise seasonal or annual concentration patterns.<br>Determine annual mean values of water quality variables. |
| Information expectations | A description of the quality of water in the water-course system before it is affected by human activities. |

| | |
|---|---|
| Sampling network required | Fixed. |
| Sampling sites | 1, 9, 10, 11, 17, 18. |

**Example 2**

| | |
|---|---|
| Objective | Detection of any signs of deterioration in water quality. |
| Key question | Are any new water quality problems arising? |
| Monitoring | Detect any short-term trends.<br>Detect any long-term trends. |
| Information expectations | Description and interpretation of water quality changes with respect to time. |
| Sampling network required | Fixed. |
| Sampling sites | 7, 12, 13, 14, 15, 19. |

**Example 3**

| | |
|---|---|
| Objective | Identification of any water bodies in the water-course system that do not meet the desired water quality standards. |
| Key questions | What are the designated uses?<br>What are the water quality standards for the designated uses? |
| Monitoring | Survey for presence of contaminants.<br>Determine extreme values of contaminating variable(s). |
| Information expectations | Description of the contaminating variable(s), the extreme values measured, when and where they occur and how they conflict with the standards.<br>Evaluation of the hazards to human health, damage to the environment and any other adverse effects.<br>Interpretation of the general state of the water bodies relative to the established standards. |
| Sampling network required | Fixed. |
| Sampling sites | 4, 5, 6, 7, 8, 15, 16, 19. |

**Example 4**

| | |
|---|---|
| Objectives | Identification of any contaminated areas. |
| Key questions | What are the indicators of contamination?<br>What are the contaminating substances?<br>Where are the possible sources? |
| Monitoring | Detect contamination.<br>Determine the concentration of the variables that indicate contamination.<br>Determine the areas affected by the contaminants and whether these areas are increasing.<br>Detect the main pollution source(s). |
| Information expectations | Maps or charts showing the distribution of contaminants.<br>Classification charts, maps and/or lists showing the contaminated areas in order of severity in relation to the source(s) of pollution.<br>Classification is usually made in three or five classes. |

| | |
|---|---|
| Sampling network required | Recurrent (short-term). |
| Sampling sites | 4, 5, 6, 7, 8, 16, 19. |

**Example 5**

| | |
|---|---|
| Objective | Determination of the extent and effects of specific waste discharges. |
| Key questions | How far away from the point of discharge does the effluent affect the receiving water?<br>What changes are being made to the ambient water quality?<br>How does the effluent affect the aquatic ecosystem? |
| Monitoring | Determine the spatial distribution of pollutants.<br>Measure the effects of pollutants on aquatic life and various water uses. |
| Information expectations | The extent to which water quality has changed compared with baseline conditions.<br>Cause-and-effect relationships at different levels of the water ecosystem, e.g. basic nutritional requirements of fishes and fish population.<br>Maps showing the distribution of pollution.<br>Comparison with water quality guidelines. |
| Sampling network required | Recurrent (short-term). |
| Sampling sites | 4, 6, 8, 12, plus others as necessary. |

**Example 6**

| | |
|---|---|
| Objective | Evaluation of the effectiveness of a water quality management intervention. |
| Key question | What are the positive and negative results of a particular water quality management intervention in terms of its effects on the concentration of polluting substances in the water? |
| Monitoring | Determine whether the particular water quality management intervention has resulted in significant changes in water quality.<br>Compare mean values of contaminant concentration before and after the intervention.<br>Detect short- and long-term trends. |
| Information expectations | Evidence that mean concentration of contaminants was higher before the management intervention.<br>Contaminant concentrations are decreasing with time. |
| Sampling network required | Fixed and recurrent (short-term). |
| Sampling sites | 4, 5, 6, 7, 14, 15, 16, 19 (for example). |

## 3.7 Selecting sampling stations

### 3.7.1 Rivers

Sampling stations on rivers should, as a general rule, be established at places where the water is sufficiently well mixed for only a single sample to be required. The lateral and vertical mixing of a wastewater effluent or a tributary stream with the main river can be rather slow, particularly if the flow in

**Table 3.2** Estimated distance for complete mixing in streams and rivers

| Average width (m) | Mean depth (m) | Estimated distance for complete mixing (km) |
|---|---|---|
| 5 | 1 | 0.08–0.7 |
| | 2 | 0.05–0.3 |
| | 3 | 0.03–0.2 |
| 10 | 1 | 0.3–2.7 |
| | 2 | 0.2–1.4 |
| | 3 | 0.1–0.9 |
| | 4 | 0.08–0.7 |
| | 5 | 0.07–0.5 |
| 20 | 1 | 1.3–11.0 |
| | 3 | 0.4–4.0 |
| | 5 | 0.3–2.0 |
| | 7 | 0.2–1.5 |
| 50 | 1 | 8.0–70.0 |
| | 3 | 3.0–20.0 |
| | 5 | 2.0–14.0 |
| | 10 | 0.8–7.0 |
| | 20 | 0.4–3.0 |

the river is laminar and the waters are at different temperatures. Complete mixing of tributary and main stream waters may not take place for a considerable distance, sometimes many kilometres, downstream of the confluence. The zone of complete mixing may be estimated from the values in Table 3.2. However, if there is any doubt, the extent of mixing should be checked by measurements of temperature or some other characteristic variable at several points across the width of the river. If there are rapids or waterfalls in the river, the mixing will be speeded up and representative samples may be obtained downstream. Sampling for the determination of dissolved oxygen, however, should take place upstream of the rapids or waterfall because the turbulence will cause the water to be saturated with oxygen. In such a case, several samples should be taken across the width of the river to allow for the possibility of incomplete mixing.

A bridge is an excellent place at which to establish a sampling station (provided that it is located at a sampling site on the river). It is easily accessible and clearly identifiable, and the station can be precisely described. Furthermore, a bridge is often a hydrological gauging station and, if so, one of the bridge piers will have a depth gauge marked on it, thus allowing the collection of stream flow information at the time of sampling (see

**Table 3.3** Suggested sampling regimes for composite samples in flowing waters

| Average discharge ($m^3 s^{-1}$) | Type of stream or river | Number of sampling points | Number of sampling depths |
|---|---|---|---|
| < 5 | Small stream | 2 | 1 |
| 5–140 | Stream | 4 | 2 |
| 150–1,000 | River | 6 | 3 |
| ≥ 1,000 | Large river | ≥ 6 | 4 |

Chapter 12). Usually, a sample taken from a bridge at mid-stream or in mid-channel, in a well mixed river, will adequately represent all of the water in the river.

To verify that there is complete mixing at a sampling station it is necessary to take several samples at points across the width and depth of the river and to analyse them. If the results do not vary significantly one from the other, a station can be established at mid-stream or some other convenient point. If the results are significantly different it will be necessary to obtain a composite sample by combining samples taken at several points in the cross-section of the stream. Generally, the more points that are sampled, the more representative the composite sample will be. Sampling at three to five points is usually sufficient and fewer points are needed for narrow and shallow streams. Suggestions are provided in Table 3.3 for the number of points from which samples should be obtained in streams or rivers of different sizes and with different flow rates.

### 3.7.2 Lakes and reservoirs

Lakes and reservoirs can be subject to several influences that cause water quality to vary from place to place and from time to time. It is, therefore, prudent to conduct preliminary investigations to ensure that sampling stations are truly representative of the water body. Where feeder streams or effluents enter lakes or reservoirs there may be local areas where the incoming water is concentrated, because it has not yet mixed with the main water body. Isolated bays and narrow inlets of lakes are frequently poorly mixed and may contain water of a different quality from that of the rest of the lake. Wind action and the shape of a lake may lead to a lack of homogeneity; for example when wind along a long, narrow lake causes a concentration of algae at one end.

If there is good horizontal mixing, a single station near the centre or at the deepest part of the lake will normally be sufficient for the monitoring of long-term trends. However, if the lake is large, has many narrow bays or contains several deep basins, more than one station will be needed. To allow for the size of a lake, it is suggested that the number of sampling stations should be the nearest whole number to the $\log_{10}$ of the area of the lake in $km^2$. Thus a lake of 10 $km^2$ requires one sampling station, 100 $km^2$ requires two stations, and so on. For lakes with irregular boundaries, it is advisable to conduct preliminary investigations to determine whether and where differences in water quality occur before deciding on the number of stations to establish.

Access to lake and reservoir sampling stations is usually by boat and returning to precisely the same locations for subsequent samples can be extremely difficult. The task is made simpler when the locations can be described in relation to local, easily identified landmarks. This, as well as the representativeness of samples taken at those locations, should be borne in mind when sampling stations are chosen.

The most important feature of water in lakes and reservoirs, especially in temperate zones, is vertical stratification (the special features of tropical lakes are described in Chapter 2), which results in differences in water quality at different depths. Stratification at a sampling station can be detected by taking a temperature reading at 1 m below the surface and another at 1 m above the bottom. If there is a significant difference (for example, more than 3 °C) between the surface and the bottom readings, there is a "thermocline" (a layer where the temperature changes rapidly with depth) and the lake or reservoir is stratified and it is likely that there will be important differences in some water quality variables above and below the thermocline. Consequently, in stratified lakes more than one sample is necessary to describe water quality.

For lakes or reservoirs of 10 m depth or more it is essential, therefore, that the position of the thermocline is first investigated by means of regularly-spaced temperature readings through the water column (e.g. metre intervals). Samples for water quality analysis should then be taken according to the position and extent (in depth) of the thermocline. As a general guide, the minimum samples should consist of:

- 1 m below the water surface,
- just above the determined depth of the thermocline,
- just below the determined depth of the thermocline, and
- 1 m above the bottom sediment (or closer if this can be achieved without disturbing the sediment).

If the thermocline extends through several metres depth, additional samples are necessary from within the thermocline in order to characterise fully the water quality variations with depth. Whilst the position of the thermocline is stable, the water quality for a given station may be monitored by fewer samples but, in practice, the position of a thermocline can vary in the short- (hours) or long-term (days) due to internal seiches (periodic oscillations of water mass) and mixing effects.

Even in warm climates, and in relatively shallow lakes and reservoirs, the possibility of a thermocline should be investigated by taking temperature profiles from the surface to the bottom. The Smir reservoir in northern Morocco, for example, although only 8 m deep had a temperature difference of 4 °C between the surface and the bottom in August 1991. A thermocline was detected between 5 m and 7 m below the surface and significant differences were reported for measured values of both pH and dissolved oxygen above and below the thermocline. In general, therefore, in tropical climates where the water depth at the sampling site is less than 10 m, the minimum sampling programme should consist of a sample taken 1 m below the water surface and another sample taken at 1 m above the bottom sediment.

### 3.7.3 Groundwater

Sampling points for groundwater monitoring are confined to places where there is access to an aquifer, and in most cases this means that samples will be obtained from existing wells. To describe such a sampling station adequately, it is essential to have certain information about the well, including depth, depth to the well screen, length of the screen and the amount by which the static water level is lowered when the well is pumped. One sample is usually sufficient to describe the water quality of the aquifer. If the water in the well is corrosive and is in contact with steel pipe or casing, the water samples may contain dissolved iron. Wells with broken or damaged casings should be avoided because surface water may leak into them and affect the water quality.

Springs can also be useful groundwater sampling points, provided that they are adequately protected against the ingress of contamination with surface water. Springs are often fed from shallow aquifers and may be subject to quality changes after heavy rainfall.

Other possible sampling stations are boreholes drilled especially to investigate the features of an aquifer, although these are expensive and would be justified only in particular circumstances.

## 3.8 Monitoring media and variables

There are three principal media which can be used for aquatic monitoring: water, particulate matter and living organisms. The quality of water and particulate matter is determined by physical and chemical analysis whereas living organisms can be used in a number of different ways (see Chapter 11). Whole communities of organisms, or the individuals that normally belong to that community, can be studied as indicators of water quality. Alternatively, the physiology, morphology or behaviour of specific organisms can be used to assess the toxicity or stress caused by adverse water quality conditions, or organisms and their tissues can be used as a medium for chemical monitoring of contaminants in the aquatic environment.

The most common biological measurement of surface water samples is the determination of phytoplankton chlorophyll pigments, which give an indication of algal biomass. Chlorophyll measurements are useful for assessing eutrophication in lakes, reservoirs and large rivers, or as an indication of productivity in most surface waters.

Each aquatic medium has its own set of characteristics for monitoring purposes, such as:

- applicability to water bodies,
- inter-comparability,
- specificity to given pollutants,
- possibility of quantification (e.g. fluxes and rates),
- sensitivity to pollution with the possibility of amplifying the pollution signal by several orders of magnitude,
- sensitivity to sample contamination,
- time-integration of the information received from instantaneous (for point sampling of the water) to integrated (for biotic indices),
- required level of field personnel, storage facility of samples, and
- length of the water quality determination process from field operation to result.

The most significant characteristics of each medium are given tentatively in Table 3.4. Water, itself, is by far the most common monitoring medium used to date, and the only one directly relevant to groundwaters. Particulate matter is widely used in lake studies, in trend monitoring and in river flux studies, whereas biological indices based on ecological methods are used more and more for long-term river and lake assessments.

Water quality can be described by a single variable or by any combination of more than 100 variables. For most purposes, however, water quality can be adequately described by fewer than 20 physical, chemical and biological characteristics. The variables chosen in a monitoring programme will depend on the programme objectives and on both existing and anticipated uses of the water. Drinking and domestic consumption, agricultural irrigation, livestock watering, industrial uses and recreational use all require water of a specific quality. Selecting the variables to include in a monitoring programme will often require a compromise between “like to know” and “need to know”. However, some variables must be measured if the basic programme objectives are to be achieved. Further information on the sources, range of

**Table 3.4** Principal characteristics of media used for water quality assessments

| Characteristics | Water | Particulate matter: Suspended | Particulate matter: Deposited | Living organisms: Tissue analyses | Living organisms: Biotests | Living organisms: Ecological surveys[1] | Living organisms: Physiol. determin.[2] |
|---|---|---|---|---|---|---|---|
| Type of analysis or observation | ←physical→ (Water to Deposited)<br>←chemical→[3] (Water to Biotests) | | | ←biological→ (Tissue analyses to Physiol. determin.) | | | |
| Applicability to water bodies | rivers, lakes, groundwater | mostly rivers | lakes, rivers | rivers, lakes | rivers, lakes | rivers, lakes | rivers, lakes |
| Intercomparability[4] | ←global→ | | | depends on species occurrence | global | ←local to regional→ | |
| Specificity to given pollutant | ←specific→ | | | | ←integrative→ | | |
| Quantification | ←complete→ quantification of concs & loads | | concen-trations only | quanti-tative | semi-quan-titative | ←relative→ | |
| Sensitivity to low levels of pollution | low | ←high→ | | | variable | medium | variable |
| Sample contam-ination risk | high[5] | medium | ←low→ | | medium | ←low→ | |
| Temporal span of information obtained | instant | short | long to very long (continuous record) | medium (1 month) to long (> 1 year) | instant. to cont-inous[6] | ←medium to long→ | |
| Levels of field operators | untrained to highly trained[7] | trained | untrained to trained | trained | ←medium to highly trained→ | | |
| Permissible sample storage duration[7] | low | high | high | high | very low | high | na |
| Minimum duration of determination | instant. (*in situ* determ.) to days | days | days to weeks | days | days to months | weeks to months | days to weeks |

na Not applicable

1 Including algal biomass estimates
2 Histologic, enzymatic, etc.
3 Including BOD determination
4 Most biological determinations depend on the natural occurrence of given species and are, therefore, specific to a given geographic region. Chemical and physical descriptors are globally representative
5 For dissolved micropollutants
6 e.g. organisms continuously exposed to water
7 Depending on water quality descriptors

Source: Meybeck *et al.*, 1996

concentrations, behaviour and transformations of water quality variables is available in the companion guidebook *Water Quality Assessments*.

The simplest combination of variables is temperature, electrical conductivity, pH, dissolved oxygen (DO) and total suspended solids (TSS). These give the bare minimum of information on which a crude assessment of overall water quality can be based. More complex programmes may analyse up to 100 variables, including a range of metals and organic micro-pollutants. Moreover, analysis of biota (plankton, benthic animals, fishes and other organisms) and of particulate matter (suspended particulates and sediments) can add valuable information.

The water quality variables for baseline stations, trend stations and global river flux stations included in the basic monitoring component of the Global Environment Monitoring System's GEMS/WATER programme are listed in Table 3.5.

In many monitoring programmes a known, or suspected, water pollution condition is the reason for deciding to analyse for specific water quality variables. Another possible reason is to determine whether a water is suitable for a planned use. Some variables may be part of a basic monitoring programme, while others will be specific to the pollution condition or planned use. Some pollution conditions and planned uses, together with the variable that should be measured, are as follows:

- *Organic wastes* are contained in municipal sewage and in the discharges from abattoirs, food-processing and similar agricultural industries. The variables that should be measured are biochemical oxygen demand (BOD), chemical oxygen demand (COD), total organic carbon (TOC), dissolved organic nitrogen, total phosphorus and faecal coliforms.
- *Eutrophication* results from nutrients entering surface water, either from a point discharge or in run-off from agricultural land. The variables that should be measured are nitrate, nitrite, ammonia, total phosphorus (filtered and unfiltered), reactive silica, transparency and chlorophyll *a*.
- *Agriculture and irrigation* can result in excessive concentrations of certain elements and compounds entering water bodies (particularly nitrates and phosphates from fertilisers, as well as pesticides and herbicides). In addition, high concentrations of some elements can cause problems for the use of water in agriculture (e.g. soil permeability may be altered, certain plants can be adversely affected and livestock can be poisoned). Agricultural practices can also influence erosion. The variables that should be measured for specific agricultural uses of water include total dissolved solids, total suspended solids, boron, selenium, sodium, calcium, magnesium and

**Table 3.5** Variables used in the GEMS/WATER programme for basic monitoring

| Measured variable | Streams: baseline and trend | Headwater lakes: baseline and trend | Groundwaters: trend only | Global river flux stations |
|---|---|---|---|---|
| Water discharge or level | x | x | x | x |
| Total suspended solids | x | – | – | x |
| Transparency | – | x | – | – |
| Temperature | x | x | x | x |
| pH | x | x | x | x |
| Electrical conductivity | x | x | x | x |
| Dissolved oxygen | x | x | x | x |
| Calcium | x | x | x | x |
| Magnesium | x | x | x | x |
| Sodium | x | x | x | x |
| Potassium | x | x | x | x |
| Chloride | x | x | x | x |
| Sulphate | x | x | x | x |
| Alkalinity | x | x | x | x |
| Nitrate | x | x | x | x |
| Nitrite | x | x | x | x |
| Ammonia | x | x | x | x |
| Total phosphorus (unfiltered) | x | x | – | x |
| Phosphorus, dissolved | x | x | – | x |
| Silica, reactive | x | x | – | x |
| Chlorophyll *a* | x | x | – | x |
| Fluoride | – | – | x | – |
| Faecal coliforms (trend stations only) | x | x | x | – |

Source: WHO, 1991

faecal coliforms (depending on the intended agricultural application). For irrigation water it is important to measure the sodium adsorption ratio (SAR) as follows:

$$SAR = \frac{Na^+}{\sqrt{\left(Ca^{2+} + Mg^{2+}\right)/2}}$$

Pesticides and herbicides used in agriculture frequently lead to the contamination of both surface water and groundwater. Monitoring programmes should take into account available information concerning the types and quantities of agricultural chemicals used in the monitoring area, their characteristics and their application patterns. Some of the variables that might have to be measured are dieldrin, aldrin, sum of DDTs, atrazine, lindane ($\gamma$-hexachlorocyclohexane or HCH), aldicarb, organophosphorus pesticides and 2,4-dichlorophenoxyacetic acid (2,4-D).

- *Industrial effluents* may contain toxic chemicals, organic or inorganic or both, depending on the industrial process. Some knowledge of industrial processes is, therefore, necessary before a rational decision can be made on the variables for which analyses should be made. Examples of the water quality variables that should be measured in industrial waters are total solids, BOD, COD, trihalomethanes, polynuclear aromatic hydrocarbons, total hydrocarbons, phenols, polychlorinated biphenyls, benzene, cyanide, arsenic, cadmium, chromium, copper, lead, iron, manganese, mercury, nickel, selenium and zinc.
- *Effluents and leachates from mining operations* affect surface water and groundwater, often very severely. The minerals being mined provide an indication of the metals for which analyses should be made. Often there will be other minerals present in non-commercial concentrations and these may be present in the mine wastes. Analysis for both the dissolved and the particulate fractions of metals should be included in the monitoring programme.
- *Acidification* of lakes, rivers and groundwater results from the long-range transport of atmospheric pollutants. Drainage water from coal mines is strongly acidic and often leads to acidification of water bodies. Water that has become acidified should be analysed for the dissolved fraction of metals such as aluminium, cadmium, copper, iron, manganese and zinc, as well as for pH and alkalinity.

## 3.9 Frequency and timing of sampling

Sampling frequency at stations where water quality varies considerably should be higher than at stations where quality remains relatively constant. A new programme, however, with no advance information on quality variation, should be preceded by a preliminary survey (see section 3.4) and then begin with a fixed sampling schedule that can be revised when the need becomes apparent.

The time interval between the collection of samples depends on the water body and its specific characteristics. An interval of one month between the collection of individual samples at a station is generally acceptable for characterising water quality over a long time period (e.g. over a year in a river) (see Table 3.6), whereas for control purposes weekly sampling may be necessary. If significant differences are suspected or detected, samples may have to be collected daily or on a continuous basis. In extreme cases, time-integrated, composite samples may have to be made up by mixing equal portions of samples taken at regular intervals over a 24-hour period, but this

**Table 3.6** Sampling frequency for GEMS/WATER stations

| Water body | Sampling frequency |
|---|---|
| *Baseline stations* | |
| Streams | Minimum: 4 per year, including high- and low-water stages<br>Optimum: 24 per year (every second week); weekly for total suspended solids |
| Headwater lakes | Minimum: 1 per year at turnover; sampling at lake outlet<br>Optimum: 1 per year at turnover, plus 1 vertical profile at end of stratification season |
| *Trend stations* | |
| Rivers | Minimum: 12 per year for large drainage areas, approximately 100,000 $km^2$<br>Maximum: 24 per year for small drainage areas, approximately 10,000 $km^2$ |
| Lakes/reservoirs | For issues other than eutrophication:<br>Minimum: 1 per year at turnover<br>Maximum: 2 per year at turnover, 1 at maximum thermal stratification<br>For eutrophication:<br>12 per year, including twice monthly during the summer |
| Groundwaters | Minimum: 1 per year for large, stable aquifers<br>Maximum: 4 per year for small, alluvial aquifers<br>Karst aquifers: same as rivers |

should be done only if they conform with the requirements of the objectives and are not to be used for the determination of unstable variables, such as dissolved oxygen. Individual samples taken at a given station should be obtained at approximately the same time of day if possible, because water quality often varies over the course of the day. However, if detection of daily quality variations or of the peak concentration of a contaminant in an effluent is of interest, sampling at regular intervals (e.g. every two or three hours throughout the day) will be necessary. Exceptional conditions of stream flow are frequently of interest because it is at maximum and minimum flow rates that extreme values of water quality are reached. For example, when flowing at its peak rate, a river usually carries its greatest load of suspended material, while pollutants will be the least diluted when a river is at minimum flow. The violation of a waste discharge regulation and the seriousness of its environmental effects will often be easier to detect during periods of minimum flow. Sampling regimes may need to take such factors into consideration (see section 12.3).

It is usual to take samples of groundwater at only one depth. Frequency of sampling is low for large, deep, confined aquifers, which typically have long residence times, but higher (perhaps monthly) for shallow, unconfined aquifers with short residence times. Sampling should be supplemented by occasional mapping to describe the aquifer fully.

Sample collection should be frequent enough to enable an accurate calculation of the mean concentrations of the variables included in the monitoring programme. The frequency of sampling required to obtain a desired level of confidence in the mean values depends on statistical measures, i.e. standard deviation and confidence interval. Sampling frequencies for GEMS/WATER baseline and trend stations are shown in Table 3.6.

## 3.10 Source literature and further reading

Chapman, D. [Ed.] 1996 *Water Quality Assessments: A Guide to the Use of Biota, Sediments and Water in Environmental Monitoring.* 2nd edition, Chapman & Hall, London.

Désilets, L. 1988 *Criteria for Basin Selection and Sampling Station Macrolocation.* Inland Waters Directorate, Water Quality Branch, Scientific Series No. 164. Environment Canada, Ottawa.

Foster, S.S.D. and Hirata, R. 1988 *Groundwater Pollution Risk Assessment: A Method Using Available Data.* Pan-American Centre for Sanitary Engineering and Environmental Science (CEPIS), Lima.

Meybeck, M., Kimstach, V. and Helmer, R. 1996 Strategies for water quality assessment. In: D. Chapman [Ed.] *Water Quality Assessments: A Guide to the Use of Biota, Sediments and Water in Environmental Monitoring.* 2nd edition, Chapman & Hall, London, 23–57.

WHO 1991 *GEMS/WATER 1991–2000: The Challenge Ahead.* Unpublished document WHO/PEP/91.2, World Health Organization, Geneva.

WHO 1992 *GEMS/WATER Operational Guide.* Third edition, Unpublished WHO document GEMS/W.92.1, World Health Organization, Geneva.

# Chapter 4

# RESOURCES FOR A MONITORING PROGRAMME

Implementing a monitoring programme requires access to resources, including an equipped laboratory, office space, equipment for field work, transport and trained personnel.

Two important points should be considered when starting a new monitoring programme:

- It is better to have a complete record of reliable data concerning water quality at a few sampling stations than to have a lot of data of questionable quality from many sampling stations.
- If reported data are not credible, the programme and its staff will lose credibility.

In the initial stages of a new monitoring programme, therefore, it is generally advisable to proceed as follows:

- Start slowly with analyses for a few variables.
- Train staff to ensure that proper procedures are followed.
- Impose quality assurance on all procedures from the beginning.
- Take samples at stations where the water quality is of major relevance to the monitoring programme.
- Prepare reports that are factual and are written so that they can be understood by persons other than scientists.
- Increase the number of variables, the number of sampling stations and the frequency of sampling as the capacities of the sampling and analysis teams increase.

## 4.1 Laboratory facilities

A number of options may be available for conducting analyses of water samples. The agency responsible for the monitoring programme may have its own laboratory or laboratories, the facilities of another agency or of a government ministry may be available, or some or all of the analytical work may be done under contract by a private laboratory. Some analytical work will inevitably be done in the field, using either field kits or a mobile laboratory. Regardless of the options chosen, the analytical services must be adequate for the volume of work expected. Some of the relevant considerations in this context are:

*This chapter was prepared by J. Bartram*

- *Variables to be analysed.* If only a few simple tests are required, analyses can be undertaken in the field using field kits. More complex testing programmes may require the services of specialised laboratories (see also section 3.8).
- *Sampling frequency and number of sampling stations.* The frequency with which samples must be taken and the number of sampling stations involved will obviously influence the volume of work necessary and, hence, the staff and facilities required (see also section 3.7).
- *Existing laboratory facilities.* Laboratory facilities may be under the direct control of the monitoring agency or be associated with another agency (e.g. the health ministry, a regional hospital, or a college or university). The major concern is that the laboratory is sufficiently close to the sampling stations to permit samples to be delivered without undue delay.
- *On-site testing.* Some analyses must be performed in the field (see Chapter 6). Modern field kits are available that permit analyses of a wide range of variables (see Appendix). This makes it possible to run a monitoring programme without the need for a fixed laboratory, but raises certain problems of analytical quality control.
- *Temporary laboratories.* If a monitoring programme is expected to be of short duration, it may be expedient to set up a temporary laboratory. Sufficient space, water and electricity supplies are essential, but equipment and supplies can be brought in and then removed after the monitoring programme is completed.
- *Mobile laboratories.* It is possible to set up a laboratory in a suitable motor vehicle, e.g. truck or van. In effect, this is a variant of on-site testing, but may provide better facilities than field kits.

In practice, the usual arrangement is for the agency responsible for water quality monitoring to establish its own central laboratory, which can be organised to provide training and supervision of staff, repair of equipment and various other services. However, if the monitoring area is large or transportation is difficult, regional laboratories may be set up or field kits used for certain analyses. Analyses that require expensive and sophisticated equipment, or that can be undertaken only by highly trained personnel, are performed only at the central laboratory, for example analysis for heavy metals using atomic absorption spectrophotometry and analysis for pesticides and herbicides using gas chromatography.

Whatever arrangements are finally chosen, it is essential that procedures are established for quality control of analytical work (see Chapter 9). This is of particular importance in relation to all aspects of field work, including sampling, sample handling and transport, as well as on-site testing.

It is beyond the scope of this handbook to make recommendations for the size, equipment and staffing of a laboratory. However, Chapters 6, 7, 8 and 10 each contain lists of instruments, equipment, glassware, reagents and other supplies necessary for various analytical procedures.

## 4.2 Transport

The types of vehicle needed for sampling expeditions depend, to a large extent, on the ease of access to the various sampling stations. If access is difficult, an "off-road" vehicle with four-wheel drive may be necessary and in some remote or rural areas a light motorcycle can be useful for transporting one person (with a minimum of equipment, e.g. portable kit), although it is important to consider the safety aspects of the latter arrangement. For some sampling programmes in large lakes and rivers, sampling stations may have to be reached by boat. It may be possible to rent a boat in the vicinity of a sampling station. Renting is generally preferable to the purchase of a boat unless there are numerous sampling stations that have to be reached regularly. If a boat must be transported by the expedition, the work of the sampling team will be slowed by the need to launch and recover it at each sampling station.

In countries or regions where reliable public transport is available, it may be possible to arrange for samples to be transported to the laboratory by bus or train. Local agents, appropriately trained and supplied with the necessary sampling equipment, can take samples at prescribed times and send them to the laboratory. This system requires careful supervision and particular attention to sample quality control.

## 4.3 Staffing

Staff on a water quality monitoring programme fall into four broad categories: programme management, field staff, laboratory staff and data processors. The numbers required in each category will depend on the size and scope of the monitoring programme.

### 4.3.1 Programme manager

The programme manager will probably require the assistance of several technical and administrative staff members during the design and planning phases of the programme. After the programme has reached the implementation stage, some of these staff members can be transferred to operations, possibly as supervisors of field and laboratory work. Others, together with the programme managers, will assume responsibility for data manipulation, preparation of reports, staff training and programme co-ordination (if other agencies are involved in the programme).

The co-ordination tasks may become quite complex if programme implementation depends on other agencies, part-time staff, temporary or rented facilities and public transport. The following description of the responsibilities of a programme manager may not be complete for any specific situation but is presented as an example of what may be expected of the manager.

The responsibilities of the programme manager will include some of the following:

- Planning of water quality monitoring activities.
- Co-ordination with regional centres, collaborating agencies, participating laboratories and others not under his or her direct control.
- Procurement of necessary equipment and consumable supplies.
- Arranging suitable transport.
- Recruitment of staff.
- Training of staff.
- Preparation of training manuals.
- Safety in the field and in the laboratory.
- Preparation of standard operating procedures (SOPs).
- Organising and managing central office facilities for the storage, handling, interpretation and distribution of data.
- Supervising and evaluating the performance of all staff.
- Reviewing and evaluating procedures.
- Preparation of reports and dissemination of the findings of the monitoring programme.

### 4.3.2 Field staff

Staff recruited for field work and sampling may not have had previous experience in water quality monitoring and new methods or procedures may be introduced from time to time. A short period of training is, therefore, appropriate. Assuming that candidates have a good general education, a well organised training session that includes practical field work will require 1–2 weeks. If field testing is also to be carried out, the training period will have to be somewhat longer. Staff should be evaluated after training and, if satisfactory, should work under fairly close supervision until they are sufficiently experienced to require only occasional supervision. Periodic short-term training sessions should be arranged for reviewing, reinforcing or extending knowledge.

Preliminary training for field staff may include, for example:

- Objectives of the water quality monitoring programme.
- The local, national and international importance of the programme.
- The importance of samples being of good quality and representative of the water body from which they are taken.
- How to ensure that samples are of good quality.
- Planning of sampling expeditions.
- Map-reading.
- Making field notes.

- Maintaining up-to-date descriptions of sampling sites and stations.
- Health and safety in the field.

Training sessions for the continuing education of field staff might include the following topics:

- Sampling for sediment.
- Sampling for biological analysis.
- Stream gauging.
- Microbiological field testing of samples (membrane filter method).
- Chemical tests in the field.

The responsibilities of field staff should include all or some of the following:

- Undertaking sampling expeditions in accordance with a planned programme.
- Obtaining samples according to SOPs.
- Labelling sample bottles, making notes and recording unusual conditions at the sampling station.
- Preparing samples for transport and delivering them to the local laboratory in accordance with SOPs.
- Routine maintenance of equipment used in the field.
- Preparing sample bottles (cleaning and addition of appropriate preservatives).
- Performing field tests for selected variables (see Chapter 6).

### 4.3.3 Laboratory staff

Two, or possibly three, types of laboratory staff may be required to undertake the required chemical, microbiological and biological analyses.

*Laboratory chiefs*. The chiefs of each type of laboratory, in liaison with the programme manager, would typically be responsible for:

- Laboratory management.
- Determining and procuring the equipment and supplies that will be needed.
- Ensuring that SOPs are being followed, which includes supervising laboratory staff and training them in the use of new equipment and procedures.
- Quality control of analytical procedures (see Chapter 9).
- Enforcing safety precautions and procedures, especially for fire, explosions and noxious fumes.

*Laboratory technician (analyst)*. Laboratory technicians will usually have had formal training and possibly practical experience in analytical work. Working under the direction of the laboratory chief, a technician will be responsible for the preparation and carrying out of analytical work in the laboratory. Their duties would typically include the following:

- Care, regular maintenance and ensuring cleanliness of laboratory and equipment, especially refrigerators, freezers, incubators, water-baths, stills, ovens and work areas.
- Ensuring the cleanliness of laboratory glassware and other reusables.

- Safe storage of all equipment and glassware.
- Preparation of reagents and media, standardising as necessary.
- Storage under proper conditions of reagents and media.
- Checking accuracy of electronic equipment used in field analyses.
- Preparation (or supervising the preparation) of sample bottles (including the addition of preservatives when appropriate) and application of correct identification labels.
- Laboratory safety.
- Training of field staff in the use of field testing equipment.
- Training of junior laboratory staff.
- Recording the results of analyses.
- Performing the tests and analyses, including those necessary for internal quality control (see Chapter 9).

*Laboratory assistant.* Laboratory assistants may have had some formal training but are usually untrained and often learn whilst in post. Their duties are performed under the direct supervision of the laboratory technician(s). These duties typically include:

- Cleaning of the laboratory, glassware and equipment.
- Preparing sample bottles, including washing, addition of preservatives and correct labelling (and sterilising if necessary).
- Storage of laboratory and sample glassware.
- Storage and stock control of chemicals and media.

In time, and with appropriate training and experience, an assistant can be promoted to a higher level of laboratory work. The technicians will teach the assistants how to use various items of laboratory equipment, prepare reagent solutions and carry out certain analyses.

### 4.3.4 Quality assurance officer

Quality assurance is dealt with in detail in Chapter 9. This section is concerned only with the responsibilities of the quality assurance officer (QAO). It is considered good practice to designate a staff member as QAO, with duties which include:

- Reporting directly to the most senior manager of the organisation on matters concerning quality assurance.
- Monitoring the quality of analytical work in the laboratory and in the field.
- Auditing reports, laboratory notebooks, field notebooks and other laboratory documentation to ensure that information is correct and complete.

Where financial or organisational constraints do not allow the appointment of a specific QAO, the responsibility for quality assurance may be delegated to a member of staff, in addition to existing duties. Quality control of work performed by the QAO and that is not itself directly concerned with quality assurance should be made the responsibility of another member of staff. In this case, senior management should ensure that conflicts of interest are minimised.

### 4.4 Human resources development and training

The quality of the data produced by a water quality monitoring programme depends on the quality of the work done by field and laboratory staff. It is, therefore, important that staff are adequately trained for the work they are expected to do. As a result, monitoring agencies often develop training programmes that are specific to their needs. The content and extent of training programmes depend on the previous training and experience of staff, the range of activities involved in the monitoring programme and whether analytical work will be done at a central laboratory or regional laboratories, and the extent to which analyses will be performed in the field.

For large, permanent monitoring programmes, a comprehensive strategy for personnel development is advisable. This should include:

- clear lines of responsibility and accountability,
- job descriptions,
- recruitment guidelines (qualifications, experience, skills requirements, etc.),
- career structures,
- mechanisms for enhancing the motivation of staff at all levels,
- systems for staff appraisal and feedback, and often
- standardised training packages, procedures manuals and training manuals as appropriate to the work of field and laboratory staff, regional and national managers.

Training is not a once-only activity but should be a continuing process. Ideally, there should be a basic framework of courses for staff at all levels, followed by short courses, seminars and workshops. Supervision of work, in both the laboratory and the field, is essential and contributes to in-service training. It is particularly valuable in water quality monitoring programmes because it permits staff to gain "hands-on" experience, thus reinforcing what was learned in formal training sessions, while taking an active part in programme operations. Laboratory staff, especially those in larger laboratories, are progressively trained and authorised to use certain items of equipment or undertake certain analyses.

Training should be flexible, responding to experience and feedback and taking account of the specific needs of individual staff members. In-house training can provide this flexibility and can be readily tailored to local requirements but it needs staff who are familiar with the necessary training techniques, usually senior laboratory and field staff (it also makes heavy demands on them). It may also make significant demands on financial resources and requires access to classrooms, field-work sites and training laboratories with appropriate equipment. For much of their staff training, therefore, many agencies will make use of courses already available at local educational establishments, supplementing these with short courses, workshops and refresher training in specific topics.

In its broadest sense, training should also be understood to include encouraging staff to join appropriate professional organisations, attend conferences and symposia, and communicate with peers in technical schools, colleges, universities and similar establishments.

### 4.4.1 Training documents

A training record should be maintained for each staff member. This should contain a detailed account of all training completed, of supervision required, and of authorisations granted for certain types of work. It may, for example, note initial training, supervised work and authorisation (for field staff) to take samples using a particular type of sampler or (for laboratory staff) to use a certain item of equipment or perform a certain analysis. Since it is a permanent record it should be in the form of a bound book and each new entry should be signed by both the trainer (or supervisor) and the trainee.

A procedures manual should be prepared, containing full details of SOPs for all programme activities. A loose-leaf format is preferable because SOPs are subject to revision and up-dating. The use of SOPs is an important element of quality assurance, which is discussed in more detail in Chapter 9.

### 4.4.2 Retaining staff

It is not unusual, after staff members have been recruited and trained, to have some of them leave for employment elsewhere, especially where environmental monitoring is a government activity and salaries are relatively poor. Staff turnover may be reduced if a career structure is developed which provides opportunities for promotion with increased expertise and experience. The contributions of staff to the programme should be adequately acknowledged in reports and similar documentation, especially when a report is published in the technical literature.

## 4.5 Communication

Good communication is important, not only for achieving programme outputs (such as data production), but also to ensure that wider aims (such as increasing awareness of environmental issues and ensuring that staff of all types see their role positively) are met. It is also indirectly important as a means of ensuring continued outside interest in, and support for, the work being undertaken.

It is good practice to ensure that responsibilities for communication are identified in every individual's job description. Communication may be aimed at a wide audience, such as writing reports or speaking at a seminar, or more specific, such as communicating results from analyst to laboratory chief or discussion of fieldwork plans between co-ordinator and field staff. In practice, much general communication should take the form of consultation and, by generating goodwill, this helps to ensure that problems are overcome through mutual interest and do not become insurmountable blocks.

Communication with external agencies (known as ESAs or External Support Agencies when they provide support to a programme) is especially important if they play a role in international assessments (as is the case in the GEMS/WATER programme) and if they provide other types of support (such as training or equipment). For consistency, liaison with such agencies, where it occurs, should be the specific responsibility of an identified member of staff. Internal communication in the form of short discussions, such as lunch time seminars or workshops, are a good means of ensuring that all staff are kept informed about general issues and are in contact with programme progress and findings.

Representatives of monitoring programmes may attend committees at both local and national levels. This representation is important as a means of communication and to maintain the profile of the programme and, as such, should also be the specific responsibility of an identified member of staff.

Communication functions, such as those noted above may demand a significant proportion of time and this should be borne in mind when preparing job descriptions.

## 4.6 Inventory of sampling stations

A listing of sampling sites will have been made and sampling stations will have been chosen during the design of the monitoring programme. An inventory of the sampling stations should be prepared that includes the following:

- A map of the general area showing the location of the sampling station.
- A narrative description of how to get to the sampling station.
- A full description of the sampling station.
- Notes concerning means of access and whether permission is needed to gain access to the station.
- Notes on any times of year when access may be difficult or impossible.
- If a boat is needed to get to the station, whether a boat can be rented locally and what other arrangements need to be made.
- Special equipment (e.g. ropes, lifebelts) and clothing (e.g. waders) that are required when sampling at this station.

- Time or times of day when samples are to be obtained.
- Tests that are to be made on-site.
- Volume of sample required and any preservative treatment needed for sample bottles.
- Travel time from the sampling station to the nearest laboratory.

Changes should be made to the inventory when, for example, the specific location of the sampling station is to be changed or some change is to be made to the sampling programme.

## 4.7 Schedules for sampling expeditions

Schedules for field work must take into account local conditions, travel time from laboratory to sampling stations and return, seasonal weather and travel problems, qualifications of field staff, availability of transport, and the maximum time permitted between collection and analysis of samples. This last point will probably determine the required length of the sampling trips and whether on-site testing is necessary. There are three principal options:

- Short trips are made and samples are returned quickly to the laboratory. This option probably requires the most travel time overall but samples would arrive at the laboratory in good condition.
- Long trips are made, samples are tested on-site with portable equipment for certain variables and separate portions of samples are "fixed" with a preservative chemical and returned to the laboratory later by the field staff or by public transport. This option reduces the distances to be travelled. However, the limitations of on-site sample preservation or field testing methods must be recognised.
- Long trips are made (perhaps for several days) and samples are tested on-site with portable equipment. No samples are returned to the laboratory. This option minimises the total distance to be travelled but, as above, the limitations of on-site testing must be recognised. This option also requires well-trained staff.

## 4.8 Source literature and further reading

WHO 1984 *Human Resources Development Handbook*. Unpublished document WHO/CWS/ETS/84.3, World Health Organization, Geneva.

WHO 1986 *Establishing and Equipping Water Laboratories in Developing Countries*. Unpublished document WHO/PEP/86.2, World Health Organization, Geneva.

# Chapter 5

# FIELD WORK AND SAMPLING

The field work associated with the collection and transport of samples will account for a substantial proportion of the total cost of a monitoring programme. Sampling expeditions should, therefore, be planned and carried out in such a way that efforts are not wasted. If, for example, an essential piece of equipment is forgotten or an inadequately described sampling station cannot be found, the value of that particular sampling expedition is seriously compromised. Similarly, if unrealistic estimates of travel time are made and an expedition takes longer than intended, samples may be held longer than the maximum allowable storage time and the results of analyses will be of questionable value.

Special aspects of field work and sampling associated with biological methods and sediment measurements are given in Chapters 11 and 13 and in the companion guidebook *Water Quality Assessments*.

The sample collection process should be co-ordinated with the laboratory so that analysts know how many samples will be arriving, the approximate time of their arrival and the analyses that are to be carried out, and can thus have appropriate quantities of reagent chemicals prepared.

It is good practice to prepare a checklist such as the one on the following page, so that nothing is missing or forgotten before a sampling expedition is undertaken. Many of the items in the list are self-explanatory; others are described more fully in later sections of this chapter.

Personnel who will collect water, biota or sediment samples must be fully trained in both sampling techniques and field test procedures. They should also be aware of the objectives of the monitoring programme since these will have some influence on the sampling procedures. Obtaining a sample that is fully representative of the whole water body is difficult and the collection and handling of samples are also frequent sources of error (often greater errors than those arising during analysis). Thus the choice of a representative sampling point and the use of appropriate sampling techniques are of fundamental importance (see Chapter 3 and the companion guidebook *Water Quality Assessments*).

*This chapter was prepared by J. Bartram, A. Mäkelä and E. Mälkki*

***Checklist for preparing for field work***

*Paperwork*
✓ Itinerary
✓ Inventory details of sampling stations; maps
✓ List of samples required at each sampling station
✓ List of stations where water level readings are to be recorded (see Chapter 12)
*Co-ordination*
✓ Local co-ordination, for example, to ensure access to sites on restricted or private land
✓ Institutional co-ordination, for example, for travel arrangements or sample transport
✓ Notify laboratories of expected date and time of sample arrival
✓ Check any available sources of information on local weather conditions and feasibility of travel
*For sampling*
✓ Sample bottles, preservatives, labels and marker pens
✓ Sample storage/transit containers and ice packs
✓ Filtering apparatus (if required)
✓ Samplers/sampling equipment
✓ Rubber boots, waders, etc.
✓ Standard operating procedures for sampling
✓ Spares of all above items if possible and when appropriate
*For documentation*
✓ Pens/wax crayons
✓ Sample labels
✓ Field notebook
✓ Report forms
*For on-site testing*
✓ List of analyses to be performed on site
✓ Check stocks of consumables (including distilled water, pH buffers, standards and blanks); replenish and refresh as appropriate
✓ Check and calibrate meters (pH, conductivity, dissolved oxygen, turbidity, thermometers)
✓ Other testing equipment according to local practice
✓ Standard operating procedures and equipment manuals
✓ Spares (e.g. batteries)
*Safety*
✓ First-aid kit
✓ Waders, gloves, etc.
✓ Fire extinguisher (if appropriate)
*Transport*
✓ Does assigned vehicle have sufficient capacity for personnel, supplies and equipment?
✓ Is vehicle road-worthy? Check battery, lubrication, coolant, windshield washer
✓ Is there sufficient fuel for the trip, either in the tank, in fuel cans, or available en route?
✓ Is the spare tyre inflated, is there a jack, wheel wrench and tool kit?
*Double-check*
✓ When was equipment last calibrated?
✓ Itinerary against travel details on inventory
✓ Accessories for equipment and meters (including cables, chargers and spare batteries) and consumables

Only if samples can be taken consistently from the same locations, can changes in the concentration of water quality variables with time be interpreted with confidence. It is advisable to carry out a pilot programme before the routine monitoring programme begins (see section 3.4). This can be used as a training exercise for new personnel and will provide the opportunity to make a final selection of sampling stations on the basis of whether they are representative of the whole water body as well as readily accessible.

Programme managers and laboratory personnel should accompany field personnel on field expeditions from time to time. This provides opportunities for field supervision as part of in-service training and for everyone working on the programme to appreciate the problems and needs of field work.

On-site testing is common for certain variables, especially those that may change (physical, chemical or biological) during transport. Dissolved oxygen, turbidity, transparency, conductivity, pH, temperature and, to a lesser extent, thermotolerant (faecal) and total coliform counts are the variables most often measured on site. Procedures for carrying out analyses in the field are covered in Chapters 6 and 10.

The special problems of analytical quality control during field testing or testing using portable equipment are outlined in section 6.7, and further information on minimising the risks associated with this approach is provided in Chapter 9.

## 5.1 Sample containers

Containers for the transportation of samples are best provided by the laboratory. This ensures that large enough samples are obtained for the planned analyses and that sample bottles have been properly prepared, including the addition of stabilising preservatives when necessary. It is essential to have enough containers to hold the samples collected during a sampling expedition. Sample containers should be used only for water samples and never for the storage of chemicals or other liquids. Glass containers are commonly used and are appropriate for samples for many analyses, but plastic containers are preferred for samples intended for certain chemical analyses or for biota or sediments. Plastic has the obvious advantage that it is less likely to break than glass.

Sample containers must be scrupulously clean so that they do not contaminate the samples placed in them. Table 5.1 provides general information on appropriate types of sample container and the recommended procedures for cleaning them when water samples are to be used for chemical analysis.

For microbiological analysis, strong, thick-walled, glass sample bottles with a minimum capacity of 300 ml should be used. They should have screw

**Table 5.1** Sample containers and their recommended washing procedures for selected water quality variables

| Variable(s) to be analysed | Recommended container[1] | Washing procedure |
|---|---|---|
| Organochlorinated pesticides and PCBs<br>Organophosphorus | 1,000 ml glass (amber) with teflon-lined cap | Rinse three times with tap water, once with chromic acid[2], three times with organic-free water, twice with washing acetone, once with special grade[3] acetone, twice with pesticide grade hexane and dry (uncapped) in a hot air oven at 360 °C |
| Pentachlorophenol<br>Phenolics<br>Phenoxy acid herbicides | 1,000 ml glass (amber) with teflon-lined cap | Rinse three times with tap water, once with chromic acid[2], three times with organic-free water, twice with washing acetone, once with special grade[3] acetone, twice with pesticide grade hexane and dry (uncapped) in a hot air oven at 360 °C for at least 1 h |
| Aluminium, Antimony, Barium, Beryllium, Cadmium, Chromium[4], Cobalt, Copper, Iron, Lead, Lithium, Manganese, Molybdenum, Nickel, Selenium, Strontium, Vanadium, Zinc | 500–1,000 ml polyethylene (depending upon number of metals to be determined) | Rinse three times with tap water, once with chromic acid[2], three times with tap water, once with 1:1 nitric acid and then three times with ultrapure distilled water[5] in that order |
| Silver | 250 ml polyethylene (amber) | Rinse three times with tap water, once with chromic acid[2], three times with tap water, once with 1:1 nitric acid and then three times with ultrapure distilled water[5] in that order |
| Mercury | 100 ml glass (Sovirel) | Rinse three times with tap water, once with chromic acid[2], three times with tap water, once with 1:1 nitric acid and then three times with ultrapure distilled water[5] in that order |
| Acidity, Alkalinity, Arsenic, Calcium, Chloride, Colour, Fluoride, Hardness, Magnesium, Non-filterable residue, pH, Potassium, Sodium, Specific conductance, Sulphate, Turbidity | 1,000 ml polyethylene | Rinse three times with tap water, once with chromic acid[2], three times with tap water, once with 1:1 nitric acid and then three times with distilled water in that order |

Continued

**Table 5.1** Continued

| Variable(s) to be analysed | Recommended container[1] | Washing procedure |
|---|---|---|
| Carbon, total organic<br>Nitrogen: ammonia<br>Nitrogen: nitrate, nitrite<br>Nitrogen: total | 250 ml polyethylene | Rinse three times with tap water, once with chromic acid[2], three times with tap water, and three times with distilled water, in that order |
| Phosphorus, total | 50 ml glass (Sovirel) | Rinse three times with tap water, once with chromic acid[2], three times with tap water, and three times with distilled water, in that order |

1 Teflon containers can also be used to replace either the recommended polyethylene or glass containers
2 Chromic acid — 35 ml saturated $Na_2Cr_2O_7$ per litre reagent grade conc. $H_2SO_4$
3 Special grade acetone — pesticide grade when GC analysis to be performed, UV grade for LC analysis
4 Chromic acid should not be used when the sample will be analysed for chromium
5 Ultrapure distilled water is obtained by passing distilled water through a Corning model AG-11 all-glass distillation unit and then through a Millipore Super Q Ultrapure Water System containing a prefilter cartridge, an activated carbon cartridge and a mixed bed deionisation cartridge

Source: After WMO, 1988

caps of a type that will maintain an effective seal, even after they have been sterilised many times in an autoclave. Some technicians fasten a Kraft paper cover over the bottle caps before autoclaving to protect them from contamination during handling. Alternatively, plastic or aluminium sleeves may be used. The neck of the bottle should not be plugged with cotton wool. To prepare sample bottles, they should be washed with a non-ionic detergent and rinsed at least three times (five is better) with distilled or deionised water before autoclaving. New bottles require the same preparation. If distilled or deionised water is not available, clean chlorine-free water may be used.

If chlorinated water is being collected for microbiological analysis, sufficient sodium thiosulphate should be added to the sample bottles to neutralise the chlorine. The recommended amount is 0.1 ml of a 1.8 per cent solution of sodium thiosulphate for each 100 ml of sample bottle volume; this should be added to the bottles before autoclaving. Sample bottles should not be rinsed with sample water or allowed to overflow because this would remove the dechlorinating chemical.

Some water quality variables are unstable and, unless an analysis can be carried out immediately after the sample is obtained, it is necessary to stabilise the sample by adding a chemical preservative. It is often convenient to add chemical preservatives to containers in the laboratory rather than in the field. When this is done, it is essential that the containers be clearly labelled with the name, concentration and quantity of the preservative chemical, the volume of the sample to be collected and the variables for which the sample is to be analysed. If preservatives are not added to containers in the laboratory, the chemicals, pipettes and directions for adding preservatives must be included in the kit of supplies and equipment taken on the sampling expedition. The subject of sample preservation is dealt with more fully in section 5.6.

## 5.2 Types of sample

### 5.2.1 Sampling surface waters

Two different types of sample can be taken from rivers, lakes and similar surface waters. The simplest, a "grab" sample, is taken at a selected location, depth and time. Normally, the quantity of water taken is sufficient for all the physical and chemical analyses that will be done on the sample. Sometimes, if the sampler is small and many analyses are to be done, two grab samples will be taken at the station and will be mixed in the same transport container. Grab samples are also known as "spot" or "snap" samples.

Composite or integrated samples, i.e. samples made up of several different parts, are often needed to fulfil some specific monitoring objectives. Composite samples may be of the following types:

- *Depth-integrated:* most commonly made up of two or more equal parts collected at predetermined depth intervals between the surface and the bottom. A piece of flexible plastic piping of several metres in length, and which is weighted at the bottom, provides a simple mechanism for collecting and integrating a water sample from the surface to the required depth in a lake. The upper end is closed before hauling up the lower (open) end by means of an attached rope. Integrated samples can also be obtained using a water pump (submersible pumps are available which allow sampling at depth) which is operated at a steady pumping rate while the water inlet is drawn upwards between the desired depths at a uniform speed.
- *Area-integrated:* made by combining a series of samples taken at various sampling points spatially distributed in the water body (but usually all at one depth or at predetermined depth intervals).
- *Time-integrated:* made by mixing equal volumes of water collected at a sampling station at regular time intervals.
- *Discharge-integrated:* It is first necessary to collect samples and to measure the rate of discharge at regular intervals over the period of interest. A common arrangement is to sample every 2 hours over a 24-hour period. The composite sample is then made by mixing portions of the

individual sample that are proportional to the rate of discharge at the time the sample was taken.

### 5.2.2 Sampling groundwater

Groundwater samples are normally obtained from existing drilled wells, dug (shallow) wells or springs. Occasionally, during the course of a hydrogeological survey, test wells may be drilled and these can be used for monitoring purposes. The usual situation, however, is that a producing well or spring will be a groundwater quality monitoring station.

If the groundwater source is a flowing spring or a well equipped with a pump, the sample can be obtained at the point of discharge. The water should flow for several minutes before sampling until it has reached constant conductivity or temperature in order to avoid any water resident in the system's piping being taken as a sample (the piping material may have contaminated the water). Samples for dissolved oxygen analysis should be taken by inserting one end of a plastic tube into the discharge pipe and the other end into a sample bottle. The water should be allowed to flow into the bottle for sufficient time to displace the contents of the bottle at least three times. Care should be taken to ensure that no air bubbles are introduced to the sample while the bottle is being filled, since this could alter the dissolved oxygen concentration.

Special care must be taken when sampling from springs that do not have an overflow and from shallow wells without pumps. The sampling container must not be allowed to touch the bottom of the well or spring catchment since this would cause settled particles to become resuspended and to contaminate the sample. Sometimes, a spring catchment is higher than the surrounding ground and this permits water to be siphoned into the sample bottle. If this is done, water should be allowed to run through the hose for 2–3 minutes to rinse it thoroughly before the sample is collected. Siphoned samples are suitable for dissolved oxygen determination provided that the sample bottle is allowed to overflow a volume of at least three times its capacity.

The depth within an aquifer from which a sample of water is collected from a well is determined by the location of the well screen (as described in section 2.2) and cannot be varied by the collector, because water enters a well at the level of the screen. Similarly, water enters a spring through fissures in the rock. Consequently, a groundwater sample can only be obtained as a grab sample. The greatest danger of getting a non-representative sample occurs when insufficient water has been pumped before the sample is collected and that the sample obtained is representative of the well rather than of the aquifer.

## 5.3 Water samplers

Several different types of sampler are available, many of them designed for specific purposes. The three types described here are those that are most useful for a general water sampling programme. Equipment required to obtain samples for biological and sediment analysis is described in the relevant sections of Chapters 11 and 13.

### 5.3.1 Dissolved oxygen sampler

A dissolved oxygen sampler is a metal tube about 10 cm in diameter and 30 cm in length, sealed at one end and with a removable cap (usually threaded) at the other. A bracket is located inside the tube in such a way that a 300-ml BOD bottle can be placed in the bracket with the top of the bottle 2–3 cm below the top of the sampler. The sampler cap has a tube extending from its underside down into the BOD bottle when the cap is in place. The upper end of this tube is open and flush with the outside face of the sampler top. A second tube in the sampler cap is flush with the inside face and extends upwards for about 8–10 cm. This second tube is sometimes incorporated into the frame to which the lowering rope is fastened. Figure 5.1 shows a typical dissolved oxygen sampler.

When the sampler is used, a BOD bottle is placed in the bracket, the sampler cap is fitted in place, and a lowering rope is fastened to the sampler which is then lowered vertically to the depth from which the sample is to be taken. Air in the sampler flows out through the highest tube and, consequently, water enters the BOD bottle through the lower tube. The volume of the sampler is about five times the volume of the bottle, therefore the incoming water flushes out the bottle at least four times and the water that finally remains in the bottle will have had no contact with the air that was originally in the sampler. Provided that the sampler is lowered quickly to the desired sampling depth, the sample obtained should be representative, in terms of its dissolved oxygen content. If a sample needs to be taken from great depth, inflow to the sampler can be prevented with a cork or similar device that can be removed when the desired depth is reached.

When the sampler is returned to the surface, the cap is removed and a ground-glass stopper is placed in the (ground-glass) neck of the BOD bottle before it is taken out of the sampler. Further handling of the dissolved oxygen sample is described in section 6.5. The water remaining in the sampler can be used for other analyses but it must be remembered that this will be water that flowed into the sampler between the surface and the depth at which the dissolved oxygen sample was taken. Moreover, its volume will be only about 1.5 litres, which may not be enough for all of the intended analyses.

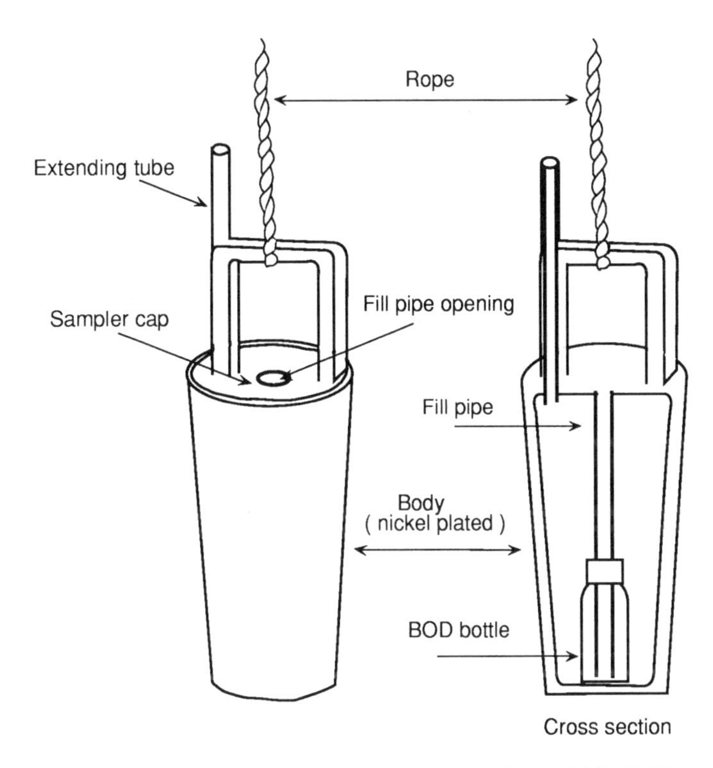


**Figure 5.1** Dissolved oxygen sampler (Adapted from WMO, 1988)

### 5.3.2 Depth sampler

The depth sampler, which is sometimes called a grab sampler, is designed in such a way that it can retrieve a sample from any predetermined depth. A typical depth sampler is shown in Figure 5.2. It consists of a tube, approximately 10 cm in diameter and 30 cm in length, fastened to a frame along which it can slide. The frame has projections at each end so that the tube can not slide off. The ends of the tube are covered by spring-loaded flaps, which can be held in the fully open position by latches. The latches can be released by applying a small amount of pressure to a lever. To accomplish this, a weight (called a "messenger") is dropped down the lowering rope, the latch is tripped and the ends of the tube close.

When the sampler is in use, the end flaps are latched into the open position. As the sampler is lowered to the required depth with the lowering rope, water passes through the open ends so that, at any depth, the water in the sampler is the water from that depth. When the desired depth is reached, the messenger weight is dropped down the rope, the latch is tripped and the end flaps close. The sampler is brought to the surface and its contents are transferred

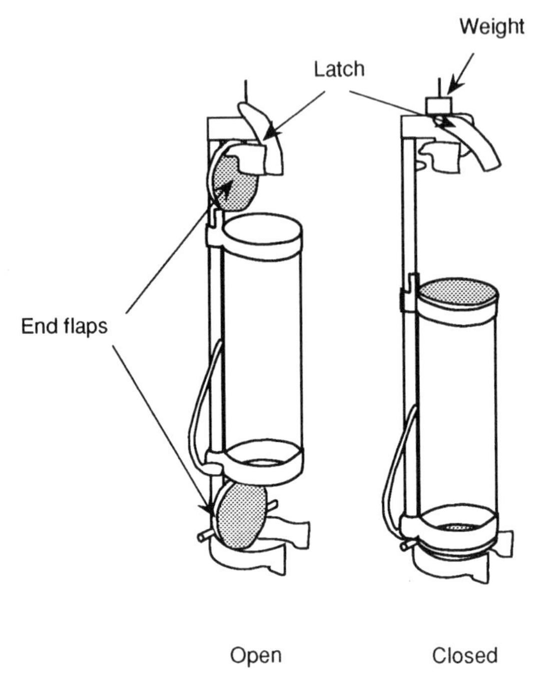


**Figure 5.2** Depth sampler

to a sample bottle. A sample obtained in this way can be used for all chemical analyses except dissolved oxygen. A simpler and less expensive model of depth sampler, suitable for moderate depths (< 30 m) is illustrated in Figure 5.3.

### 5.3.3 Multi-purpose sampler

A multi-purpose sampler, also called a sampling iron, is most frequently used for taking samples in flowing streams or rivers. It consists of a weighted platform equipped with clamps or similar means of holding a sample bottle, a rudder to maintain its position in the flowing water, and rings at the top and bottom to which lowering ropes can be attached as shown in Figure 5.4. One end of the rope may be attached to the top ring and a friction release device, connected between the rope and the bottom ring, holds the bottle in an inverted position during lowering. An alternative arrangement is to use two ropes, one fastened to the lower ring and one to the upper. Both arrangements permit the collection of samples from a deep location by allowing the sampler to be lowered in an inverted position and then restored to the upright position when the required depth is reached.

The multi-purpose sampler is very easy to use for sampling near the surface. It is simply immersed in the water and allowed to fill up. For samples

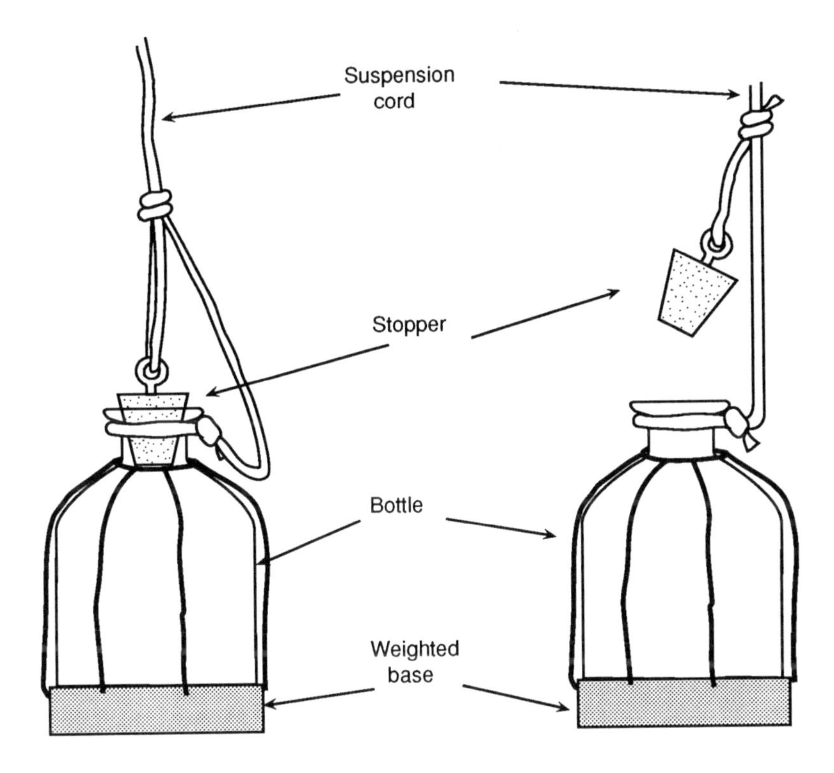


**Figure 5.3** Depth sampler suitable for moderate depths

from greater depths, it must be lowered in the inverted position and then, when the desired depth is reached, righted either by a sharp tug on the rope (for the one-rope configuration) or by transferring restraint to the rope connected to the upper ring. Although some water may enter the sampler during its descent, this type of sampler has the advantage that the sample does not need to be transferred to another container for shipment because it can remain in the container in which it was collected. Samples taken with the multi-purpose sampler cannot be used for dissolved oxygen determination.

When samples are taken for chemical and physical analysis from rivers and lakes, it is often sufficient merely to immerse an open-mouthed vessel, such as a bucket, below the water surface. The contents can then be poured into an appropriate set of sample bottles. Alternatively, the sample bottle can be immersed in the water and allowed to fill up. Care should be taken to avoid the entry of water from the surface since this will often contain very fine floating material that cannot be easily seen. If the water is flowing, the open mouth of the bottle should point upstream.

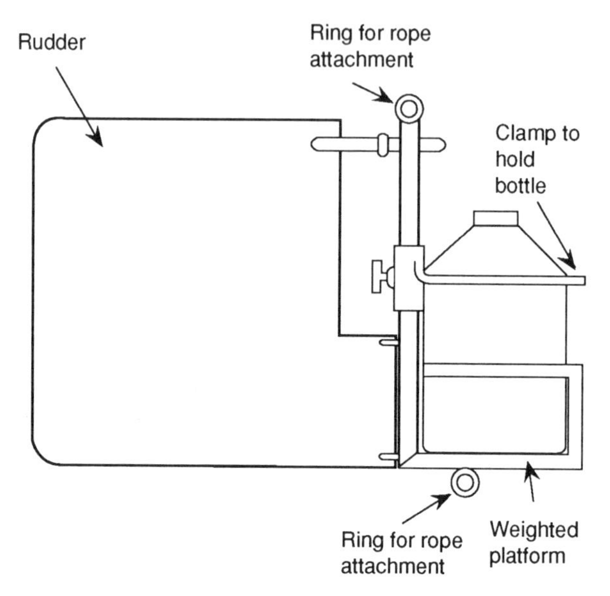


**Figure 5.4** Multi-purpose sampler (Adapted from WMO, 1988)

## 5.4 Manual sampling procedures

This section deals principally with sampling for water analysis or bacteriological analysis. Details of procedures for other biological methods and for sediment measurements are given in Chapters 11 and 13 respectively.

### 5.4.1 Guidelines

*Samples for physical and chemical analyses*

The minimum sample size varies widely depending on the range of variables to be considered and the analytical methods to be employed, but it is commonly between 1 and 5 litres. The volumes required for individual analyses are summarised in Table 5.2.

The following general guidelines can be applied to the collection of water samples (to be analysed for physical or chemical variables) from rivers and streams, lakes or reservoirs and groundwater.

- Before collecting any sample, make sure that you are at the right place. This can be determined by the description of the station, from the position of landmarks and, in lakes, by checking the depth. If samples must be taken from a boat, a sampling station may be marked by placing a buoy at the desired location; otherwise it is necessary to identify the sampling station by the intersection of lines between landmarks on the shore.

**Table 5.2** Sample volumes required for individual physico-chemical analyses

| Analysis | Sample volume (ml) | Analysis | Sample volume (ml) |
|---|---|---|---|
| Alkalinity | 100 | Kjeldahl nitrogen | 400 |
| Aluminium | 25 | Nitrate nitrogen | 200 |
| BOD | 1,000 | Nitrite nitrogen | 50 |
| Boron | 1 | Phosphorus | 100 |
| Calcium | 50 | Potassium | 100 |
| Chloride | 100 | Selenium | 1,000 |
| Fluoride | 50 | Silica | 50 |
| Iron | 50 | Sodium | 100 |
| Magnesium | 75 | Sulphate | 200 |
| Manganese | 90 | TOC | 200 |
| Ammonia nitrogen | 400 | TSS | 1,000 |

BOD Biochemical oxygen demand
TOC Total organic carbon
TSS Total suspended solids

- Do not include large, non-homogeneous pieces of detritus, such as leaves, in the sample. Avoid touching and disturbing the bottom of a water body when taking a depth sample, because this will cause particles to become suspended. As an example, the GEMS/WATER monitoring programme sets the upper size limit of particulate matter at 0.063 mm. To remove larger material pass the water sample through a sieve and collect it in a bottle for transport.
- Sampling depth is measured from the water surface to the middle of the sampler.
- Samples taken to describe the vertical profile should be taken in a sequence that starts at the surface and finishes at the bottom. When taking the sample at the maximum depth it is important to ensure that the bottom of the sampler is at least 1 m above the bottom.
- Do not lower a depth sampler too rapidly. Let it remain at the required depth for about 15 seconds before releasing the messenger (or whatever other device closes the sampler). The lowering rope should be vertical at the time of sampling. In flowing water, however, this will not be possible and the additional lowering necessary to reach the required depth should be calculated.
- A bottle that is to be used for transport or storage of the sample should be rinsed three times with portions of the sample before being filled. This does not apply, however, if the storage/transport bottle already contains a preservative chemical.
- The temperature of the sample should be measured and recorded immediately after the sample is taken.

- The sample to be used for dissolved oxygen determination should be prepared immediately after the temperature is measured. If an electronic technique is being used, a portion of the sample is carefully poured into a beaker for measurement. If the Winkler method is being used, the chemical reagents are added to the bottle in accordance with the directions contained in section 6.5.
- Separate portions of the sample should be set aside for pH and conductivity determinations. The same portion must not be used for both determinations because of the possibility of potassium chloride diffusing from the pH probe.
- At any time that the sample bottles are not closed, their tops must be kept in a clean place.
- A small air space should be left in the sample bottle to allow the sample to be mixed before analysis.
- All measurements taken in the field must be recorded in the field notebook before leaving the sampling station.
- All supporting information should be recorded in the field notebook before leaving the sampling station. Such conditions as the ambient air temperature, the weather, the presence of dead fish floating in the water or of oil slicks, growth of algae, or any unusual sights or smells should be noted, no matter how trivial they may seem at the time. These notes and observations will be of great help when interpreting analytical results.
- Samples should be transferred to sample bottles immediately after collection if they are to be transported. If analysis is to be carried out in the field, it should be started as soon as possible.

*Samples for bacteriological analysis*

Most of the guidelines for sampling for physical and chemical analyses apply equally to the collection of samples for bacteriological analyses. Additional considerations are:

- Samples for bacteriological analyses should be taken in a sterile sampling cup and should be obtained before samples for other analyses.
- Care must be exercised to prevent contamination of the inside of the sampling cup and sampling containers by touching with the fingers or any non-sterile tools or other objects.
- Bottles in which samples for bacteriological analyses are to be collected (or transported) should be reserved exclusively for that purpose.

### 5.4.2 Procedures

*Sampling from a tap or pump outlet*

1. Clean the tap. Remove any attachments that may cause splashing from the tap. These attachments are a frequent source of contamination that may influence the

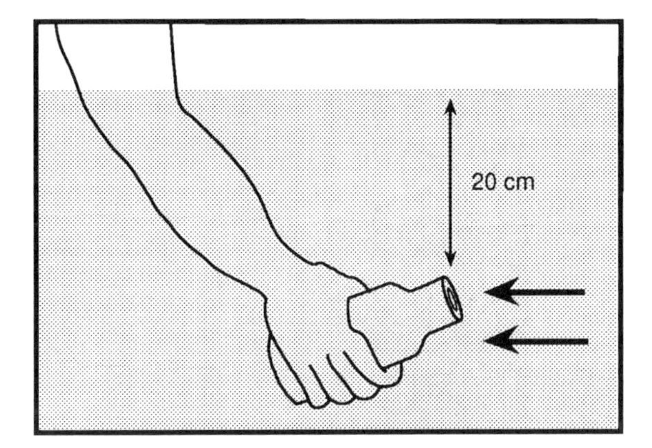


**Figure 5.5** Collecting a sample from surface water

perceived quality of the water supply. Use a clean cloth to wipe the outlet and to remove any dirt.

2. Open the tap. Turn on the tap to maximum flow and let the water run for 1–2 minutes. Turn off the tap.

*Note:* Some people omit the next two steps and take the samples at this stage, in which case the tap should not be adjusted or turned off, but left to run at maximum flow.

3. Sterilise the tap for 1 minute with a flame (from a gas burner, cigarette lighter or an alcohol-soaked cotton wool swab).
4. Open the tap before sampling. Carefully turn on the tap and allow water to flow at medium rate for 1–2 minutes. Do not adjust the flow after it has been set.
5. Fill the bottle. Carefully remove the cap and protective cover from the bottle, taking care to prevent entry of dust that may contaminate the sample. Hold the bottle immediately under the water jet to fill it. A small air space should be left to allow mixing before analysis. Replace the bottle cap.

*Sampling water from a water-course or reservoir*

Open the sterilised bottle as described in step 5 above.

1. Fill the bottle (see Figure 5.5). Hold the bottle near its bottom and submerge it to a depth of about 20 cm, with the mouth facing slightly downwards. If there is a current, the bottle mouth should face towards the current. Turn the bottle upright to fill it. Replace the bottle cap.

*Sampling from dug wells and similar sources*

1. Prepare the bottle. With a length of string, attach a weight to the sterilised sample bottle (Figure 5.6).
2. Attach the bottle to the string. Take a 20 m length of string, rolled around a stick, and tie it to the bottle string. Open the bottle as described above.
3. Lower the bottle. Lower the weighted bottle into the well, unwinding the string slowly. Do not allow the bottle to touch the sides of the well (see Figure 5.6).

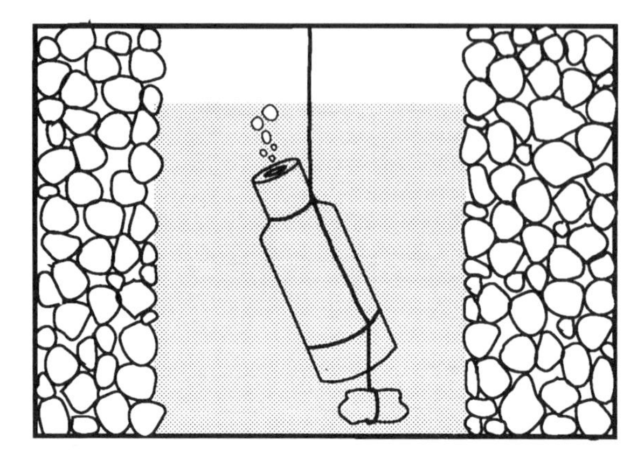

**Figure 5.6** Lowering a weighted bottle into a well

4. Fill the bottle. Immerse the bottle completely in the water and continue to lower it to some distance below the surface (see Figure 5.6). Do not allow the bottle to touch the bottom of the well or disturb any sediment.
5. Raise the bottle. Once the bottle is judged to be full, bring it up by rewinding the string around the stick. If the bottle is completely full, discard a little water to provide an air space. Cap the bottle as described previously.

## 5.5 Recording field observations

Sampling officers should have a field notebook in which all details of relevance are recorded at the time. The field book should be hard-bound and not loose-leaf. Full books should not be discarded but stored for future reference because they represent data in original form and are sometimes invaluable for reference purposes.

Details recorded should include:

- those noted on the sample bottle (see section 5.7),
- what samples were collected, and
- what measurements were made, how they were made, and the results obtained (including blanks, standards, etc., and the units employed).

All supporting information (any unusual local features at the site and time of sampling) should also be noted. If there has been any variation from the agreed sampling station, this should be noted, with reasons. Any need for a permanent change in sampling station should be brought to the attention of the programme co-ordinator and the inventory should be changed if necessary.

If a standard field record layout is used in place of a plain notebook, adequate space should be available for comments and observations. To facilitate field work, the layout and content of the pages should reflect the sequence in which the various procedures will be carried out (see Figure 5.7).

***Site:*** No./code Description
***Station:*** No./code Description
***Date:*** ***Time:***

***Weather conditions:***

---

***Samples collected:*** Standard chemistry Yes/No Sample no.
Microbiology Yes/No Sample no.

***Sampling depth:***

***Problems encountered/adaptations made during sampling:***

***Sample preservation and storage:***

***Sample transport:***

---

***Analyses undertaken on site:***

| Variable | Method used | Equipment no. | Sample/blank | Reading value | Units |
|---|---|---|---|---|---|

***Notes on on-site analyses:***

---

***General remarks:***

***Collector:*** Name Signature Date
***Samples received by:*** Name Signature Date
***Data received by:*** Name Signature Date

**Figure 5.7** Example page from a field notebook

## 5.6 Sample preservation

There is little clear consensus on the best means of preserving samples for specific analyses, the practicality of preservation in the field and the length of time for which samples may be stored without deteriorating. In general, sample bottles should be resealed and stored in a clean, cool, dark environment and protected from recontamination. Additional methods of preservation include freezing, solvent extraction and the addition of chemical preservatives.

Storage may vary according to the method employed for analyses and correct information may, therefore, be available in standard methods. If not, stability testing may have to be carried out before sampling is initiated and should, in any case, be done as part of method validation.

Suggested chemical preservatives and recommended maximum storage times for samples for various analyses are summarised in Table 5.3.

The preservative treatment may be applied immediately on sampling or may already be contained in the sample bottle. For neutralising chlorine in samples for microbiological analysis, however, sodium thiosulphate must be in the sample bottles before they are autoclaved. Where pretreatment by filtration is required, the preservative should be added immediately after filtration, as some chemicals may alter the physical characteristics of the sample. Acidification, for example, may lead to dissolution of colloidal and particulate metals.

## 5.7 Transportation and storage of samples

The sample collection process should be co-ordinated with the laboratory. Analysts need to know how many samples will be arriving, the approximate time of arrival and the analyses that are to be carried out, so that appropriate quantities of reagent chemicals can be prepared. If sample bottles are provided by the laboratory, this ensures that they are of adequate volume and have been properly prepared (with added chemical preservatives where necessary).

Each sample bottle must be provided with an identification label on which the following information is legibly and indelibly written:

- Name of the study.
- Sample station identification and/or number.
- Sampling depth.
- Date and time of sampling.
- Name of the individual who collected the sample.
- Brief details of weather and any unusual conditions prevailing at the time of sampling.

**Table 5.3** Suggested preservative treatments and maximum permissible storage times

| Variable | Recommended container[1] | Preservative | Max. permissible storage time |
|---|---|---|---|
| Alkalinity | Polyethylene | Cool 4 °C | 24 h |
| Aluminium | Polyethylene | 2 ml Conc. $HNO_3$ $l^{-1}$ sample | 6 months |
| Arsenic | Polyethylene | Cool 4 °C | 6 months |
| BOD | Polyethylene | Cool 4 °C | 4 h |
| Boron | Polyethylene | Cool 4 °C | 6 months |
| Cadmium | Polyethylene | 2 ml Conc. $HNO_3$ $l^{-1}$ sample | 6 months |
| Calcium | Polyethylene | Cool 4 °C | 7 days |
| Carbamate pesticides | Glass | $H_2SO_4$ to pH < 4, 10g $Na_2SO_4$ $l^{-1}$ | Extract immediately |
| Carbon | | | |
| inorganic/organic | Polyethylene | Cool 4 °C | 24 h |
| particulate | Plastic Petri dish | Filter using GF/C filter; Cool, 4 °C | 6 months |
| Chloride | Polyethylene | Cool 4 °C | 7 days |
| Chlorinated hydrocarbon | Glass | Cool 4 °C | Extract immediately |
| Chlorophyll | Plastic Petri dish | Filter on GF/C filter; freeze –20 °C | 7 days |
| Chromium | Polyethylene | 2 ml Conc. $HNO_3$ $l^{-1}$ sample | 6 months |
| COD | Polyethylene | Cool 4 °C | 24 h |
| Copper | Polyethylene | 2 ml Conc. $HNO_3$ $l^{-1}$ sample | 6 months |
| Dissolved oxygen (Winkler) | Glass | Fix on site | 6 h |
| Fluoride | Polyethylene | Cool 4 °C | 7 days |
| Iron | Polyethylene | 2 ml Conc. $HNO_3$ $l^{-1}$ sample | 6 months |
| Lead | Polyethylene | 2 ml Conc. $HNO_3$ $l^{-1}$ sample | 6 months |
| Magnesium | Polyethylene | Cool 4 °C | 7 days |
| Manganese | Polyethylene | 2 ml Conc. $HNO_3$ $l^{-1}$ sample | 6 months |
| Mercury | Glass or teflon | 1 ml Conc. $H_2SO_4$ + 1 ml 5% $K_2Cr_2O_7$ | 1 month |
| Nickel | Polyethylene | 2 ml Conc. $HNO_3$ $l^{-1}$ sample | 6 months |
| Nitrogen | | | |
| Ammonia | Polyethylene | Cool 4 °C, 2 ml 40% $H_2SO_4$ $l^{-1}$ | 24 h |
| Kjeldahl | Polyethylene | Cool 4 °C | 24 h |
| Nitrate + Nitrite | Polyethylene | Cool 4 °C | 24 h |
| Organic nitrogen | Polyethylene | Cool 4 °C | 24 h |
| Organic particulates | Plastic Petri dish | Filter using GF/C filter, Cool 4 °C | 6 months |
| Organophosphorus pesticides | Glass | Cool, 4 °C, 10% HCl to pH 4.4 | No holding, extraction on site |
| Pentachlorophenol | Glass | $H_2SO_4$ to pH < 4, 0.5 g $CuSO_4$ $l^{-1}$ sample; Cool 4 °C | 24 h |
| pH | Polyethylene | None | 6 h |
| Phenolics | Glass | $H_3PO_4$ to pH < 4, 1.0 g $CuSO_4$ $l^{-1}$ sample; Cool 4 °C | 24 h |
| Phenoxy acid herbicides | Glass | Cool 4 °C | Extract immediately |
| Phosphorus | | | |
| Dissolved | Glass | Filter on site using 0.45 µm filter | 24 h |
| Inorganic | Glass | Cool 4 °C | 24 h |
| Total | Glass | Cool 4 °C | 1 month |

Continued

**Table 5.3** Continued

| Variable | Recommended container[1] | Preservative | Max. permissible storage time |
|---|---|---|---|
| Potassium | Polyethylene | Cool, 4 °C | 7 days |
| Residue | Polyethylene | Cool, 4 °C | 7 days |
| Selenium | Polyethylene | 1.5 ml Conc. $HNO_3$ $l^{-1}$ sample | 6 months |
| Silica | Polyethylene | Cool, 4 °C | 7 days |
| Sodium | Polyethylene | Cool, 4 °C | 7 days |
| Electrical conductivity | Polyethylene | Cool, 4 °C | 24 h |
| Sulphate | Polyethylene | Cool, 4 °C | 7 days |
| Zinc | Polyethylene | 2 ml Conc. $HNO_3$ $l^{-1}$ sample | 6 months |

[1] Teflon containers can also be used to replace either the polyethylene or glass containers shown in the table.

Source: Adapted from Environment Canada, 1981

- Record of any stabilising preservative treatment.
- Results of any measurements completed in the field.

This information, as discussed earlier, will also be recorded in the field notebook.

Sample bottles should be placed in a box for transport to the laboratory. Sturdy, insulated wooden or plastic boxes will protect samples from sunlight, prevent the breakage of sample bottles, and should allow a temperature of 4 °C to be attained and maintained during transport. Figure 5.8 shows a suitable transport box. Rapid cooling of samples for BOD and/or microbiological analyses requires that the transport box should contain cold water in addition to ice or an "ice pack". The use of a solid coolant alone is inadequate because heat transfer and sample cooling are too slow. Bottles containing samples for bacteriological analysis should ideally be placed in clear plastic bags to protect them from external contamination.

If the delay between sample collection and bacteriological analysis will be less than 2 hours, samples should simply be kept in a cool, dark place. When more than 2 hours will elapse, samples should be chilled rapidly to about 4 °C by placing them in a cold water/ice mixture (see above) in an insulated container, where they should remain during shipment. If the time between collection and analysis exceeds 6 hours, the report of the analysis should include information on the conditions and duration of sample transport.

In practice, it is difficult to ensure the transport of samples under conditions that do not affect their bacteriological quality and equipment designed for conducting analyses in the field is, therefore, becoming

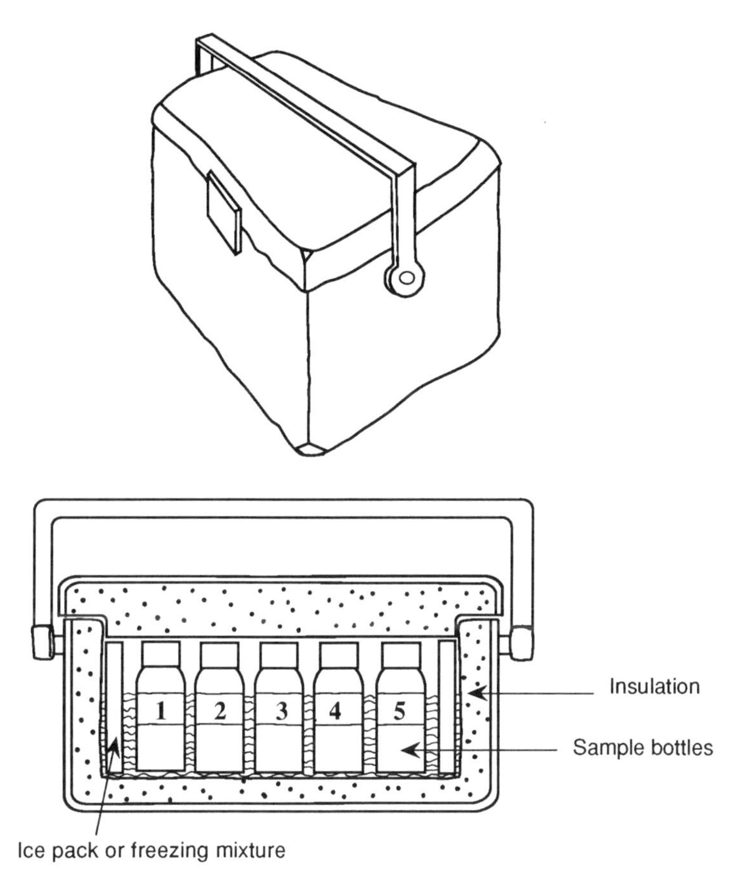


**Figure 5.8** Sample transport box

increasingly popular. It is also possible to filter samples in the field and to place the filters on a holding medium for later treatment in the laboratory.

On arrival at the laboratory, samples for bacteriological analysis should be placed in a refrigerator and analysis should be started within 2 hours. Any samples arriving more than 24 hours after they were collected, or arriving unchilled more than 2 hours after they were collected, should be discarded. Analysis of such samples is unlikely to reflect the bacteriological condition of the water at the time of sampling.

Samples for chemical analysis should arrive at the analytical laboratory and be analysed within 24 hours of collection, since some variables are subject to change during storage (although others, such as hardness, fluoride, chloride and sulphate, are stable for 2–3 weeks).

## 5.8 Reception of samples by the laboratory

It is good practice for a member of the laboratory staff to sign for the receipt of samples and to make the following checks at the same time:

- All of the necessary details are recorded on the labels of sample bottles.
- The samples are contained in appropriate bottles.
- Samples have arrived in time for subsequent analysis to provide a reliable picture of water quality at the time of sampling.
- Samples have been treated with any necessary preservatives.
- Samples have been stored at appropriate temperatures, maintained throughout transport.

Samples should be logged into the laboratory system as soon as they arrive and (generally) transferred to a refrigerator at about 4 °C. If someone other than laboratory staff is to receive the samples, they should be instructed to transfer them directly to a refrigerator, noting the time and the condition of the samples, and then inform laboratory staff accordingly.

Where sample storage times and/or conditions have been such as to make it unlikely that analyses will yield reliable results, laboratory staff should decline to accept the affected samples. Supervisory staff should lend their support to the decision to reject samples on this basis.

## 5.9 Safety during field work

Field staff will encounter a wide range of hazards in the course of their work. To give just a few examples, water-courses may be highly contaminated with sewage or chemicals, access to sampling stations may involve crossing dangerous terrain, and wading in streams inevitably carries the possibility of slipping and personal injury. Where there is a risk of infection from contact with water (as in the case of schistosomiasis, for example), suitable protective clothing, such as rubber gloves, should be provided and its use by staff strongly encouraged.

Field staff should be trained to recognise and deal with as many as possible of the hazards they are likely to encounter. As a minimum, training should include water safety and first-aid. A basic first-aid kit should be carried at all times, and should not be left in the transport vehicle if staff are obliged to move any significant distance from it.

## 5.10 Source literature and further reading

APHA 1992 *Standard Methods for the Examination of Water and Wastewater*. 18th edition. American Public Health Association (APHA), American Water Works Association (AWWA) and Water Pollution Control Federation (WPCF), Washington, DC.

Bartram, J.B. 1990 *Drinking Water Supply Surveillance*. Robens Institute, University of Surrey, Guildford.

Chapman, D. [Ed.] 1996 *Water Quality Assessments. A Guide to the Use of Biota, Sediments and Water in Environmental Monitoring*. 2nd edition. Chapman & Hall, London.

Environment Canada 1981 *Analytical Methods Manual*. Water Quality Branch, Inland Waters Directorate, Environment Canada, Ottawa.

ISO 1990 *Water Quality Sampling. Part 1: Guidelines on the Design of Sampling Programmes*. International Standard ISO 5667-1, International Organization for Standardization, Geneva.

ISO 1990 *Water Quality Sampling. Part 2: Guidelines on Sampling Techniques*. International Standard ISO 5667-2, International Organization for Standardization, Geneva.

ISO 1990 *Water Quality Sampling. Part 3: Guidelines on the Preservation and Handling of Samples*. International Standard ISO 5667-3, International Organization for Standardization, Geneva.

ISO 1990 *Water Quality Sampling. Part 4: Guidelines for Sampling from Lakes, Natural and Man-Made*. International Standard ISO 5667-4, International Organization for Standardization, Geneva.

ISO 1990 *Water Quality Sampling. Part 6: Guidelines for Sampling of Rivers and Streams*. International Standard ISO 5667-6, International Organization for Standardization, Geneva.

Lloyd, B. and Helmer, R. 1991 *Surveillance of Drinking-Water Quality in Rural Areas*. Longman, Harlow.

WHO 1985 *Guidelines for Drinking-Water Quality. Volume 3, Drinking-Water Quality Control in Small-Community Supplies*. World Health Organization, Geneva.

WHO 1986 *Establishing and Equipping Water Laboratories in Developing Countries*. Unpublished document WHO/PEP/86.2, World Health Organization, Geneva.

WMO 1988 *Manual on Water Quality Monitoring: Planning and Implementation of Sampling and Field Testing*. Operational Hydrology Report No. 27, World Meteorological Organization, Geneva.

# Chapter 6

# FIELD TESTING METHODS

Analyses for many important physical, chemical and microbiological variables can be carried out in the field using apparatus made specifically for field use. A significant advantage of field analysis is that tests are carried out on fresh samples whose characteristics have not been contaminated or otherwise changed as a result of storage in a container. This is of special importance for samples that are to undergo microbiological analysis but cannot be transported to a laboratory within the time limits or under the conditions described in Chapter 5. Some variables must be measured in the field, either *in situ* or very soon after the sample has been collected. Field analysis is necessary for temperature, transparency and pH. Dissolved oxygen may be determined in the field or the sample may be treated (fixed) in the field and the remainder of the analysis completed in a laboratory. If samples are to be chemically preserved before being transported to the laboratory, conductivity (if required) must be measured before preservative chemicals are added.

Another advantage of field analysis is that samples are highly unlikely to lose the labels that identify the time and place of sampling. Loss of such identification would be disastrous if, for example, many samples had been collected to determine the water quality profile of a river.

Where there are no laboratories within a reasonable distance of the sampling stations, field analysis may be the only feasible way to obtain water quality information. Mobile laboratories are expensive to set up and maintain, while a temporary laboratory is justified only if a large sampling and analysis programme is to be carried out within a relatively compact sampling area.

The limitations of field analysis must, however, be recognised. Some of the methods used in field analysis produce less accurate results than those that can be used in a well-equipped laboratory. In addition, the limits of detection and the reproducibility of field analyses will often be less than is possible with laboratory methods. Furthermore, it is difficult to implement an analytical quality assurance programme when analyses are done in the field, although some attempt should be made to control the quality of field results.

*This chapter was prepared by R. Ballance*

Several manufacturers produce equipment designed specifically for use in the field (see Appendix 1 for examples). This equipment, often in the form of a "kit", almost invariably contains instruments, glassware, reagent packages and other consumables that are unique to the kit. Replacement of broken components and replenishment of reagents and consumables therefore requires re-supply from the manufacturer or an approved agent, and many developing countries experience difficulties with this. Extensive delays may be experienced in replacement and re-supply and, for various reasons, these replacements may prove expensive. While the same problems apply to standard laboratory supplies and reagents, there is always the possibility of borrowing from another laboratory while waiting for an order to be filled.

The ease with which the procedure involved in a field analysis may be followed will depend on the complexity of the procedure and the climatic and other conditions at the time of analysis. For example, a field measurement that can be made with an electrode and a meter (temperature, pH, conductivity, dissolved oxygen) is normally very easily accomplished provided that little or no meter calibration is necessary. On the other hand, photometric or titrimetric procedures that involve the addition of two or more reagents, a reaction time of several minutes, and the observation of a colour change or a photometer reading might present difficulties under some field conditions.

In using any field kit it is essential to follow exactly the procedures specified in the manufacturer's instruction manual. Carefully measured quantities of reagents may be supplied in pre-packaged form, and use of a different concentration of a chemical from that recommended would distort the result of an analysis. Even when methods are followed exactly it is vital that they are validated before use. Method validation is discussed in Chapter 9 and is especially important where on-site testing is to be undertaken because analytical quality assurance is more difficult, as noted above. Care must also be taken to ensure that the batteries in any battery-powered apparatus (such as a photometer or a pH meter) are fresh and are supplying the correct voltage.

As an alternative to purchasing a field kit, it is possible for an analyst to assemble and package the glassware and chemicals needed to analyse for some of the variables. If this is done, plastic (as opposed to glass) bottles, burettes, pipettes, flasks and beakers should be used whenever possible, and the quantity of reagent chemicals should be no greater than is needed for one sampling expedition. The field kit thus made should be "field-tested" close to the laboratory where it is assembled. Staff should practise using the kit to identify any problems associated with its use; they should also make realistic estimates of the supplies required and the time that will be needed to conduct the planned series of analyses under field conditions. For example, they should establish how much distilled water is needed for rinsing glassware, whether paper towels or wiping cloths are needed, whether a table is needed to hold glassware during the analyses. This sort of trial run before analytical work is attempted in the field must be part of the training course for all field staff who will carry out on-site testing, regardless of the type of field test kit or portable equipment that is being used.

Appendix 1 lists some suppliers of water testing equipment, although their inclusion in this handbook is not necessarily an endorsement of these suppliers. It is worth noting that many laboratory suppliers market other companies' products under their own name. Sometimes the general laboratory suppliers are less expensive than the manufacturer because they are able to buy in bulk. Price quotations should therefore be sought from several sources. It is important to be aware of the full extent of the package when a kit is purchased for water testing: many suppliers do not mention consumable costs and there may be expensive calibration standards that must be purchased at extra cost.

The fact that field testing assays are performed away from the laboratory should not mean that less care is taken to ensure reliability of results. The use of quality assessment and internal quality control techniques to ensure and monitor quality is still possible (see Chapter 9).

## 6.1 Temperature

Temperature must be measured *in situ* because a water sample will gradually reach the same temperature as the surrounding air. If it is not possible to measure the temperature *in situ*, a sample must be taken from the correct location and depth of the sampling station and its temperature measured immediately it is brought to the surface.

Temperature is measured with a glass thermometer, either alcohol/toluene-filled or mercury-filled, with 0.1 °C graduations, or an electronic thermometer of the type that is usually an integral part of a dissolved oxygen meter or a conductivity meter.

*Procedure*

The procedure to follow depends on the type of thermometer being used and on whether direct access to the point at which the temperature is to be measured is impossible (as, for example, when the water to be tested is in a deep well or when a water sample can be taken only from a bridge).

1. When a glass thermometer is used and the testing point can be reached, immerse the thermometer in the water until the liquid column in the thermometer stops moving (approximately 1 minute, or longer if necessary). For a pumping well, immerse the thermometer in a container with water flowing through until the temperature stabilises. Record the reading to the nearest 0.1 °C.
2. When either a glass thermometer or an electronic thermometer is used and the measurement point is inaccessible, obtain a water sample of at least 1 litre. Rinse

the thermometer (or the probe) with a portion of the sample and discard the rinse water. Immerse the thermometer (or the probe) in the sample. Hold it there for approximately 1 minute (longer if the temperature reading has not become constant). Record the reading to the nearest 0.1 °C.

3. When an electronic thermometer having a probe with long leads is used, lower the probe to the required depth. Hold it at that depth until the reading on the meter is constant. Record the temperature to the nearest 0.1 °C and the depth to the nearest 10cm. Lower (or raise) the probe to the next measurement point for the next reading.

## 6.2 Transparency

Transparency is a water quality characteristic of lakes and reservoirs and can be measured quickly and easily using simple equipment. This characteristic varies with the combined effects of colour and turbidity. Some variation may also occur with light intensity and with the apparatus used.

The apparatus used for transparency measurement is called a Secchi disc — named after Secchi, who first used it in 1865 to measure the transparency of the Mediterranean Sea. The disc is made of rigid plastic or metal, but the details of its design are variable. It may be 20 to 30 cm or even larger in diameter and is usually painted white. Alternatively, it may be painted with black and white quadrants.

The disc is suspended on a light rope or chain so that it remains horizontal when it is lowered into the water. The suspension rope is graduated at intervals of 0.1 and 1 metre from the level of the disc itself and usually does not need to be more than 30 m in length. A weight fastened below the disc helps to keep the suspension rope vertical while a measurement is being made. Figure 6.1 shows a typical Secchi disc.

The same size and pattern of disc should be used at any given sampling station so that a series of measurements made over a number of years will be as free as possible from distortions arising from differences in apparatus.

Use of a boat to reach the measurement site is essential.

*Procedure*

The observation should not be made early in the morning or late in the afternoon.

1. Lower the Secchi disc, where possible, through a shaded area of water surface (glare on the water surface can distort the observation).
2. As the disc is lowered, note the depth at which it just disappears from view.
3. Lower the disc a little further, then raise it and note the depth at which it reappears.

*Reporting*

The average of the two depth readings is reported as the Secchi disc transparency. The report must also state the diameter of the disc and the pattern, if any, on the upper surface of the disc.

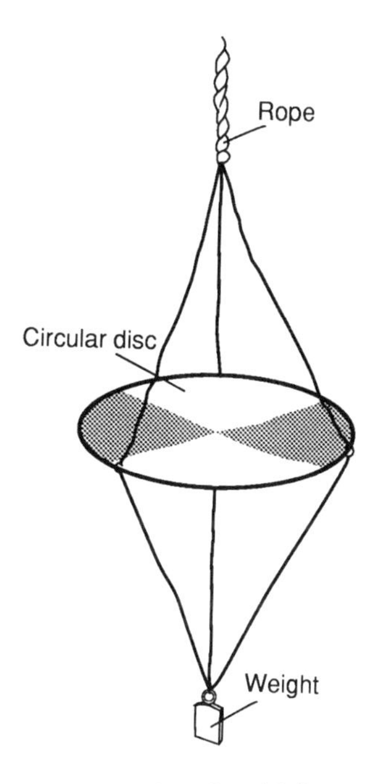


**Figure 6.1** The Secchi disc

## 6.3 pH

Determination of the pH of water should, if possible, be made *in situ*. If this is not possible, for example with well water or when access to a lake or river is very difficult, the measurement should be made immediately after the sample has been obtained.

There are three different methods of pH measurement: pH indicator paper, liquid colorimetric indicators and electronic meters. The use of pH indicator paper is simple and inexpensive, but the method is not very accurate and requires a subjective assessment of colour by the user. Liquid colorimetric indicators change colour in accordance with the pH of the water with which they are mixed. The colour that develops can then be compared with a printed card, with coloured glass standards, or with a set of prepared liquid standards. Colorimetric methods are reasonably simple and accurate to about 0.2 pH units. Their main disadvantage is that standards for comparison or a comparator instrument must be transported to the sampling station. Moreover, physical or chemical characteristics of the water may interfere with the colour developed by the indicator and lead to an incorrect measurement. The third method, electrometric pH measurement, is accurate and free from interferences. Pocket-sized, battery-powered, portable meters that give readings with an accuracy of ± 0.05 pH units are suitable for field use. Larger, more sophisticated models of portable meter can attain an accuracy of ± 0.01 pH units. Care must be taken when handling such equipment. The electrodes used for measurement generally need replacing periodically (e.g. yearly). Old or poor quality electrodes often show a slow drift in the readings.

### 6.3.1 Measurement of pH using colour indicators

A comparator and colour discs are required for this method of measuring pH. The instrument made by one manufacturer uses colour discs for the pH ranges and indicators listed below.

| Indicator | pH range |
|---|---|
| Universal | 4.0–11.0 |
| Bromocresol green | 3.6–5.2 |
| Methyl red | 4.4–6.0 |
| Bromocresol purple | 5.2–6.8 |
| Bromothymol blue | 6.0–7.6 |
| Phenol red | 6.8–8.4 |
| Thymol blue | 8.0–9.6 |
| Phenolphthalein | 8.6–10.2 |

*Procedure*

1. Fill three comparator cells to the 10-ml mark with portions of the water sample and place one of the cells in the left-hand compartment of the comparator.
2. Add 1 ml of universal indicator to one of the cells and mix well; place the cell in the right-hand compartment. Compare the colour in the right-hand cell with the glass standards of the universal disc.
3. From the above list, choose an indicator that has the mid-point of its range near to the approximate pH determined with the universal indicator. Add 0.5 ml of this indicator to the third comparator cell, mix, and place the cell in the right-hand compartment of the comparator.
4. Put the appropriate standard disc in the comparator and compare the colour of the sample with the glass standards on the disc. Record the result to the closest 0.2 pH unit.

### 6.3.2 Measurement of pH using a pH meter

There are many models of pH meter and it is beyond the scope of this manual to describe them all. The common features are a sensing electrode and a reference electrode connected to an electronic circuit that amplifies the voltages produced when the electrodes are immersed in a solution or water sample. The amplified voltage is displayed on a meter graduated in pH units. Sensing and reference electrodes designed for field use are often combined in one element. The electronic circuitry in a portable meter is powered by either disposable or rechargeable batteries, depending on the design of the meter.

It is possible to purchase a more complex instrument that is designed for measurement of conductivity and temperature as well as pH. Some instruments offer additional features and are equipped with circuitry and probes that allow the measurement of dissolved oxygen, redox potential and/or turbidity. It is not possible to provide detailed operating instructions for all of the many makes and models of pH meter. Operating and maintenance instructions are supplied by the manufacturer. There is, however, a general procedure that should be followed. The sensing, or glass, electrode must be soaked in distilled water for several hours before use when it is new or if it has dried out during storage of more than a day. When a glass electrode is not in use for more than a few hours, its tip (the lower 1–2 cm) should be kept immersed in distilled water. The tips of glass electrodes should be carefully protected against abrasion and breakage.

The reference electrode is usually supplied with a rubber cap that protects the tip against breakage as well as preventing the crystallisation of dissolved salts on the tip. A hole in the side of the electrode is provided for filling the body of the electrode with saturated potassium chloride (KCl) solution. The correct liquid level is approximately 5 mm below the bottom edge of the hole. When the electrode is not in use the hole should be covered with a rubber sleeve that slides over the body of the electrode.

One or more buffer solutions are necessary for standardising the meter. It is usual for the manufacturer to provide a container of buffer solution with the meter, and the supply may be replenished with purchases from either the manufacturer or a chemicals supplier. Buffer solutions may also be prepared in an analytical laboratory where volumetric glassware and an analytical balance (accurate to ± 1 mg) are available.

*Procedure*

*Standardising the meter*

1. Remove the protective rubber cap and slide the rubber sleeve up to expose the hole in the side of the reference electrode.
2. Rinse both electrodes with distilled water and blot them dry with soft absorbent paper.
3. Pour sufficient buffer solution into a beaker to allow the tips of the electrodes to be immersed to a depth of about 2 cm. The electrodes should be at least 1 cm away from the sides and the bottom of the beaker.
4. Measure the temperature of the buffer solution with a thermometer and set this on the temperature adjustment dial of the meter (if the meter is so equipped). Some meters have an automatic temperature adjustment feature.
5. Turn on the pH meter.

6. Adjust the needle on the pH dial to the known pH of the buffer. If the needle keeps jumping, check that the leads from the electrodes are firmly connected to the meter. When the needle stops moving, make the fine adjustment.
7. Turn the instrument to stand-by (if it is equipped for this).
8. Raise the electrodes clear of the buffer solution. Remove the buffer and rinse the electrodes with distilled water.
9. Proceed to determination of pH of the sample. If the sample is not ready, place the electrodes in distilled water.

*Determination of pH of sample*

1. The electrodes are either immersed in, or have been rinsed with, distilled water. Remove them from the water and blot dry.
2. Rinse the electrodes and a small beaker with a portion of the sample.
3. Pour sufficient of the sample into the small beaker to allow the tips of the electrodes to be immersed to a depth of about 2 cm. The electrodes should be at least 1 cm away from the sides and the bottom of the beaker.
4. Measure the temperature of the water sample and set the temperature adjustment dial accordingly (if the instrument does not have automatic temperature compensation).
5. Turn on the pH meter.
6. Read the pH of the water sample on the dial of the meter. Make sure that the needle has stopped moving before the pH is recorded.
7. Turn the pH meter to stand-by and raise the electrodes out of the sample. Remove the sample and discard it. Rinse the electrodes and the beaker with distilled water, and blot the electrodes with soft tissue.
8. If other samples are to be tested, repeat steps 2 to 7.
9. If no other samples are to be tested, slide the rubber sleeve down to cover the hole in the side of the reference electrode and replace the protective rubber cap on the tip.
10. Switch the meter off and pack it in its carrying case for transport.

*Reporting*

Report the pH and the temperature of the water at the time the measurement was made.

## 6.4 Conductivity (or specific conductance)

The ability of water to conduct an electric current is known as conductivity or specific conductance and depends on the concentration of ions in solution. Conductivity is measured in millisiemens per metre ($1\ mS\ m^{-1} = 10\ \mu S\ cm^{-1} = 10\ \mu mhos\ cm^{-1}$). The measurement should be made *in situ*, or in the field immediately after a water sample has been obtained, because conductivity changes with storage time. Conductivity is also temperature-dependent; thus, if the meter used for measuring conductivity is not equipped with automatic temperature correction, the temperature of the sample should be measured and recorded.

Many manufacturers of scientific equipment produce conductivity meters. The apparatus consists of a conductivity cell containing two rigidly attached electrodes, which are connected by cables to the body of the meter. The meter contains a source of electric current (a battery in the case of portable models), a Wheatstone bridge (a device for measuring electrical resistance) and a small indicator (usually a galvanometer). Some meters are arranged to provide a reading in units of conductance (mhos), while others are graduated in units of resistance (ohms). The conductivity cell forms one arm of the Wheatstone bridge. The design of the electrodes, i.e. shape, size and relative position, determines the value of the cell constant, $K_c$, which is usually in the range 0.1 to 2.0. A cell with a constant of 2.0 is suitable for measuring conductivities from 20 to 1,000 mS $m^{-1}$.

The cell constant $K_c$ can be determined by using the apparatus to measure the conductivity of a standard solution (0.0100 mol $l^{-1}$) of potassium chloride and dividing the true conductivity of the solution (127.8 mS $m^{-1}$ at 20 °C) by the measured conductivity, $K_c = C_t / C_m$. Care must be taken to ensure that measured and true conductivities are expressed in the same units. The temperature of the solution is critical because electrolytic conductivity increases with temperature at a rate of approximately 1.9 per cent per °C. Some meters provide a reading of resistance, which is the reciprocal of conductance. When resistance is measured, the cell constant is calculated by dividing the measured resistance by the true resistance, $K_c = R_m / R_t$.

Platinised electrodes require replating if readings become erratic or when the platinum black coating peels or flakes off. The replating procedure is not difficult but it should be done in the laboratory. Stainless steel electrodes are more appropriate for field use but they must be kept clean. If they become contaminated, for example with oily wastewater, they must be cleaned with a solvent, then with alcohol and finally be well rinsed with distilled water. When not in use the cell should be wiped dry and stored in its carrying case.

*Reagents*

✓ Distilled water for preparing standard potassium chloride solution should have a very low conductivity. It must not contain $CO_2$. Use redistilled water and boil it immediately before use. Allow to cool in a hard-glass bottle fitted with a $CO_2$ trap.

✓ Standard potassium chloride solution, 0.0100 mol $l^{-1}$, for the calibration of electrodes and determination of the cell constant. Dissolve 0.7456 g of anhydrous KCl (dried at 105 °C and cooled in a desiccator) in $CO_2$-free distilled water. Make up to 1,000 ml at 20 °C. Store in a hard-glass bottle fitted with a $CO_2$ trap. The conductivity of this solution is 127.8 mS $m^{-1}$ at 20 °C.

*Procedure*

*Determination of cell constant*

1. Rinse out the conductivity cell with at least three portions of standard KCl solution.
2. Adjust the temperature of a fourth portion of the solution to 20 ± 0.1 °C (or as near as possible to that temperature).

3. Immerse the conductivity cell in a sufficient volume of the KCl solution for the liquid level to be above the vent holes in the cell. There should be no air bubbles clinging to the electrodes and the cell should not be closer than 2 cm to the sides and bottom of the container.

4. Observe and record the temperature of the KCl solution to the nearest 0.1 °C. Some meters have built in thermometers and/or automatic temperature compensation.

5. Turn the meter on. Follow the manufacturer's operating instructions and record the meter reading.

6. Calculate the cell constant. The formula includes a factor that compensates for the difference in temperature if the reading was taken at a temperature other than 20.0 °C. The value of the temperature correction factor [0.019($t$ – 20) + 1] can be determined from the graph in Figure 6.2.

   If *conductivity* was measured, the calculation is:

$$K_c = (127.8/C_{KCl}) \times [0.019(t - 20) + 1]$$

   where $C_{KCl}$ = measured conductivity (µmhos)
   $t$ = observed temperature (°C)
   $K_c$ = the cell constant ($cm^{-1}$).

   If *resistance* was measured, the calculation is:

$$K_c = R_{KCl} \times 0.001278 \times [0.019(t - 20) + 1]$$

   where $R_{KCl}$ = measured resistance (ohms)
   $t$ = observed temperature (°C)
   $K_c$ = the cell constant ($cm^{-1}$)

*Measurement of sample conductivity*

1. Rinse the conductivity cell with at least three portions of the sample.

2. Adjust the temperature of a portion of the sample to 20 ± 0.1 °C (or as close as possible to that temperature).

3. Immerse the conductivity cell containing the electrodes in a sufficient volume of the sample for the liquid level to be above the vent holes in the cell. There should be no air bubbles clinging to the electrodes and the cell should not be closer than 2 cm to the sides and bottom of the container.

4. Observe and record the temperature of the sample to the nearest 0.1 °C. Some meters have built in thermometers and/or automatic temperature compensation.

5. Turn the meter on. Follow the manufacturer's operating instructions and record the meter reading.

6. Turn the meter off and pack it and the electrode in the carrying case for transport.

*Calculation*

Electrolytic conductivity increases with temperature at a rate of 1.9 per cent per °C. Conductivity measurements will therefore be the most accurate when made at the same temperature as that at which the cell constant is determined. While a temperature correction factor is included in the calculation, the temperature coefficient, 0.019, is for standard (0.0100 mol $l^{-1}$) KCl solution: the temperature coefficient for water will usually be different. Thus, if the temperature is much higher or much lower than 20 °C, an error will be introduced into the calculation.

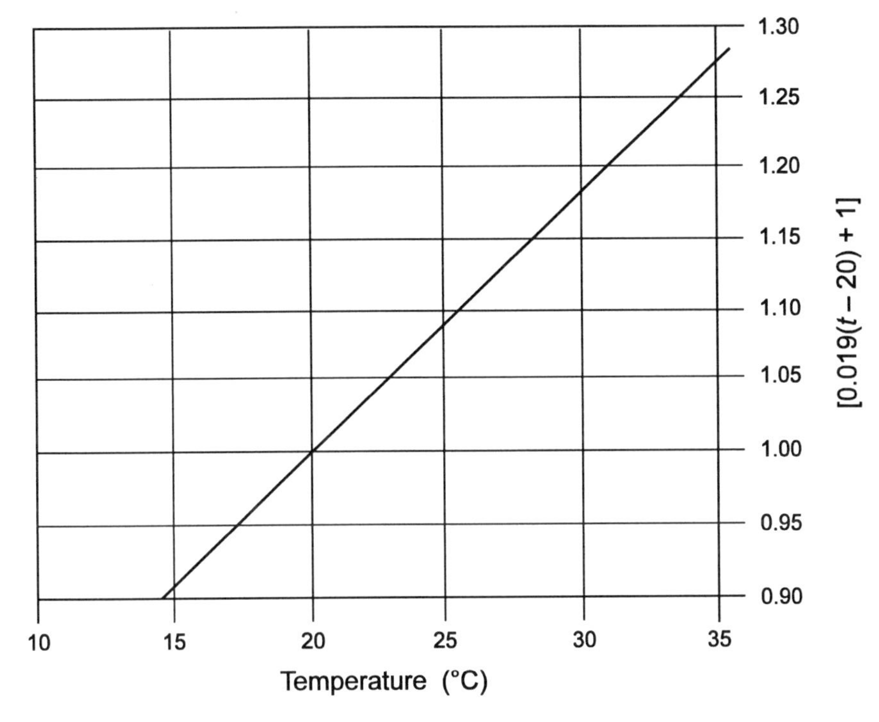


**Figure 6.2** Values of [0.019($t$ – 20) + 1]

When *conductivity* of the sample has been measured, the calculation is:

$$\text{Conductivity} = \frac{C_m \times K_c}{0.019(t-20)+1}\ \mu\text{mhos cm}^{-1}$$

When *resistance* of the sample has been measured, the calculation is:

$$\text{Conductivity} = \frac{10^6 \times K_c}{R_m\left[0.019(t-20)+1\right]}\ \mu\text{mhos cm}^{-1}$$

where $K_c$ = cell constant ($cm^{-1}$)
$C_m$ = measured conductivity of sample at $t$ °C (µmhos $cm^{-1}$)
$R_m$ = measured resistance of sample at $t$ °C (ohms)
$t$ = temperature of sample (°C).

*Note:* The value for [0.019($t$ – 20) + 1] may be taken from the graph in Figure 6.2.

*Reporting*
Record the meter reading, the units of measurement, and the temperature of the sample at the time of reading. Report conductivity at 20 °C.

## 6.5 Dissolved oxygen

The dissolved oxygen concentration depends on the physical, chemical and biochemical activities in the water body, and its measurement provides a good indication of water quality. Changes in dissolved oxygen concentrations can be an early indication of changing conditions in the water body.

Two main methods are available for the determination of dissolved oxygen: the Winkler method and the electrometric method using membrane electrodes. Use of the Winkler method requires the addition of three chemical reagents to the sample very soon after it is obtained. The dissolved oxygen concentration (in mg $l^{-1}$) is then determined by titration with sodium thiosulphate solution, which may be done in the field or up to 6 hours later in a laboratory. The electrometric method is suitable for the field determination of dissolved oxygen and is simple to perform. It requires an electrically powered meter and an appropriate electrode. The result it gives requires the application of correction factors to compensate for salinity and temperature; some meters have built in temperature compensation.

### 6.5.1 Winkler method

*Apparatus*

- ✓ BOD bottle, capacity 250 to 300 ml
- ✓ Graduated cylinder
- ✓ Flask
- ✓ Burette (or other device for dispensing and measuring liquid)
- ✓ Pipettes (or similar means of adding reagents)

*Reagents*

- ✓ Distilled water in a rinse bottle.
- ✓ Manganous sulphate solution. Dissolve 500 g manganous sulphate pentahydrate, $MnSO_4.5H_2O$, in distilled water. Filter if there is any undissolved salt, and make up to 1 litre.
- ✓ Alkaline-iodide-azide solution. Dissolve 500 g sodium hydroxide, NaOH (or 700 g potassium hydroxide, KOH), and 135 g sodium iodide, NaI (or 150 g potassium iodide, KI), in distilled water and dilute to 1 litre. Sodium and potassium salts may be used interchangeably. Dissolve 10 g sodium azide, $NaN_3$, in 40 ml distilled water and add to the NaOH/NaI mixture. This reagent should not give a colour with starch when diluted 1:25 and acidified.
- ✓ Concentrated sulphuric acid.
- ✓ Starch indicator solution. Make a smooth paste by blending 1 g of soluble starch with a little cold distilled water in a beaker of capacity at least 200 ml. Add 200 ml of boiling distilled water while stirring constantly. Boil for 1 minute and allow to cool. Store in a refrigerator or at a cool temperature. Alternatively, thiodene powder may be used as an indicator.
- ✓ Sodium thiosulphate solution (0.025 mol $l^{-1}$ for 200 ml sample). Dissolve 6.3 g sodium thiosulphate pentahydrate, $Na_2S_2O_3.5H_2O$, in distilled water and make up to 1 litre. Standardise against KI. Add either 1 ml chloroform or 10 mg mercuric iodide to stabilise the solution. Store in a brown bottle.

*Procedure*

The procedure described here assumes that the sample has already been properly collected in a dissolved oxygen sampler and is contained in a BOD bottle.

1. Remove the BOD bottle containing the sample from the dissolved oxygen sampler and insert the matching ground-glass stopper in the neck of the bottle. Be sure that no air bubbles have been trapped under the stopper and maintain a water seal around the stopper until ready for the next step of the procedure.
2. Pour off the water seal and remove the ground glass stopper. Add 1 ml of $MnSO_4$ solution, then 1 ml of alkaline-iodide-azide solution. For both additions, hold the tip of the pipette against the inside of the bottle neck to prevent splashing.
3. Replace the ground-glass stopper, being careful to avoid trapping air bubbles under it.
4. Mix the contents by inverting the bottle several times. Keep a finger over the stopper during mixing to make sure that it does not fall out. A brown floc will form in the bottle before and during the mixing. If there was no dissolved oxygen in the sample, the floc will be white. When the bottle is set down the floc will settle, leaving a clear liquid above it.
5. Allow the floc to settle between a half and two-thirds of the way down the bottle, then mix again as in step 4 (above). Allow to settle once more, until all of the floc is in the lower third of the bottle.
6. Remove the stopper, add 1 ml of $H_2SO_4$ without splashing, replace the stopper and mix the contents of the bottle by inverting it several times. The floc will disappear and the liquid in the bottle will be a yellowish-brown colour. If there was no dissolved oxygen in the sample the liquid will be colourless.

   *Note:* The dissolved oxygen in the sample is now "fixed". The amount of iodine that has been released from the reagent (causing the yellow-brown colour) is proportional to the amount of oxygen that was in the sample. If the bottle is kept tightly stoppered it may be stored for up to 6 hours before step 8, titration with sodium thiosulphate solution.
7. Transfer a volume, $V_t$, corresponding to 200 ml of the original sample to the flask. Adjustment should be made to compensate for the amount by which the sample was diluted when 1 ml of $MnSO_4$ and 1 ml of the alkaline-iodide-azide solution were added. If, for example, a 300-ml BOD bottle is used, the volume would be:

$$V_t = 200 \times \frac{300}{300-2} = 201.3 \text{ ml}$$

8. Titrate with sodium thiosulphate solution (0.025 mol $l^{-1}$), stirring the contents of the flask until the yellow-brown colour fades to a pale straw colour. Add a few drops of starch solution and a blue colour will develop. Continue titrating a drop at a time until the blue colour disappears.

*Calculation*

For titration of 201.3 ml (200 ml of sample plus 1.3 ml allowance for reagents) with 0.025 mol $l^{-1}$ sodium thiosulphate:

$$1 \text{ ml } Na_2S_2O_3 \text{ solution} = 1 \text{ mg } l^{-1} \text{ dissolved oxygen}$$

If sodium thiosulphate is used at a strength other than 0.025 mol $l^{-1}$, and if the sample volume titrated is other than 200 ml (excluding the volume added to compensate for the chemical reagents as described in step 7), the dissolved oxygen in the sample may be calculated from the following formula:

$$\text{Dissolved oxygen} = \frac{\text{ml titrant} \times \text{mol } l^{-1} \text{ titrant} \times 8{,}000}{\text{volume of sample titrated}} \text{ mg } l^{-1}$$

### 6.5.2 Electrometric method

*Apparatus*

✓ Battery-powered meter. This is a meter designed specifically for dissolved oxygen measurement. Other meters, such as a specific ion meter or an expanded scale pH meter, may also be used.

✓ Oxygen-sensitive membrane electrode.

*Procedure*

1. Follow exactly the calibration procedure described in the manufacturer's operating instructions. Generally, electrodes are calibrated by reading against air or against a sample of known dissolved oxygen content. This "known" sample could be one for which dissolved oxygen concentration has been determined by the Winkler method or one that has been saturated with oxygen by bubbling air through it. The zero end of a calibration curve can be determined by reading against a sample containing no dissolved oxygen, prepared by adding excess sodium sulphite, $Na_2SO_3$, and a trace of cobalt chloride, $CoCl_2$, to the sample.
2. Rinse the electrode in a portion of the sample which is to be analysed for dissolved oxygen.
3. Immerse the electrode in the water, ensuring a continuous flow of water past the membrane to obtain a steady response on the meter.
4. Record the meter reading and the temperature, and the make and model of the meter.
5. Switch the meter off and pack it and the electrode in the carrying case for transport.

## 6.6 Thermotolerant (faecal) coliforms

Samples for microbiological testing are very prone to changes during transport and storage and there is therefore considerable advantage in field testing for variables such as thermotolerant (faecal) coliforms.

Almost all kits for field testing for thermotolerant coliforms are based on the membrane filtration method. This method is described in detail in Chapter 10 and is not repeated here.

An example of a kit originally developed for testing drinking water samples is provided in Figure 6.3. The kit includes equipment to test for thermotolerant coliforms, turbidity, pH, residual chlorine and (optionally) conductivity. This and some similar kits include an integral incubator and rechargeable battery which provide independence from mains electricity for up to a week.

### 6.6.1 Disinfecting equipment in the field

Disinfection of equipment is essential for many microbiological analytical procedures, but some of the devices used for this purpose in a laboratory are unsuitable for transporting to the field. Some of the methods for disinfecting equipment in the field are:

- *Dry heat:* The flame from a gas cigarette-lighter, for example, can be used to disinfect forceps for manipulation of membrane filters. It must be a

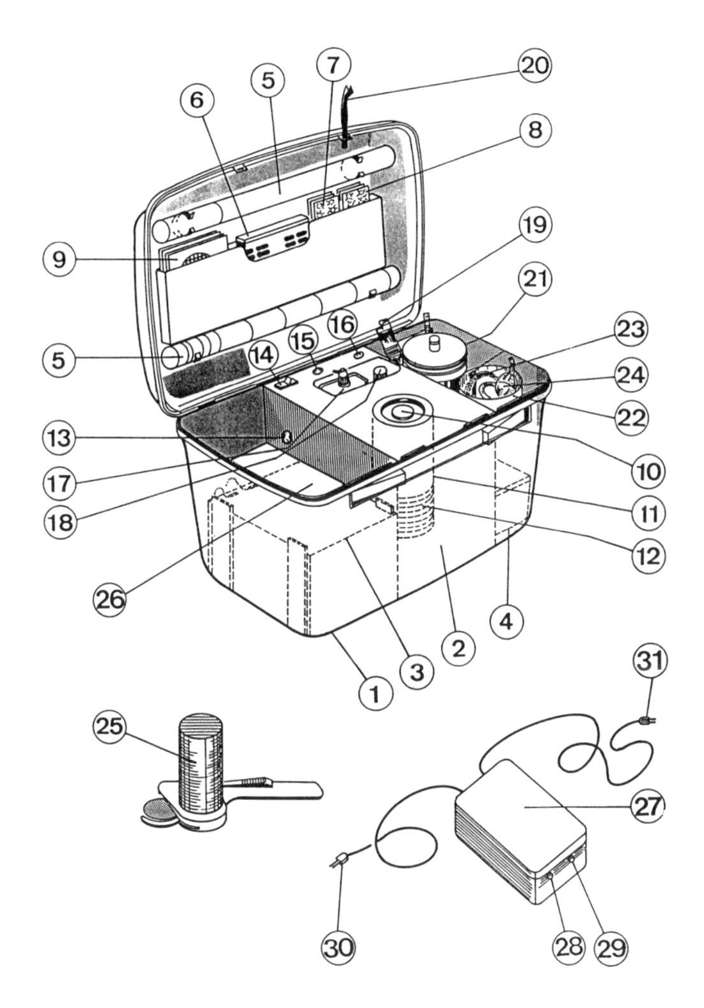

1. Case
2. Incubator
3. Battery
4. Spares case
5. Turbidity tubes (pair)
6. Chlorine and pH comparator
7. Chlorine test tablets
8. pH test tablets
9. Membrane filters
10. Incubator lid
11. Incubator pot
12. Petri dishes
13. Power socket
14. On/Off switch
15. Power “On” indicator
16. Heater “On” indicator
17. Methanol dispenser
18. Culture medium bottle
19. Lighter
20. Tweezers
21. Filtration assembly with sample cup
22. Vacuum cup
23. Sample cable
24. Vacuum pump
25. Absorbent pad dispenser
26. Storage space

*Charger unit*

27. Battery charger/mains power unit
28. Power “On” indicator
29. Charged battery indicator
30. Incubator plug
31. Mains supply plug

**Figure 6.3** Example of a field kit that includes the facility to test for thermotolerant coliforms. Reproduced with the permission of the Robens Institute, Guildford, UK

butane or propane gas lighter, not one that uses gasoline or similar liquid fuel, which would blacken the forceps.

- *Formaldehyde:* This gas is a powerful bactericide. It may be generated by the combustion of methanol (but no other alcohol) in a closed space where oxygen becomes depleted. In the field, this is a convenient way to disinfect the filtration apparatus between uses. A minimum contact time of 10 minutes is recommended.
- *Disinfecting reusables:* The two main items of reusable equipment, Petri dishes (glass or aluminium) and bottles, may be disinfected by immersion in boiling water for a few minutes, by dry heat sterilisation in an oven or by heating in a pressure cooker for at least 20 minutes.
- *Disposal of contaminated material:* Autoclaving (or pressure cooking) of contaminated material is impractical in the field. Contaminated materials such as membrane filters and pads may be burned.

## 6.7 Quality assurance in the field

Quality control of analytical procedures is as important in field testing as it is in traditional laboratory analysis. This topic is covered in detail in Chapter9 but certain problems are peculiar to field testing. Such problems may include one or more of the following:

- Staff performing the analyses may not be fully trained in chemical or microbiological analyses.
- The conditions of sampling and analysis may vary considerably between sampling sites.
- Analytical methodology may be of limited accuracy and precision because of the need for equipment to be portable and easy to use on site.

Such problems can lead to the generation of erroneous data.

All field testing should be conducted to specified standards. Training in the proper use of the equipment and a programme of regular equipment maintenance and calibration are essential. In addition, there should be an internal quality control system, similar to those described in Chapter 9, for each analytical technique.

Ideally, field analysis should be performed under the supervision of a central laboratory. This allows the independent checking of quality control data before they are accepted into any formal data base. Quality assurance of field assays can be integrated with the quality system used by the central laboratory.

When a link with an analytical laboratory is not possible, appropriate quality control procedures should be implemented and monitored by the management of the field testing service, in particular by personnel who are not directly involved with generation of data. This again ensures independent checking of data. Often the expertise required to conduct monitoring is available only from field testing staff themselves, in which case it is advisable for management to institute some form of audit system to monitor the validity of reported data.

## 6.8 Source literature and further reading

APHA 1992 *Standard Methods for the Examination of Water and Wastewater.* 18th edition. American Public Health Association (APHA), American Water Works Association (AWWA) and Water Pollution Control Federation (WPCF), Washington, DC.

AWWA 1978 *Simplified Methods for the Examination of Drinking Water.* American Water Works Association, Denver, CO.

Hach Company 1989 *Hach Water Analysis Handbook.* Hach Company, Loveland, CO.

Hutton, L.G. 1983 *Field Testing of Water in Developing Countries.* Water Research Centre, Medmenham.

ISO 1990 *Water Quality—Detection and Enumeration of Coliform Organisms, Thermotolerant Coliform Organisms and Presumptive Escherichia coli. Part 1: Membrane Filtration Method.* International Standard ISO 9308-1, International Organization for Standardization, Geneva.

WHO 1992 *GEMS/WATER Operational Guide.* Third edition, Unpublished WHO document GEMS/W.92.1, World Health Organization, Geneva.

# Chapter 7

# PHYSICAL AND CHEMICAL ANALYSES

In compiling this chapter, care has been taken to avoid procedures that require delicate or sophisticated equipment. For many of the variables for which methods of analysis are presented here, further information relating to their selection and inclusion in water quality monitoring and assessment programmes (such as their environmental significance, normal ranges of concentrations, and behaviour in the aquatic environment) can be found in the companion guidebook *Water Quality Assessments*.

## 7.1 Preparation and use of chemical reagents

The following general rules should be followed in the preparation and use of chemical reagents. The best quality chemical reagents available should be used — normally "analytical reagent grade". For most laboratory purposes, water distilled in a borosilicate glass still or a tin still will be satisfactory. For preparing some reagents, dilution water requires special treatment, such as a second distillation, boiling to drive off $CO_2$ or passing through a mixed bed ion exchanger. Where such special treatment is necessary, this is stated.

Recipes for the preparation of reagents usually give directions for the preparation of a 1-litre volume. For those reagents that are not used often, smaller volumes should be prepared by mixing proportionally smaller quantities than those given in the recipe. Where a working standard or working solution is to be made by dilution of a stock solution, no more of the stock solution should be prepared than will be used within the next six months. Furthermore, only the amount of stock solution necessary to meet the immediate need for a working or standard solution should be diluted at one time.

Reagent solutions should be kept in tightly stoppered glass bottles (except where they are incompatible with glass, as with silica solutions). Rubber or neoprene stoppers or screw tops with gaskets are suitable, provided that the reagents do not react with these materials. For short-term storage, for example during a field trip of a week or two, small quantities of reagent may be transported in plastic bottles with plastic screw caps.

Reagent containers should always be accurately labelled with the name of the reagent, its concentration, the date that it was prepared and the name or initials of the person who prepared it.

*This chapter was prepared by R. Ballance*

**Table 7.1** Characteristics of three common acids

| Characteristic | Hydrochloric acid (HCl) | Sulphuric acid ($H_2SO_4$) | Nitric acid ($HNO_3$) |
|---|---|---|---|
| Relative density of reagent grade concentrated acid | 1.174–1.189 | 1.834–1.836 | 1.409–1.418 |
| Percentage of active ingredient in concentrated acid | 36–37 | 96–98 | 69–70 |
| Molarity of concentrated acid (mol $l^{-1}$) | 11–12 | 18 | 15–16 |

**Table 7.2** Volume (ml) of concentrated acid needed to prepare 1 litre of dilute acid

| Desired strength (mol $l^{-1}$) | HCl | $H_2SO_4$ | $HNO_3$ |
|---|---|---|---|
| 6 mol $l^{-1}$ | 500 | 333.3 | 380 |
| 1 mol $l^{-1}$ | 83 | 55.5 | 63.3 |
| 0.1 mol $l^{-1}$ | 8.3 | 5.6 | 6.3 |

To prepare a 0.1 mol $l^{-1}$ solution, measure 16.6 ml of 6 mol $l^{-1}$ solution and dilute it to 1 litre.

To prepare a 0.02 mol $l^{-1}$ solution, measure 20.0 ml of 1 mol $l^{-1}$ solution and dilute it to 1 litre.

Hydrochloric, sulphuric and nitric acids and sodium hydroxide are in common use in the analytical laboratory. The characteristics of the acids are given in Table 7.1, and directions for preparing dilutions that are frequently needed in Table 7.2. Preparation of four different concentrations of sodium hydroxide is detailed in Table 7.3. Other concentrations may be made by appropriate dilution with distilled water.

## 7.2 Alkalinity

The alkalinity of water is its capacity to neutralise acid. The amount of a strong acid needed to neutralise the alkalinity is called the total alkalinity, *T*, and is reported in mg $l^{-1}$ as $CaCO_3$. The alkalinity of some waters is due only to the bicarbonates of calcium and magnesium. The pH of such water does not exceed 8.3 and its total alkalinity is practically identical with its bicarbonate alkalinity.

**Table 7.3** Preparation of uniform solutions of sodium hydroxide

| Desired concentration of NaOH solution ($mol\ l^{-1}$) | Weight (g) of NaOH to prepare 1 litre of solution | Volume (ml) of NaOH (15 $mol\ l^{-1}$) to prepare 1 litre of solution |
|---|---|---|
| 15 | 600 | 1,000 |
| 6 | 240 | 400 |
| 1 | 40 | 67 |
| 0.1 | 4 | 6.7 |

Water having a pH above 8.3 contains carbonates and possibly hydroxides in addition to bicarbonates. The alkalinity fraction equivalent to the amount of acid needed to lower the pH value of the sample to 8.3 is called phenolphthalein alkalinity, *P*. This fraction is contributed by the hydroxide, if present, and half of the carbonate (the pH range of 8.3 is approximately that of a dilute bicarbonate solution).

The stoichiometric relationships between hydroxide, carbonate and bicarbonate are valid only in the absence of significant concentrations of other weak anions. This applies especially to the alkalinity (and acidity) of polluted waters and wastewaters.

*Principle*

Alkalinity is determined by titration of the sample with a standard solution of a strong mineral acid. The procedure given uses two colour indicators to determine the end-points in a titration. It is satisfactory for most routine applications. If high levels of accuracy are essential, electrometric titration is preferred, and must also be used when the colour, turbidity or suspended matter in a sample interferes seriously with the determination by the indicator method. Low alkalinities (below approximately 10 $mg\ l^{-1}$) are also best determined by electrometric titration.

Titration to the end-point of pH 8.3 determines the phenolphthalein alkalinity and to the end-point of pH 4.5 the total alkalinity. The pH to which the titration for total alkalinity should be taken lies between 4 and 5, depending on the amount of the alkalinity and free carbon dioxide in the sample. For practical purposes the end-point of pH 4.5 (indicated by methyl orange) gives sufficiently accurate results.

Wherever possible, the titration should be carried out on filtered water at the point of sampling. If this is not possible, the sampling bottle must be completely filled and the alkalinity determined within 24 hours.

*Interferences*
Colour, turbidity and suspended matter may interfere with the visual titration by masking the colour change of an indicator. Turbidity and suspended matter can be eliminated by filtration. The colour of the sample can be reduced by activated carbon and filtration. Free chlorine may affect the indicator colour response and should be removed by the addition of a small amount (usually one drop) of 0.1 mol $l^{-1}$ sodium thiosulphate solution.

The presence of finely divided calcium carbonate suspensions in some natural waters may cause a fading end-point and should be removed by filtration. Silicate, phosphate, borate, sulphide and other anions of weak inorganic and organic acids (e.g. humic acids) will be included in the total alkalinity estimate. They do not interfere with the titration but can influence the validity of stoichiometric relationships.

*Apparatus*

- ✓ White porcelain dish, 200-ml capacity, or conical flask.
- ✓ Burette, 25 ml or 50 ml.

*Reagents*

- ✓ Carbon dioxide-free distilled water must be used for the preparation of all stock and standard solutions. If the distilled water has a pH lower than 6.0, it should be freshly boiled for 15 minutes and cooled to room temperature. Deionised water may be substituted for distilled water provided that it has a conductance of less than 0.2 mS $m^{-1}$ and a pH greater than 6.0.
- ✓ Sodium carbonate, 0.05 mol $l^{-1}$. Dissolve in water 5.300 g anhydrous sodium carbonate previously oven-dried for 1 hour at 250–300 °C and make up to 1 litre.
- ✓ Sulphuric acid, 0.05 mol $l^{-1}$. Dilute 3.1 ml sulphuric acid (density 1.84) to 1 litre. Standardise against 0.05 mol $l^{-1}$ sodium carbonate using methyl orange indicator. If required, this solution may be diluted to 0.01 mol $l^{-1}$.
- ✓ Phenolphthalein indicator. Dissolve 0.5 g of phenolphthalein in 50 ml of 95 per cent ethanol, and add 50 ml of distilled water. Add a dilute (e.g. 0.01 to 0.05 mol $l^{-1}$) carbon dioxide-free solution of sodium hydroxide one drop at a time, until the indicator turns faintly pink.
- ✓ Methyl orange indicator. Dissolve 0.05 g of methyl orange in 100 ml water.
- ✓ Mixed indicator. Dissolve 0.02 g of methyl red and 0.1 g of bromocresol green in 100 ml of 95 per cent ethanol. This indicator is suitable over the pH range 4.6–5.2.

*Procedure*

1. Mix 100 ml of the sample with two or three drops of phenolphthalein indicator in the porcelain basin (or in a conical flask over a white surface). If no colour is produced, the phenolphthalein alkalinity is zero. If the sample turns pink or red, determine the alkalinity by titrating with standard acid until the pink colour just disappears. In either

case, continue the determination using the sample to which phenolphthalein has been added.

2a. Add a few drops of methyl orange indicator. If the sample is orange without the addition of acid, the total alkalinity is zero. If the sample turns yellow, titrate with standard acid until the first perceptible colour change towards orange is observed.

2b. The determination by means of mixed indicator is done in the same way as with methyl orange. The mixed indicator yields the following colour responses: above pH 5.2, greenish blue; pH 5.0, light blue with lavender grey; pH 4.8, light pink-grey with a bluish cast; pH 4.6, light pink.

Any difficulty experienced in detecting the end-point may be reduced by placing a second 100-ml sample with the same amount of indicator (phenolphthalein, methyl orange or mixed indicator) in a similar container alongside that in which the titration is being carried out. Another way to provide a standard end-point is to prepare buffer solutions to which are added indicators in the same amount as in an alkalinity titration.

*Calculation*

Phenolphthalein alkalinity as $CaCO_3$

$$P = \frac{100{,}000 \times A \times M}{V} \text{ mg l}^{-1}$$

Total alkalinity as $CaCO_3$

$$T = \frac{100{,}000 \times B \times M}{V} \text{ mg l}^{-1}$$

where $A$ = volume of standard acid solution (ml) to reach the phenolphthalein end-point of pH 8.3
$B$ = volume of standard acid solution (ml) to reach the end-point of methyl orange or mixed indicator
$M$ = concentration of acid (mol $l^{-1}$)
$V$ = volume of sample (ml)

Using 100 ml of sample and 0.01 mol $l^{-1}$ standard acid solution, the numerical value of alkalinity as mg $l^{-1}$ $CaCO_3$ is 10 times the number of millilitres of titrant consumed.

*Precision*

The precision of visual titration is estimated at 2–10 per cent for alkalinity between 50 and 5 mg $l^{-1}$.

## 7.3 Aluminium

Although aluminium is among the most abundant elements in the earth's crust, it is present in only trace concentrations in natural waters. Because it occurs in many rocks, minerals and clays, aluminium is present in practically all surface waters, but its concentration in waters at nearly neutral pH rarely exceeds a few tenths of a milligram per litre. In addition, in treated water or wastewater, it may be present as a residual from the alum coagulation process. The median concentration of aluminium in river water is reported to be 0.24 mg $l^{-1}$ with a range of 0.01 to 2.5 mg $l^{-1}$.

*Sample handling*

Because aluminium may be lost from solution to the walls of sample containers, samples should be acidified with 1.5 ml of concentrated nitric acid per litre of sample before storage in plastic containers. If the pH is not less than 2 after the addition of acid, more nitric acid should be added. If only soluble aluminium is to be determined, filter a portion of unacidified sample through a 0.45 μm membrane filter, discard the first 50 ml of filtrate and use the succeeding filtrate, after acidification, for the determination. Do not use filter paper, absorbent cotton or glass wool for filtering any solution that is to be tested for aluminium because these materials will remove most of the soluble aluminium.

*Principle*

Dilute aluminium solutions, buffered to a pH of 6.0 and with Eriochrome cyanine R dye added, produce a red to pink complex with a maximum absorption at 535 nm. The intensity of the developed colour is influenced by the aluminium concentration, reaction time, temperature, pH, alkalinity and the concentration of other colours in the sample. To compensate for colour and turbidity, the aluminium in one portion of the sample is complexed with EDTA to provide a blank. Interference by iron and manganese is eliminated by adding ascorbic acid. The limit of detection in the absence of fluoride and polyphosphates is approximately 6 mg $l^{-1}$.

*Interferences*

Negative errors are caused by both fluoride and polyphosphates because of their complexation with aluminium. When the fluoride concentration is constant, the percentage error decreases with increasing amounts of aluminium. The fluoride concentration is often known or can be readily determined, and fairly accurate results can therefore be obtained by adding the known amount of fluoride to a set of standards. A procedure is given for the removal of complex phosphate interference. Orthophosphate under 10 mg $l^{-1}$ does not interfere. The interference caused by even small amounts of alkalinity is removed by acidifying the sample just beyond the neutral point of methyl orange. Sulphate does not interfere up to a concentration of 2,000 mg $l^{-1}$.

*Apparatus*

✓ Colorimetric equipment. One of the following is required:
- —Spectrophotometer: for use at 535 nm with a light path of 1 cm or longer.
- —Filter photometer: equipped with a green filter, with maximum transmittance between 525 and 535 nm and with a light path of 1 cm or longer.
- —Nessler tubes: matched set, tall form, 50-ml capacity.

- ✓ Glassware: all glassware should be treated with warm 1+1 HCl and rinsed with aluminium-free distilled water to avoid errors due to materials adsorbed on the glass. The glassware should be well rinsed to remove all traces of the acid.

*Reagents*

- ✓ Stock aluminium solution. Use either metal or salt to prepare a solution in which 1 ml contains 500 µg Al. Dissolve 500.0 mg aluminium metal in 10 ml concentrated HCl and dilute to 1,000 ml with distilled water. Alternatively, dissolve 8.791 g aluminium potassium sulphate, $AlK(SO_4).12H_2O$, in water and dilute to 1,000 ml. Adjust the weight of the chemical (8.791 g) by dividing it by the decimal fraction of assayed aluminium potassium sulphate in the reagent used.
- ✓ Standard aluminium solution. Dilute 10.00 ml stock aluminium solution to 1,000 ml with distilled water (1.00 ml = 5.00 µg Al). Prepare fresh daily.
- ✓ Sulphuric acid, $H_2SO_4$, 3 mol $l^{-1}$ and 0.01 mol $l^{-1}$.
- ✓ Ascorbic acid solution. Dissolve 0.1 g ascorbic acid in water and make up to 100 ml in a volumetric flask. Prepare fresh daily.
- ✓ Buffer reagent. Dissolve 136 g sodium acetate, $NaC_2H_3O_2.3H_2O$, in water, add 40 ml of 1 mol $l^{-1}$ acetic acid and make up to 1 litre.
- ✓ Stock dye solution. The stock dye solution is stable for at least a year and can be prepared from any one of several commercially available dyes. Suitable dyes, their suppliers and directions for preparing a solution are:
  Solochrome cyanine R 200 (*Arnold Hoffman and Co, Providence, RI, USA*), or Eriochrome cyanine (*K & K Laboratories, Plainview, NY, USA*). Dissolve 100 mg in water and dilute to 100 ml in a volumetric flask. This solution should have a pH of about 2.9.
  Eriochrome cyanine R (*Pfaltz & Bauer Inc., Stamford, CT, USA*). Dissolve 300 mg dye in about 50 ml water. Adjust pH from about 9 to about 2.9 with 1+1 acetic acid (approximately 3 ml will be required). Dilute with water to 100 ml.
  Eriochrome cyanine R (*EM Science, Gibbstown, NJ, USA*). Dissolve 150 mg dye in about 50 ml water. Adjust pH from about 9 to about 2.9 with 1+1 acetic acid (approximately 2 ml will be required). Dilute with water to 100 ml.
- ✓ Working dye solution. Dilute 10.0 ml of stock dye solution to 100 ml with distilled water in a volumetric flask. Working solution is stable for at least six months.
- ✓ Methyl orange indicator solution. Dissolve 50 mg methyl orange powder in distilled water and dilute to 100 ml.
- ✓ EDTA. Dissolve 3.7 g of the sodium salt of ethylenediaminetetraacetic acid dihydrate in water and dilute to 1 litre.
- ✓ Sodium hydroxide, NaOH, 1 mol $l^{-1}$ and 0.1 mol $l^{-1}$.

*Procedure*

*Preparation of calibration graph*

1. Prepare standards and a blank by diluting 0 ml to 7.0 ml portions (0 to 7.0 µg Al) of the aluminium working standard to approximately 25 ml in 50-ml volumetric flasks. Add 1 ml of 0.01 mol $l^{-1}$ $H_2SO_4$, and mix. Add 1 ml ascorbic acid solution and mix. Add 10 ml buffer solution and mix.

2. With a volumetric pipette add 5.00 ml working dye solution and mix. Immediately make up to 50 ml with distilled water. Mix and let stand for 5 to 10 minutes. The colour begins to fade after 15 minutes.

3. Read transmittance or absorbance on a spectrophotometer using a wavelength of 535 nm or a green filter providing maximum transmittance between 525 and 535 nm.

Adjust the instrument to zero absorbance with the standard containing no aluminium. Plot the concentration of aluminium (µg Al in 50 ml final volume) against absorbance.

*Sample treatment when there is no interference by fluoride or phosphate*

4. Pour 25.0 ml of sample or a measured portion of sample diluted to 25 ml into a porcelain dish or flask, add a few drops of methyl orange indicator and titrate with 0.01 mol $l^{-1}$ $H_2SO_4$ to a faint pink colour. Record the amount of acid used and discard the sample.

5. Pour 25 ml of sample into each of two 50-ml volumetric flasks. To each of these, add the amount of 0.01 mol $l^{-1}$ sulphuric acid that was used in the titration plus 1 ml excess. To one of the samples add 1 ml EDTA solution; this will serve as a blank by complexing any aluminium present and compensating for colour and turbidity. To both samples add 1 ml ascorbic acid solution and mix. Add 10 ml buffer solution and mix.

6. With a volumetric pipette add 5.00 ml working dye solution and mix. Immediately make up to 50 ml with distilled water. Mix and let stand for 5–10 minutes. Set the instrument to zero absorbance or 100 per cent transmittance using the EDTA blank. Read transmittance or absorbance of the sample and determine aluminium concentration from the calibration curve.

*Visual comparison*

7. If photometric equipment is not available, prepare and treat standards and a sample, as described above, in 50-ml Nessler tubes. Make up to the mark with water and compare sample colour with the standards after a contact time of 5–10 minutes. A sample treated with EDTA is not needed when Nessler tubes are used. If the sample contains turbidity or colour, the Nessler tube method may result in considerable error.

*Removal of phosphate interference*

8. Add 1.7 ml of 3 mol $l^{-1}$ $H_2SO_4$ to 100 ml of sample in a 200-ml Erlenmeyer flask. Heat on a hotplate for at least 90 minutes, keeping the temperature of the solution just below the boiling point. At the end of the heating period the volume of the solution should be about 25 ml. Add distilled water if necessary to keep it at, or slightly above, that volume.

9. Cool the solution and then bring the pH to 4.3 to 4.5 with NaOH (use 1 mol $l^{-1}$ NaOH and then 0.1 mol $l^{-1}$ as the end-point is approached). Monitor with a pH meter. Make up to 100 ml with distilled water, mix, and use a 25-ml portion for the test. Treat a blank in the same manner using 100 ml distilled water and 1.7 ml of 3 mol $l^{-1}$ $H_2SO_4$. Subtract the blank reading from the sample reading or use it to set the instrument to zero absorbance before taking the sample reading.

*Correction for samples containing fluoride*

10. Measure the fluoride concentration in the sample by either the SPADNS or electrode method (see section 7.10, Fluoride). Add the measured amount of fluoride to each of the samples used for preparing the calibration curve or used in the visual comparison.

*Calculation*

$$\text{Aluminium} = \frac{\mu\text{g Al (in 50 ml final volume)}}{\text{ml sample}} \text{ mg l}^{-1}$$

## 7.4 Biochemical oxygen demand

The biochemical oxygen demand (BOD) is an empirical test, in which standardised laboratory procedures are used to estimate the relative oxygen requirements of wastewaters, effluents and polluted waters. Microorganisms use the atmospheric oxygen dissolved in the water for biochemical oxidation of organic matter, which is their source of carbon. The BOD is used as an approximate measure of the amount of biochemically degradable organic matter present in a sample. The 5-day incubation period has been accepted as the standard for this test (although other incubation periods are occasionally used).

The BOD test was originally devised by the United Kingdom Royal Commission on Sewage Disposal as a means of assessing the rate of biochemical oxidation that would occur in a natural water body to which a polluting effluent was discharged. Predicting the effect of pollution on a water body is by no means straightforward, however, and requires the consideration of many factors not involved in the determination of BOD, such as the actual temperature of the water body, water movements, sunlight, oxygen concentrations, biological populations (including planktonic algae and rooted plants) and the effects of bottom deposits. As determined experimentally by incubation in the dark, BOD includes oxygen consumed by the respiration of algae. The polluting effect of an effluent on a water body may be considerably altered by the photosynthetic action of plants and algae present, but it is impossible to determine this effect quantitatively in 5-day BOD experiments. Consequently, no general ruling can be given on the BOD of samples containing algae, and each case should be considered on its merits. Suspended organic matter in an effluent is frequently deposited over a short distance immediately downstream of an outfall, where it may result in a very considerable decrease in the local dissolved oxygen concentration.

A further complication in the BOD test is that much of the oxygen-consuming capacity of samples may be due to ammonia and organically bound nitrogen, which will eventually be oxidised to nitrite and nitrate if nitrifying bacteria are present. Furthermore, the ammonia added in the dilution water used for the method presented here may also be nitrified so that, to this extent, the BOD value is not representative of the sample alone. Nitrifying bacteria are extremely sensitive to trace elements that may be present, and the occurrence of nitrification is sporadic and unpredictable even with samples known to contain nitrifying bacteria. Moreover, because of the slow growth of nitrifying bacteria, the degree of nitrification will depend on the number of these organisms initially present. Nitrification does not occur to any detectable extent during the 5-day BOD determination of crude and settled sewage and almost all industrial effluents. The BOD test is thus useful for determining the relative waste loadings to treatment plants and the degree of oxygen demand removal provided by primary treatment. Occurrence of nitrification during the 5-day incubation is almost always confined to treated effluents and river waters, which have already been partially nitrified. Only these cases need special attention, presenting the question of whether or not to use the method incorporating an inhibitor of nitrification. Determination of the degree of nitrification is tedious but, unless it is known, the BOD value may be misleading in assessing treatment plant performance or in calculating the effect of an effluent on a river. In some instances, nitrification has been shown to account for more than 70 per cent of the BOD of a well purified sewage effluent. Nevertheless, procedures in which nitrification may occur have been in use for many years and no attempt is made in the following method to eliminate nitrification.

The BOD determined by the dilution method presented here has come to be used as an approximate measure of the amount of biochemically degradable organic matter in a sample. For this purpose the dilution test, applied skilfully to samples in which nitrification does not occur, remains probably the most suitable single test, although manometric methods may warrant consideration in some cases. The analyst should also consider whether the information required could be obtained in some other way. For example, the chemical oxygen demand (COD) test will result in virtually complete oxidation of most organic substances, thereby indicating the amount of oxygen required for complete oxidation of the sample. In other circumstances, and particularly in research work, determination of the organic carbon content may be more appropriate. In any case, results obtained by the BOD test should never be considered in isolation but only in the context of local conditions and the results of other tests.

Complete oxidation of some wastes may require too long a period of incubation for practical purposes. For certain industrial wastes, and for waters polluted by them, it may be advisable to determine the oxidation curve obtained. Calculations of ultimate BOD from 5-day BOD values (e.g. based on calculations using exponential first-order rate expressions) are not correct. Conversion of data from one incubation period to another can be made only if the course of the oxidation curve has been determined for the individual case by a series of BOD tests carried out for different incubation periods.

The dilution method of determining BOD described below is the one most generally used. The dissolved oxygen content of the liquid is determined before and after incubation for 5 days at 20 °C. The difference gives the BOD of the sample after allowance has been made for the dilution, if any, of the sample.

*Sample handling*

The test should be carried out as soon as possible after samples have been taken. If samples are kept at room temperature for several hours, the BOD may change significantly, depending on the character of the samples. In some instances it may decrease and in others it may increase. The decrease at room temperature has occasionally been found to be as much as 40 per cent during the first 8 hours of storage. If samples cannot be dealt with at once they should, whenever practicable, be stored at about 5 °C. In the case of individual samples collected over a long period, it is desirable to keep all the samples at about 5 °C until the composite sample can be made up for the BOD determination.

Samples must be free from all added preservatives and stored in glass bottles.

It is necessary that excess dissolved oxygen be present during the whole period of incubation, and desirable that at least 30 per cent of the saturation value remains after 5 days. Since the solubility of atmospheric oxygen at the temperature of incubation is only 9 mg $l^{-1}$, samples that absorb more than about 6 mg $l^{-1}$ during incubation for 5 days will not fulfil this condition. This is the case with sewage, nearly all sewage effluents, and many other waste liquids. The additional oxygen is supplied by diluting the sample with clean, well aerated water. The amount of dilution depends upon the nature of the sample.

*Interferences*

If the pH of the sample is not between 6.5 and 8.5, add sufficient alkali or acid to bring it within that range. Determine the amount of acid and alkali to be added by neutralising a separate portion of the sample to about pH 7.0 with a 1 mol $l^{-1}$ solution of acid or alkali, using an appropriate indicator (e.g. bromothymol blue), or pH meter. Add a calculated aliquot volume of acid or alkali to the sample for the BOD test.

Some samples may be sterile, and will need seeding. The purpose of seeding is to introduce into the sample a biological population capable of oxidising the organic matter in the wastewater. Where such micro-organisms are already present, as in domestic sewage or unchlorinated effluents and surface waters, seeding is unnecessary and should not be carried out.

When there is reason to believe that the sample contains very few micro-organisms, for example as a result of chlorination, high temperature, extreme pH or the specific composition of some industrial wastes, the dilution water should be seeded.

For seeding, to each litre of dilution water add 5 ml of a fresh sewage effluent of good quality obtained from a settling tank following an aerobic biological process of purification. If necessary, settle (not filter) the effluent in a glass cylinder for about 30 minutes.

If such effluent is not available, use settled domestic sewage that has been stored at 20 °C for 24 hours; for seeding, add 1–2 ml of the supernatant to each litre of dilution water.

The special difficulties in choosing a seed for industrial effluents that are toxic, or that are not broken down by sewage bacteria, are dealt with in the following sub-section on "Seeding samples of industrial effluents". If the samples are analysed in different laboratories, better agreement between test results will be achieved by using the same type of seed or, preferably, the same seed.

Some samples may be supersaturated with dissolved oxygen, especially waters containing algae. If such samples are to be incubated without dilution, the dissolved oxygen concentration should be lowered to saturation to prevent loss of oxygen during incubation. The sample should be brought to about 20 °C in a partly filled bottle and well shaken.

A few sewage effluents and certain industrial effluents contain either residual chlorine or the products of the action of chlorine on certain constituents. Such liquids cannot be used directly for the determination of BOD because of the bactericidal effect of the chlorine or of its products and also because chlorine would introduce an error into the determination of dissolved oxygen. If the samples are allowed to stand for 1 to 2 hours, the residual chlorine will often be dissipated. Dilutions for BOD can then be prepared with properly seeded standard dilution water.

Higher concentrations of chlorine, and of many compounds containing available chlorine, may be removed by treating a portion of the sample with sodium bisulphite. The treated portion is then used for the BOD test. This procedure will probably give reasonably good results for domestic sewage effluents that have been chlorinated, since the chlorine will be present chiefly as chloramines formed by combination of chlorine with the ammonia present. However, in the case of other effluents consisting of, or containing, industrial wastes, the chlorine may have combined with organic compounds present to produce substances which, although giving no reaction for chlorine with the starch–iodide test described below, are inhibitory to biochemical oxidation or are even bactericidal. The BOD, as determined in these circumstances, is generally lower than would be expected for the organic content as measured by other tests.

Should a value for BOD of a chlorinated effluent be required, notwithstanding the uncertainty of the interpretation of the test, the following procedure should be used:

1. If the sample is alkaline to phenolphthalein bring it to a pH of 5.0 by the addition of dilute sulphuric acid. Add a crystal of potassium iodide to a convenient measured

volume of sample (e.g. 100 ml) and titrate it with approximately 0.0125 mol $l^{-1}$ or 0.025 mol $l^{-1}$ sodium bisulphite (or sulphite) solution, using a few drops of starch solution as an indicator.

2. To another portion of sample, sufficient to carry out the BOD test, add the requisite amount of dilute sulphuric acid to adjust the pH to 5.0, followed by the volume of sodium bisulphite solution determined by the previous titration. After thorough mixing allow to stand for several minutes, then check the absence of chlorine by testing a small portion of the treated sample with neutral starch–iodide.
3. Confirm the absence or excess of bisulphite on another portion by means of starch solution and a drop of 0.0125 mol $l^{-1}$ iodine, which should develop a blue colour. Adjust the pH to about 7.3 before proceeding with the test.
4. Make up the dilution with seeded dilution water and proceed as for unchlorinated samples.

*Note:* Some wastewaters contain substances reacting with iodine, which precludes the determination of dissolved oxygen by iodometric titration. An instrumental method should be used (see determination of dissolved oxygen, section 6.5.)

*Seeding samples of industrial effluents*

A seed of sewage effluent, as described above, is satisfactory for many industrial effluents. However, if the BOD of such effluents as found by the standard test is substantially less than the chemical oxygen demand (COD) it may be for one of the following reasons:

(i) the sample contains compounds resistant to biochemical breakdown,
(ii) the seeding organisms are of an unsuitable type or require acclimatisation, or
(iii) toxic or bacteriostatic compounds are present, exerting an inhibiting effect at the concentration employed for the test.

Compounds constitutionally resistant to breakdown will not exert an oxygen demand on the receiving waters, but substances amenable to breakdown will generally contribute to the pollution load, even if the BOD test fails for reasons (ii) and (iii) above. Before embarking on the tedious, and sometimes impossible, task of preparing a seed by the method described below, the analyst should decide whether sufficient information about the sample may be given by alternative methods such as determinations of COD and organic carbon. Sometimes, if the difficulty is the result of condition (iii), it is possible to obtain reliable BOD values merely by increasing the dilution until the toxic constituents of the sample are below the inhibitory threshold concentrations. If this procedure fails, or if condition (ii) applies, the following method should be used:

1. Neutralise the sample if necessary, then add about 10 per cent of the threshold toxic concentration of the sample (if known; otherwise add a concentration that is thought unlikely to kill activated sludge organisms) to a mixture of settled sewage and activated sludge (2,000 mg $l^{-1}$ suspended solids) and aerate by diffused air or by stirring.

2. After one day, allow the sludge to settle and decant the supernatant liquid, top up to the same volume with sewage and sample as before. Repeat daily. After 3 or 4 days measure the BOD of the sample using dilution water seeded with the settled mixture, then increase the proportion of sample in the mixture by a factor of 2.
3. Continue the procedure, doubling the proportion in 3- or 4-day intervals, until a maximum BOD has been reached.

   If a laboratory-scale, continuously-fed, activated sludge unit is available, this can be used in a similar way to produce a seed acclimatised to the sample. Sometimes, adapted seed is available from the effluent of a biological treatment process receiving the waste in question, or the seed might be developed from the receiving water below the point of discharge of this waste, if it is not being treated.

*Apparatus*

✓ Incubation bottles. It is recommended that narrow-mouthed, glass-stoppered bottles of a nominal capacity of 250 ml be used, and it is essential that the bottles are clean. New bottles should be cleaned with either 5 mol $l^{-1}$ hydrochloric or sulphuric acid and thoroughly rinsed. In normal use, bottles are kept clean by the acidic iodine solution of the Winkler procedure and require no treatment apart from thorough rinsing with tap water and distilled water. Special cleaning may be necessary in some cases, but the use of chromic acid is not recommended because traces of chromium may remain in the bottle.
Some analysts prefer to use bottles of about 125 ml capacity, thus reducing the incubator space required. There is evidence, however, that with samples of some types the size of bottles (i.e. the ratio of the glass surface to the volume of liquid) may influence the result. The analyst wishing to use small bottles must, therefore, be satisfied that such a procedure gives results similar to those obtained by use of bottles of standard size.
As a precaution against drawing air into the dilution bottle during incubation, a water seal is recommended. Satisfactory water seals are obtained by inverting the bottles in a water-bath or adding water to the flared mouth of special BOD bottles.

✓ Incubator or water-bath. The temperature of incubation should be 20 ± 0.5 °C. A water-bath, or constant temperature room is usually employed. Incubation must be carried out in the dark. Some samples may contain algae which, if incubated in the light, would give off oxygen by photosynthetic action, and thus interfere with the BOD determination.

*Reagents*

✓ Dilution water. The logical diluent for a sewage effluent is the river water into which the effluent is discharged, but this method can be adopted only in special cases and is obviously unsuitable where effluents from widely differing localities are dealt with in one laboratory. Moreover, the river water may itself have a considerable BOD.
Distilled water alone is unsatisfactory as a diluent, and it is recommended that a synthetic dilution water be employed. This is prepared by adding reagents to good quality distilled water. Water from copper stills should not be used since copper inhibits biochemical oxidation (0.01 mg $l^{-1}$ is the maximum safe concentration). Some commercial vapour-compression stills have also been shown to produce water containing copper.
Deionised water produced in some commercial units has been found satisfactory, but deionising columns in hard-water areas require frequent regeneration. It may be convenient, however, to run two deionising columns in series, or to deionise the water from a vapour compression still. Water from a new or freshly regenerated column should always be shown to give similar BOD values to distilled water, bearing in mind that the resins may introduce or fail to remove undesirable organic matter.

Stock solutions of the following pure chemicals are required; any solutions showing signs of precipitates or growths should be discarded.

- ✓ Ferric chloride solution: dissolve 0.125 g ferric chloride, $FeCl_3.6H_2O$, in 1 litre water.
- ✓ Calcium chloride solution: dissolve 27.5 g anhydrous calcium chloride, $CaCl_2$, (or equivalent if hydrated calcium chloride is used), in 1 litre water.
- ✓ Magnesium sulphate solution: dissolve 25 g magnesium sulphate, $MgSO_4.7H_2O$, in 1 litre water.
- ✓ Phosphate buffer stock solution: dissolve 42.5 g potassium dihydrogen phosphate, $KH_2PO_4$, in 700 ml water and add 8.8 g sodium hydroxide. This should give a solution of pH 7.2 which should be checked. Add 2 g ammonium sulphate, and dilute to 1 litre.

Add 1 ml of each reagent to each litre of freshly distilled (or deionised) water. Bring the water to incubation temperature 20 ± 1 °C and saturate with oxygen by bubbling air through it or by shaking the partially filled bottle, and use as soon as possible. Discard any dilution water remaining unused and clean the bottle, preferably with a sterilising agent. Thoroughly wash and rinse free from residual traces of the agent, and store out of direct sunlight. Stocks of dilution water should never be "topped up" with fresh solution.

A satisfactory dilution water, when incubated with or without a seed under standard conditions should not absorb more than 0.2 mg $l^{-1}$ of oxygen, and in any case must not absorb more than 0.5 mg $l^{-1}$. A high oxygen uptake may sometimes be associated with the presence of water-soluble organic vapours in the laboratory atmosphere. Water for dilution should therefore be distilled (or deionised) and used in a room from which volatile organic compounds are excluded. Air used for aeration must be as clean as possible.

*Procedure*

1. Pretreatment of dilution water by seeding is sometimes necessary (see above). Pretreatment of sample (see "Interferences") is needed if the sample is supersaturated with oxygen or if the sample contains residual chlorine. If the pH of the sample is not between 6.5 and 8.5, it should be brought within this range.

2. Samples that have been stored in a refrigerator should be allowed to reach room temperature before dilutions are made. All samples must be well mixed just before dilution.

3. In some wastes, suspended matter may cause difficulty because the distribution of the solids may be uneven when the sample is made up into dilutions. This may cause discrepancies in the results from different dilutions or duplicate dilutions. In such cases, shake the sample vigorously immediately before the dilutions are made. Artificial homogenising procedures may cause an increased oxygen demand, and cannot be recommended.

4. Sometimes, the BOD determination in settled or filtered samples is needed. In such cases a settling time of 30 minutes is usually applied. For the BOD test of filterable substances, membrane filter, glass-fibre filter or paper filter may be used. The type of filter should be indicated in reporting the result. If determinations other than the BOD test are carried out on the filtered sample (e.g. residue, COD), it is recommended that filters of the same type and porosity be used for all of those procedures.

*Dilution*

5. Unless the approximate BOD of the sample is already known, the required degree of dilution will not be known and more than one dilution will have to be made. Recommended dilutions are given in Table 7.4. With experience, the analyst will often be able to use the COD as a guide to the dilution required. As low a dilution as possible should be used consistent with at least 30 per cent of the oxygen remaining after 5 days. It should be noted that some metals, e.g. copper, chromium, lead, will partially inhibit oxygen consumption.

**Table 7.4** Recommended dilutions for the BOD test

| Range of BOD values to be determined (mg $l^{-1}$) | Sample volume (ml) | Dilution water volume (ml) | Dilution factor "d" | Report to nearest mg $l^{-1}$ | Source of sample |
|---|---|---|---|---|---|
| 0 to 6 | undiluted | 0 | 1 | 0.1 | R |
| 4 to 12 | 500 | 500 | 2 | 0.2 | R,E |
| 10 to 30 | 200 | 800 | 5 | 0.5 | R,E |
| 20 to 60 | 100 | 900 | 10 | 1 | E,S |
| 40 to 120 | 50 | 950 | 20 | 2 | S |
| 100 to 300 | 20 | 980 | 50 | 5 | S,C |
| 200 to 600 | 10 | 990 | 100 | 10 | S,C |
| 400 to 1,200 | 5 | 995 | 200 | 20 | I,C |
| 1,000 to 3,000 | 2 | 998 | 500 | 50 | I |
| 2,000 to 6,000 | 1 | 999 | 1,000 | 100 | I |

R River water
E Biologically purified sewage effluent
S Settled sewage or weak industrial wastewater
C Crude (raw) sewage
I Strong industrial wastewater

6. In preparing dilutions for the BOD test, siphon or pour carefully the standard dilution water (seeded if necessary) into a graduated cylinder of capacity 1,000–2,000 ml, filling the cylinder half-way without entrainment of air. Add the quantity of carefully mixed sample to make the desired dilution and dilute to the desired level with dilution water. Mix well. Each analyst will have a preferred detailed procedure for preparing dilutions. Nevertheless, the following principles must be strictly adhered to:

   (i) The sample and dilution water must be mixed thoroughly, but violent agitation leading to the formation of minute air bubbles must be avoided. Mixing may be accomplished by careful repeated inversion of a bottle or stoppered measuring cylinder containing the sample and dilution water, or by use of a magnetic stirrer in a completely filled bottle.

   (ii) Dilutions involving the measurement of less than 5 ml of sample should be made by first diluting the sample in a volumetric flask (e.g. × 10 dilution) and then using the appropriate volume of this mixture for final dilution to the required strength.

   (iii) The diluted mixture is transferred to two incubation bottles (more if replicate results are required) by siphoning or by careful pouring. If a siphon is used, at least 50 ml of mixture must be discarded before the first bottle is filled. Bottles must be filled completely, allowed to stand for a few minutes and then tapped gently to remove bubbles. The stoppers are then inserted firmly without trapping air bubbles in the bottle.

   (iv) On any one occasion, exactly the same mixing and transfer techniques must be used for all dilutions and samples.

(v) Bottles of the dilution water used in the test must be prepared at the same time as the sample dilutions to permit a determination of the blank.

*Determination of dissolved oxygen and incubation*

7. Determine the initial concentration of dissolved oxygen in one bottle of the mixture of sample and dilution water, and in one of the bottles containing only dilution water. Place the other bottles in the incubator (those containing the sample, or the mixture of sample and dilution water, and that containing the plain dilution water to act as a blank, unseeded or seeded in accord with previous steps).

8. Incubate the blank dilution water and the diluted samples for 5 days in the dark at 20 °C. The BOD bottles should be water-sealed by inversion in a tray of water in the incubator or by use of a special water-seal bottle. Although it is known that the BOD of some samples is increased if the liquid is agitated during the incubation, it is not at present suggested that agitation should be provided.

9. After 5 days determine the dissolved oxygen in the diluted samples and the blank using the azide modification of the iodometric method or an electrometric method. (for particulars see "Dissolved oxygen", section 6.5.) Those dilutions showing a residual dissolved oxygen of at least 30 per cent of the initial value and a depletion of at least 2 mg $l^{-1}$ should be considered the most reliable.

*Independent check of the technique*

10. It might be thought desirable, from time to time, to check the technique. For this purpose, pure organic compounds of known or determinable BOD are used. If a particular organic compound is known to be present in a given waste, it may well serve as a control on the seed used. A number of organic compounds have been proposed, such as glucose and glutamic acid. In exceptional cases, a given component of a particular waste may be the best choice to test the efficacy of a particular seed. For general use, a mixture of glucose and glutamic acid has certain advantages. Glucose has an exceptionally high and variable oxidation rate with relatively simple seeds. When glucose is used with glutamic acid, the oxidation rate is stabilised and is similar to that obtained with many municipal wastes.

11. For the check, dissolve 150 mg each of glucose and glutamic acid (both dried at 103 °C for 1 hour) in 1 litre of water. This solution should be freshly prepared.

12. Make up a 1 in 50 dilution using seeded dilution water and determine the BOD in the usual way. The BOD should be approximately 220 mg $l^{-1}$. If the result obtained is less than 200 mg $l^{-1}$ or more than 240 mg $l^{-1}$, some defect in the seed, dilution water or experimental techniques should be suspected.

### *Immediate dissolved oxygen demand*

Substances oxidisable by molecular oxygen, such as ferrous iron, sulphite, sulphide and aldehyde, impose a load on the receiving water and must be taken into consideration. The total oxygen demand of such a substrate may be determined by using a calculated initial dissolved oxygen (DO) or by using the sum of the immediate dissolved oxygen demand (IDOD) and the 5-day BOD.

Where a differentiation of the two components is desired, the IDOD should be determined. It should be understood that the IDOD does not necessarily represent the immediate oxidation by molecular dissolved oxygen but may represent an oxidation by the iodine liberated in the acidification step of the iodometric method.

The depletion of dissolved oxygen in a standard water dilution of the sample in 15 minutes has been arbitrarily selected as the IDOD. To determine the IDOD, the dissolved oxygen of the sample (which in most cases is zero) and of the dilution water are determined separately. An appropriate dilution of the sample and dilution water is prepared, and the dissolved oxygen of the sample dilution minus the observed dissolved oxygen after 15 minutes is the IDOD (mg $l^{-1}$) of the sample dilution.

*Calculation*

(1) When BOD has been determined in an undiluted sample

BOD (mg $l^{-1}$) = DO before incubation (mg $l^{-1}$) – DO after incubation (mg $l^{-1}$)

(2) When BOD has been determined in a diluted sample

A. *Without correction for blank (i.e. for the BOD of the dilution water itself)*

When seeding is not required:

$$\text{BOD} = \frac{D_1 - D_2}{P} \text{ mg l}^{-1}$$

When using seeded dilution water:

$$\text{BOD} = \frac{(D_1 - D_2) - (B_1 - B_2)f}{P}$$

Including IDOD if small or not determined:

$$\text{BOD} = \frac{D_c - D_2}{P} \text{ mg l}^{-1}$$

$$\text{IDOD} = \frac{D_c - D_1}{P} \text{ mg l}^{-1}$$

where:
$D_o$ = DO of original dilution water
$D_1$ = DO of diluted sample immediately after preparation (mg $l^{-1}$)
$D_2$ = DO of diluted sample after 5 days' incubation
$D_c$ = DO available in dilution at zero time = $D_o p + D_s P$
$D_s$ = DO of original undiluted sample
$p$ = decimal fraction of dilution water used
$P$ = decimal fraction of sample used: ($P + p = 1.00$)
$B_1$ = DO of dilution of seed control* before incubation;
$B_2$ = DO of dilution of seed control* after incubation;
$f$ = ratio of seed in sample to seed in control*

$$f = \frac{\text{\% of seed in } D_1}{\text{\% of seed in } B_1}$$

Seed correction = $(B_1 - B_2)f$

*The seed control refers to a separate test to check the BOD attributable to the seed added to the sample. For this purpose, measure the oxygen depletion of a series of seed dilutions and use the one giving 40–70 per cent oxygen depletion.

The DO determined on the unseeded dilution water after incubation is not used in the BOD calculations because this practice would overcorrect for the dilution water. In all the above calculations, corrections are not made for small losses of DO in the dilution water during incubation. If the dilution water is unsatisfactory, proper corrections are difficult and the results are questionable.

*B. With correction for the BOD of the dilution water*

If the BOD of the dilution water reaches the limit of 0.5 mg $l^{-1}$ or approaches it, the correction may be of importance, especially for samples of water having a low BOD but requiring a dilution. In such cases, correction for BOD may be used. The calculation is then:

$$\mathrm{BOD} = \frac{1{,}000}{V} \times (\mathrm{BOD_m} - \mathrm{BOD_d}) + \mathrm{BOD_d} \quad \mathrm{mg\ l^{-1}}$$

$$\mathrm{BOD} = \frac{1{,}000}{V}\left[(S_\mathrm{m} - S_t) - (D_\mathrm{m} - D_t)\right] \quad \mathrm{mg\ l^{-1}}$$

where: BOD = BOD of the sample
$\mathrm{BOD_m}$ = BOD of the mixture (sample + dilution water)
$\mathrm{BOD_d}$ = BOD of the dilution water (blank)
$V$ = volume (ml) of sample in 1 litre of the mixture
$S_\mathrm{m}$ = DO of the mixture before incubation
$S_t$ = DO of the mixture after incubation (after *t* days)
$D_\mathrm{m}$ = DO of the dilution water before incubation
$D_t$ = DO of the dilution water after incubation for *t* days

*Expression of results*

$\mathrm{BOD}_t$ in mg $l^{-1}$, where *t* indicates the number of days in incubation.

*Precision and accuracy*

Using a procedure very similar to the above, 78 analysts in 55 laboratories analysed natural water samples plus an exact increment of biodegradable organic compounds. At a mean value of 2.1 and 175 mg $l^{-1}$ BOD, the standard deviations were $\pm$ 0.7 and $\pm$ 26 mg $l^{-1}$ respectively. There is no acceptable procedure for determining the accuracy of the BOD test.

## 7.5 Chemical oxygen demand

The chemical oxygen demand (COD) is the amount of oxygen consumed by organic matter from boiling acid potassium dichromate solution. It provides a measure of the oxygen equivalent of that portion of the organic matter in a water sample that is susceptible to oxidation under the conditions of the test. It is an important and rapidly measured variable for characterising water bodies, sewage, industrial wastes and treatment plant effluents.

In the absence of a catalyst, however, the method fails to include some organic compounds, such as acetic acid, that are biologically available to the aquatic organisms but does include some biological compounds, such as cellulose, that are not part of the immediate biochemical demand on the available oxygen of the receiving water. The carbonaceous portion of nitrogen compounds can be determined but the dichromate is not reduced by any ammonia in a waste or by any ammonia liberated from the proteinaceous matter. With certain wastes containing toxic substances, COD or a total organic carbon determination may be the only method for determining the organic load. It should be noted that the COD is not a measure of organic carbon, although the same chemical reactions are involved. Where wastes contain only readily available organic bacterial nutrients and no toxic matter, the results can be used to obtain an approximate estimate of the ultimate carbonaceous BOD values.

The use of exactly the same technique each time is important because only a part of the organic matter is included, the proportion depending on the chemical oxidant used, the structure of the organic compounds and the manipulative procedure.

The dichromate method has been selected as a reference method for the COD determination because it has advantages over other oxidants owing to its oxidising power, its applicability to a wide variety of samples and its ease of manipulation. The test will have most value for monitoring and control of effluents and receiving waters after correlation with other variables has been established.

*Principle*

The sample is boiled under reflux with potassium dichromate and silver sulphate catalyst in strong sulphuric acid. Part of the dichromate is reduced by organic matter and the remainder is titrated with ferrous ammonium sulphate.

*Interferences*

Straight-chain aliphatic compounds, aromatic hydrocarbons and pyridine are not oxidised to any appreciable extent, although this method gives more nearly complete oxidation than a permanganate method. The straight-chain compounds are more effectively oxidised when silver sulphate is added as a catalyst. However, silver sulphate reacts with chlorides, bromides or iodides to produce precipitates that are only partially oxidised. There is no advantage in using the catalyst in the oxidation of aromatic hydrocarbons, but it is essential to the oxidation of straight-chain alcohols and acids.

The oxidation and other difficulties caused by the presence of chlorides in the sample may be overcome by adding mercuric sulphate before refluxing, in order to bind the chloride ion as a soluble mercuric chloride complex, which greatly reduces its ability to react further.

Nitrite nitrogen exerts a COD of 1.14 mg $mg^{-1}$ of nitrite nitrogen. To eliminate significant interference due to nitrites, 10 mg of sulphamic acid for every 1 mg of nitrite nitrogen in the refluxing flask may be added. If a series of samples containing nitrite is analysed, the sulphamic acid may be added to the standard dichromate solution, since it must be included in the distilled water blank. Thus, 120 mg of sulphamic acid per litre of dichromate solution will eliminate the interference of up to 6 mg of nitrite nitrogen per litre in a 20-ml sample. An aliquot volume of the sample diluted to 20 ml should be used to eliminate the interference of higher concentrations of nitrite.

Ferrous iron and hydrogen sulphide exert COD of 0.14 mg $mg^{-1}$ $Fe^{2+}$ and 0.47 mg $mg^{-1}$ $H_2S$ respectively. Appropriate corrections can be calculated and subtracted from the result or both interferences can be removed by bubbling air through the sample, if easily volatile organic matter is not present.

The procedure can be used to determine COD values of 50 mg $l^{-1}$ with the standard dichromate solution (0.0417 mol $l^{-1}$). With the dilute dichromate, values are less accurate, especially below 10 mg $l^{-1}$, but may be used to indicate an order of magnitude.

*Sample handling*

Samples should be taken with bottles that do not release organic substances into water; glass-stoppered glass bottles are satisfactory. Unstable samples should be tested without delay, especially wastewater and polluted water samples. Natural, not heavily polluted, water should be analysed on the same day or at least within 24 hours and the sample should be kept cold before analysis.

If there is to be a delay before analysis the sample may be preserved by adding sulphuric acid, about 2 ml $H_2SO_4$ ($d$ = 1.84) diluted 1+2 to each 100 ml of sample. If samples are to be stored for longer than 24 hours, deep freezing is recommended.

Depending on the aim of the analysis, COD can be determined on unfiltered and/or filtered samples. When both determinations are carried out, the difference gives the COD of the particulate matter. Samples containing settleable solids should be homogenised sufficiently by means of a blender to permit representative sampling for the COD determination in unfiltered samples. For the analysis of filtrate, the original (not homogenised) sample is used. Filtration through glass-fibre filters is recommended, but hard paper filters may be used if the sample has a high COD. The filters should be pre-rinsed with water.

*Apparatus*

✓ A reflux apparatus consisting of a 250-ml Erlenmeyer flask (500 ml if large samples are used) with ground-glass neck, and a 300-mm double surface condenser (Liebig, Friedrichs, West or equivalent) with a ground-glass joint. Since absolute cleanliness is essential, flasks and condensers should be protected from dust by inverted cups when not in use. The glassware must be used exclusively for COD determinations.

✓ A heating mantle or hotplate.
✓ A hotplate producing at least 1.5 W $cm^{-2}$ of heating surface to ensure adequate boiling of the liquid in the flask. Heating mantles are preferred because they prevent the problem of overheating.

*Reagents*
✓ Sulphuric acid ($d$ =1.84).
✓ Standard potassium dichromate solution, 0.0417 mol $l^{-1}$. Dissolve 12.259 g of $K_2Cr_2O_7$ primary standard grade, dried at 103 °C for 2 hours, in distilled water and dilute to 1.000 litre.
✓ Dilute standard potassium dichromate solution, 0.00417 mol $l^{-1}$. Dilute 100 ml of the standard potassium dichromate solution to 1.000 litre.
✓ Standard ferrous ammonium sulphate solution, 0.250 mol $l^{-1}$. Dissolve 98 g of $Fe(NH_4)_2(SO_4)_2.6H_2O$ analytical grade crystals in distilled water. Add 20 ml of $H_2SO_4$ ($d$ = 1.84), cool and dilute to 1.000 litre. This solution may be standardised against the standard potassium dichromate solution as follows:
Dilute 10.0 ml of standard potassium dichromate solution, 0.0417 mol $l^{-1}$, to about 100 ml. Add 30 ml $H_2SO_4$ ($d$ = 1.84) and allow to cool. Titrate with the ferrous ammonium titrant, using 2 or 3 drops of ferroin indicator.

$$\text{Concentration (mol l}^{-1}) = \frac{V_1 \times 0.25}{V_2}$$

where: $V_1$ = volume (ml) of $K_2Cr_2O_7$
$V_2$ = volume (ml) of $Fe(NH_4)_2(SO_4)_2$

✓ Dilute standard ferrous ammonium sulphate solution, 0.025 mol $l^{-1}$. Dilute 100 ml of the standard ferrous ammonium sulphate solution to 1.000 litre. Standardise daily against the dilute standard potassium dichromate, 0.00417 mol $l^{-1}$.
✓ Silver sulphate, reagent powder. This reagent may be used either directly in powder form or in saturated solution, or it may be added to the sulphuric acid (about 5 g of $Ag_2SO_4$ to 1 litre of $H_2SO_4$; 1–2 days are required for dissolution).
✓ Mercuric sulphate, analytical grade crystals.
✓ Ferroin indicator solution. Dissolve 0.695 g of ferrous sulphate, $FeSO_4.7H_2O$, in water. Add 1.485 g of 1,10-phenanthroline monohydrate, shaking until dissolved. Dilute to 100 ml. This solution is also commercially available.
✓ Sulphamic acid, analytical grade (required only if the interference of nitrites is to be eliminated).
✓ Anti-bumping granules that have been previously heated to 600 °C for 1 hour.

*Procedure*
Initial dilutions in volumetric flasks should be made on waste with a high COD value to reduce the error that is inherent in measuring small samples.

*A. Samples with low chloride concentrations*
If the sample contains less than 100 mg $l^{-1}$ chloride after evaporation, proceed as follows:

1. Place in an Erlenmeyer flask 20.0 ml of the sample or an aliquot diluted to 20.0 ml with distilled water.
2. Add 10.0 ml of standard potassium dichromate solution, 0.0417 mol $l^{-1}$, and a few anti-bumping granules. Mix well.
3. Add slowly, with caution, 30 ml of concentrated $H_2SO_4$ containing silver sulphate, mixing thoroughly by swirling while adding the acid. If $H_2SO_4$ containing silver sulphate is not used, add 0.15 g of dry silver sulphate and then, slowly, 30 ml of concentrated $H_2SO_4$.

*Note:* If the liquid has not been well mixed local heating may occur on the bottom of the flask and the mixture may be blown out of the flask.

4. Attach the condenser to the flask and reflux the mixture for 2 hours. Allow to cool and then wash the condenser with distilled water.

5. Dilute the mixture to about 150 ml with distilled water, cool to room temperature, and titrate the excess dichromate with standard ammonium ferrous sulphate using 2–3 drops of ferroin indicator. Although the quantity of ferroin is not critical, do not vary it among different samples even when analysed at different times. The end-point is when the colour changes sharply from blue-green to reddish-brown, even though the blue-green may reappear within several minutes.

6. Reflux in the same manner a blank consisting of 20 ml of distilled water together with the reagents and titrate as in step 5, above.

*B. Samples with high chloride concentration*
If the sample contains more than 100 mg $l^{-1}$ chloride after evaporation or dilution, proceed as follows:

To 20.0 ml of sample or aliquot in the flask add 0.5 g of mercuric sulphate and shake thoroughly. This addition is sufficient to complex 40 mg of chloride ion or 2000 mg $l^{-1}$ when 20.0 ml of sample are used. If more chloride is present, add more $HgSO_4$ to maintain a $HgSO_4$: $Cl^-$ ratio of 10:1. It is not important if a slight precipitate develops because it will not affect the determination. Continue with steps 2 to 6, as above.

*Adjustments for other sample sizes*
If a water is expected to have a higher or lower than normal COD, a sample ranging in size from 10.0 ml to 50.0 ml may be used with the volumes, weights and concentrations adjusted accordingly. Table 7.5 gives the appropriate reagent quantities for different sample sizes. Use these quantities when following the procedure given above. When using large samples, increase the size of the Erlenmeyer flask to 500 ml to permit titration within the refluxing flask.

*C. Samples with low COD*
Follow one of the procedures given above for high and low chloride concentrations with the following differences:

1. Use dilute standard potassium dichromate, 0.00417 mol $l^{-1}$.
2. Perform the back titration with either 0.025 mol $l^{-1}$ or 0.01 mol $l^{-1}$ ferrous ammonium sulphate.
3. Use redistilled water for the preparation of all reagents and blanks.

Exercise extreme care with this procedure because a trace of organic matter in the glassware or the atmosphere may cause a gross error. If a further increase in sensitivity is required, reduce a larger sample to 20 ml (final total volume 60 ml) by boiling in the refluxing flask on a hotplate in the presence of all the reagents. Carry a blank through the same procedure. The ratio of water to sulphuric acid must not fall much below 1.0 or a high blank will result because of the thermal decomposition of potassium dichromate. This technique has the advantage of concentrating without significant loss of easily digested volatile materials. Hard-to-digest volatile materials, such as volatile acids, are lost but an improvement is gained over ordinary evaporative concentration methods. Moreover, as sample volume increases, correspondingly lower concentrations of chlorides will be complexed by 0.4 g of $HgSO_4$.

**Table 7.5** Reagent quantities for different sample sizes

| Sample size (ml) | Standard potassium dichromate (ml) | $H_2SO_4$ with $Ag_2SO_4$ (ml) | $HgSO_4$ (g) | Ferrous ammonium sulphate (mol $l^{-1}$) | Final volume before titration (ml) |
|---|---|---|---|---|---|
| 10.0 | 5.0 | 15 | 0.2 | 0.05 | 70 |
| 20.0 | 10.0 | 30 | 0.4 | 0.10 | 140 |
| 30.0 | 15.0 | 45 | 0.6 | 0.15 | 210 |
| 40.0 | 20.0 | 60 | 0.8 | 0.20 | 280 |
| 50.0 | 25.0 | 75 | 1.0 | 0.25 | 350 |

*Test of the technique and reagents*
The technique and the quality of the reagents may be tested using potassium acid phthalate as a standard substance. Potassium acid phthalate has a theoretical COD of 1.176 g $g^{-1}$. Dissolve 425.1 mg of potassium acid phthalate in distilled water and dilute to 1,000 ml for a 500 mg $l^{-1}$ COD solution. A recovery of 98–100 per cent of the theoretical oxygen demand can be expected.

*Calculation*

$$\text{Concentration of COD} = \frac{(a-b) \times c \times 8{,}000}{v} \text{ mg l}^{-1}$$

where: $a$ = ferrous ammonium sulphate (ml) used for blank
$b$ = ferrous ammonium sulphate (ml) used for sample
$c$ = molarity (mol $l^{-1}$) of ferrous ammonium sulphate
$v$ = volume of sample (ml)

*Precision and accuracy*
A set of synthetic unknown samples containing potassium acid phthalate and sodium chloride was tested by 74 laboratories. At a COD of 200 mg $l^{-1}$ in the absence of chloride, the standard deviation was 6.5 per cent. At 160 mg $l^{-1}$ COD and 100 mg $l^{-1}$ chloride the standard deviation was 6.3 per cent, while at 150 mg $l^{-1}$ COD and 1,000 mg $l^{-1}$ chloride it was 9.3 per cent. These standard deviations refer to the distribution of results from all laboratories.

For most organic compounds the oxidation is 95–100 per cent of the theoretical value. Benzene, toluene and pyridine are not completely oxidised.

## 7.6 Boron

In most natural waters boron is rarely found in concentrations greater than 1 mg $l^{-1}$, but even this low concentration can have deleterious effects on certain agricultural products, including citrus fruits, walnuts and beans. Water having boron concentrations in excess of 2 mg $l^{-1}$ can adversely affect many of the more common crops. Groundwater may have greater concentrations of boron, particularly in areas where the water comes in contact with igneous rocks or other boron-containing strata. The boron content in many waters has been increasing as a result of the introduction of industrial waste and of the use of boric acid and its salts in cleaning compounds.

Ingestion of boron at concentrations usually found in natural water will have no adverse effects on humans. Ingestion of large quantities of boron, however, can affect the central nervous system, while extended consumption of water containing boron can lead to a condition known as borism.

The photometric curcumin method described below for determining boron concentrations is applicable in the 0.10–1.00 mg $l^{-1}$ range and can be extended by selection of an appropriate sample volume. The curcumin method of analysis is a reference method.

*Sample handling*

Many types of glass contain boron and their use should therefore be avoided. Samples should be stored in polyethylene bottles or alkali-resistant, boron-free glassware.

*Principle*

When a sample of water containing boron is acidified and evaporated in the presence of curcumin, a red-coloured product called rosocyanine is formed. The rosocyanine is taken up in ethanol, and the red colour is compared photometrically with standards. The minimum detectable amount of boron is 0.2 μg.

*Interferences*

The method is not applicable if more than 20 mg of $NO_3$-N per litre is present. When the total "hardness" of the sample exceeds 100 mg $l^{-1}$ as $CaCO_3$, the colour measurement may be affected by the formation of an opalescence in the alcoholic solution of rosocyanine. This interference may be prevented by passage of the sample through a cation exchange column.

*Apparatus*

- ✓ Colorimetric equipment. Either spectrophotometer or absorptiometer for measurement at 540 nm, providing a light path of at least 1 cm.
- ✓ Evaporating dishes: capacity 100–150 ml, of low boron content. Glass, platinum or other material found by experience to be suitable may be used. All dishes must be of a similar size, shape and composition.
- ✓ Water-bath: set at 55 ± 2 °C.
- ✓ Glass-stoppered volumetric flasks: capacity 25 and 50 ml.
- ✓ Ion-exchange column: 50 cm long x 1.3 cm in diameter.

*Reagents*

- ✓ Curcumin reagent. Dissolve 40 mg of finely ground curcumin and 5.0 g of oxalic acid in 80 ml of 950 ml $l^{-1}$ ethanol. Add 4.2 ml of concentrated HCl and make the solution up to 100 ml with ethanol in a 100-ml volumetric flask. This reagent will be stable for several days if stored in a refrigerator.
- ✓ Strongly acidic cation exchange resin.

- ✓ Hydrochloric acid: 1+5.
- ✓ Ethanol: 95 per cent.
- ✓ Boron stock standard. Dissolve 571.6 mg of anhydrous boric acid, $H_3BO_3$, in distilled water and dilute to 1 litre (1.00 ml ≡ 100 µg B).
- ✓ Boron working standard. Dilute 10.00 ml of stock boron solution to 1 litre with distilled water (1.00 ml ≡ 1.00 µg B).

*Procedure*

*Preparation of calibration graph*

1. Pipette 0 (blank), 0.25, 0.50, 0.75 and 1.00 ml of the boron working standard into evaporating dishes of the same shape, size and composition. Add distilled water to each standard to bring the total volume to 1.0 ml. Treat these standards as described in step 4, below, beginning with the addition of the 4.0 ml of curcumin.
2. Subtract the absorbance of the blank from the absorbances of the standards and prepare a calibration graph relating the net absorbance to the amount of boron (µg).

*Ion-exchange removal of calcium and magnesium*

3. If the "hardness" is greater than 100 mg $l^{-1}$, as $CaCO_3$, the sample should be treated by ion exchange before analysis as follows. Charge the column with 20 g of a strongly acidic cation exchange resin. Backwash the column with distilled water to remove the entrained air bubbles. Thereafter, make certain that the resin remains covered with liquid at all times. Pass 50 ml of 1+5 hydrochloric acid through the column and then wash the column free of chloride ion with distilled water. Pipette a 25 ml sample, or a smaller aliquot of a sample of known high boron content, onto the resin column. Adjust the rate of flow through the column to about 2 drops per second and collect the effluent in a 50-ml volumetric flask. Wash the column with small portions of distilled water until the flask is full to the mark. Mix the contents of the flask and transfer 2.00 ml into the evaporating dish. Add 4.0 ml of curcumin reagent and complete the analysis as described in the following section.
4. For water containing 0.10–1.00 mg of boron per litre, use 1.00 ml of sample. For water containing more than 1.00 ml of boron per litre, make an appropriate dilution with boron-free distilled water, so that a 1.00-ml aliquot contains approximately 0.50 µg of boron. Pipette the 1.00 ml of sample or dilution into an evaporating dish.
5. Add 4.0 ml of curcumin reagent to each evaporating dish and swirl the dish gently to mix the contents thoroughly. Float the dishes on a water-bath set at 55 ± 2 °C and evaporate the contents to complete dryness. Remove each dish from the water-bath 15 minutes after the contents appear dry and the odour of HCl has disappeared.
6. After the dishes cool to room temperature, add 10.0 ml of 95 per cent ethanol to each dish, stirring gently with a polyethylene rod to ensure complete dissolution of the red-coloured product. Wash the contents of each dish into a 25-ml volumetric flask, using 95 per cent ethanol. Make up to the mark with 95 per cent ethanol and mix thoroughly by inverting the flask.
7. Measure the absorbance at 540 nm within 1 hour of drying the samples. Subtract the absorbance of a boron-free blank to obtain the net absorbance. If the final solution is turbid, it should be filtered through filter paper before the absorbance is measured.

*Precautions*
Variables such as volume and concentration of reagents, as well as the time and temperature of drying, must be carefully controlled for a successful determination. Evaporating dishes must be identical in shape, size and composition to ensure equal evaporation times. Increasing the time of evaporation results in intensification of the resulting colour.

*Calculation*
Determine the amount of boron (µg) equivalent to the net absorbance from the calibration graph. The concentration of boron in mg $l^{-1}$ is obtained by dividing this weight by the volume of the sample in millilitres.

If hardness ions were removed by ion exchange, the volume of sample applied to the resin column should be divided by 25, i.e.:

$$\text{Concentration of boron} = \frac{\mu\text{g B}}{\text{ml sample}/25}\ \text{mg l}^{-1}$$

*Precision and accuracy*
A synthetic unknown sample containing B, 240 µg $l^{-1}$; As, 40 µg $l^{-1}$; Be, 250 µg $l^{-1}$; Se, 20 µg $l^{-1}$; and V, 6 µg $l^{-1}$ in distilled water was analysed in 30 laboratories by the curcumin method with an overall relative standard deviation of 22.8 per cent and a difference of 0 per cent between the overall mean and the true value.

## 7.7 Calcium

Calcium dissolves out of almost all rocks and is, consequently, detected in many waters. Waters associated with granite or siliceous sand will usually contain less than 10 mg of calcium per litre. Many waters from limestone areas may contain 30–100 mg $l^{-1}$ and those associated with gypsiferous shale may contain several hundred milligrams per litre.

Calcium contributes to the total hardness of water. On heating, calcium salts precipitate to cause boiler scale. Some calcium carbonate is desirable for domestic waters because it provides a coating in the pipes which protects them against corrosion.

The method described here is the EDTA titrimetric method.

*Sample handling*
Samples should be collected in plastic or borosilicate glass bottles without the addition of preservative. If any calcium carbonate is formed during sample storage, it must be redissolved by the addition of nitric acid before analysis.

*Principle*
When EDTA is added to water containing calcium and magnesium ions, it reacts with the calcium before the magnesium. Calcium can be determined in the presence of magnesium by EDTA titration; the indicator used is one that reacts with calcium only. Murexide indicator gives a colour change when all of the calcium has been complexed by EDTA at a pH of 12–13.

*Interferences*

Orthophosphate precipitates calcium at the pH of the test. Strontium and barium interfere with the calcium determination, and alkalinity in excess of 300 mg $l^{-1}$ may cause an indistinct end-point with hard waters. Under the conditions of the test, normal concentrations of the following ions cause no interference with the calcium determination: $Cu^{2+}$, $Fe^{2+}$, $Fe^{3+}$, $Mn^{2+}$, $Zn^{2+}$, $Al^{3+}$, $Pb^{2+}$ and $Sn^{4+}$.

*Apparatus*

✓ Porcelain dishes, 100-ml capacity.
✓ Burette, 25 or 50 ml.
✓ Pipettes.
✓ Stirring rods.
✓ Graduated cylinder, 50 ml.

*Reagents*

✓ Sodium hydroxide, 1 mol $l^{-1}$. Cautiously dissolve 40 g of NaOH in 600 ml of distilled water. Make up to 1 litre in a volumetric flask and allow to stand for at least 48 hours to permit sodium carbonate to precipitate. Decant the supernatant and store in a rigid polyethylene bottle, tightly sealed. When needed, withdraw solution by siphoning to avoid opening the bottle.
✓ Murexide indicator. The indicator changes from pink to purple at the end-point. Prepare by dissolving 150 mg of the dye in 100 g of absolute ethylene glycol. Alternatively, a dry mixture of the dye with sodium chloride is a stable form of the indicator. Prepare by mixing 200 mg of dye with 100 g of solid NaCl and grinding the mixture to a fine powder (i.e. one that passes through a 40-mesh sieve).
✓ Standard EDTA titrant, 0.01 mol $l^{-1}$. Weigh 3.723 g of disodium ethylenediaminetetraacetate dihydrate, dissolve in distilled water and make up to 1 litre in a volumetric flask. Standardise against standard calcium solution, 1,000 mg $l^{-1}$, and adjust so that 1 ml of standard EDTA is equivalent to 1 ml of standard calcium solution.
✓ Standard calcium solution. Weigh 1.000 g of $CaCO_3$ (primary standard grade) that has been dried at 105 °C and place it in a 500-ml Erlenmeyer flask. Place a funnel in the neck of the flask and add small volumes of 6 mol $l^{-1}$ HCl until all the $CaCO_3$ has dissolved. Add 200 ml of distilled water and boil for a few minutes to expel $CO_2$. Cool, add a few drops of methyl red indicator, and adjust to intermediate orange colour by adding 3 mol $l^{-1}$ $NH_4OH$ or 6 mol $l^{-1}$ HCl as necessary. Transfer the contents of the flask and rinsings to a 1-litre volumetric flask and make up to the mark with distilled water (1.0 ml is equivalent to 1 mg $CaCO_3$, 0.4008 mg $Ca^{2+}$, 0.2431 mg $Mg^{2+}$ and 1 ml standard EDTA titrant).

*Sample pretreatment*

Heavily polluted water containing considerable organic matter should be treated by nitric acid digestion as follows. Mix the sample and transfer 50 ml to a 125-ml Erlenmeyer flask. Add 5 ml of concentrated $HNO_3$ and a few boiling chips or glass beads. Bring to a slow boil and evaporate until about 10 ml remain. Continue heating and add $HNO_3$ in 1-ml portions. The solution will be clear and light in colour when digestion is complete. Do not let any dry areas appear on the bottom of the flask during digestion.

Remove from heat and cool. Wash down the walls of the flask with distilled water and transfer the contents to a 50-ml volumetric flask. Rinse the Erlenmeyer flask two or three times with a little distilled water and add the rinsings to the volumetric flask. Cool to ambient temperature before filling the volumetric flask to the mark with distilled water.

*Procedure*

1. Prepare a colour comparison blank by placing 50 ml of distilled water in a white porcelain dish.
2. Prepare the sample for titration by placing 50 ml of sample in a white porcelain dish. If the sample is high in calcium, use a measured portion of the sample containing 5–10 mg of calcium and make up to 50 ml with distilled water.
3. Add 2 ml of NaOH solution to both the sample and the comparison blank and stir. The resulting pH should be between 12 and 13.
4. Add 0.1–0.2 mg of Murexide indicator mixture (or 1–2 drops of indicator solution) to the blank. Stir, then add 1–2 drops of EDTA titrant from the burette. Stir until the colour turns from red to an orchid purple. Record the burette reading. Keep the blank for a colour reference comparison.
5. Add 0.1–0.2 mg of indicator mixture (or 1–2 drops of indicator solution) to the sample.
6. If the sample turns red, add EDTA titrant slowly, with constant stirring. Continue to add EDTA until the colour turns from red to faint purple. Add EDTA drop by drop until the colour matches the colour comparison blank. The change from faint purple to orchid purple usually takes place with the addition of 5 or 6 drops of EDTA titrant.
7. Read the burette and determine the volume of EDTA titrant used by subtracting the burette reading obtained at step 4.

*Calculation*

$$\text{Concentration of Ca} = \frac{A \times C \times 400.8}{\text{ml sample}} \quad \text{mg l}^{-1}$$

where $A$ = volume of EDTA titrant used for titration of sample (ml)
$C$ is calculated from the standardisation of the EDTA titrant:

$$C = \frac{\text{ml of standard calcium solution}}{\text{ml of EDTA titrant}}$$

## 7.8 Chloride

Chloride anions are usually present in natural waters. A high concentration occurs in waters that have been in contact with chloride-containing geological formations. Otherwise, a high chloride content may indicate pollution by sewage or industrial wastes or by the intrusion of seawater or saline water into a freshwater body or aquifer.

A salty taste in water depends on the ions with which the chlorides are associated. With sodium ions the taste is detectable at about 250 mg $l^{-1}$ $Cl^-$, but with calcium or magnesium the taste may be undetectable at 1,000 mg $l^{-1}$.

A high chloride content has a corrosive effect on metal pipes and structures and is harmful to most trees and plants.

### 7.8.1 Silver nitrate method

*Principle*

Chloride is determined in a neutral or slightly alkaline solution by titration with standard silver nitrate, using potassium chromate as indicator. Silver chloride is quantitatively precipitated before red silver chromate is formed.

*Interferences*

Bromide, iodide and cyanide are measured as equivalents of chloride. Thiosulphate, sulphite and sulphide interfere and the end-point may be difficult to detect in highly coloured or very turbid samples.

*Apparatus*

- ✓ Porcelain dish, 200-ml capacity.
- ✓ Graduated cylinder, 100 ml.
- ✓ Stirring rods.
- ✓ Burette, 25 or 50 ml.
- ✓ pH meter.

*Reagents*

- ✓ Silver nitrate standard solution. Dissolve 4.791 g silver nitrate, $AgNO_3$, in distilled water and dilute to 1 litre. Store in a brown glass bottle (1.0 ml ≡ 1.0 mg $Cl^-$).
- ✓ Sodium chloride standard solution. Dissolve 1.648 g dried sodium chloride, NaCl, in about 200 ml of distilled water in a beaker. Transfer to a 1-litre volumetric flask. Rinse the beaker twice with distilled water and pour the rinsings into the volumetric flask. Make up to the mark with distilled water (1.0 ml ≡ 1.0 mg $Cl^-$).
- ✓ Potassium chromate indicator. Dissolve 5 g potassium chromate, $K_2CrO_4$, in 100 ml distilled water. Add silver nitrate solution drop by drop to produce a slight red precipitate of silver chromate, and filter.

*Procedure*

1. Measure 100 ml of sample into a porcelain dish. Check the pH; it must be between 5.0 and 9.5 in this procedure. If the pH of the sample is below 5.0, add a small amount of calcium carbonate and stir. If the pH is above 9.5, add 0.1 mol $l^{-1}$ nitric acid drop by drop to bring the pH to about 8. Stir, and add a small amount of calcium carbonate.
2. Add 1 ml potassium chromate indicator solution and stir. The solution should turn a reddish colour.
3. Titrate with silver nitrate solution with constant stirring until only the slightest perceptible reddish coloration persists. If more than 25 ml is required, take 50 ml of the sample and dilute it to 100 ml before titration.
4. Repeat steps 1 to 3 on a 100-ml distilled water blank to allow for the presence of chloride in any of the reagents and for the solubility of silver chromate.

*Calculation*

$$\text{Chloride as } Cl^- = \frac{1{,}000(V_1 - V_2)}{\text{Volume of sample}} \text{ mg l}^{-1}$$

where $V_1$ = volume of silver nitrate required by the sample (ml)
$V_2$ = volume of silver nitrate required by the blank (ml)

When the sample contains less chloride than 50 mg $l^{-1}$, and the amount of silver nitrate required is consequently small, an appreciable error may be introduced because of the difficulty of determining the exact point at which the reddish colour becomes perceptible. It is helpful to compare the colour with that of a similar solution that has not been taken to the end-point of the titration and, therefore, contains no reddish component in its colour. Alternatively, the titration can be taken to match the colour of a liquid that has been titrated to a small, but known, amount beyond its end-point, subtracting this amount from that required by the sample.

*Reporting*
Record the $Cl^-$ results in mg $l^{-1}$. The results should be rounded off as follows:

| Concentration range (mg $l^{-1}$) | < 50 | 50–100 | 100–200 | 200–500 |
|---|---|---|---|---|
| Record to nearest (mg $l^{-1}$) | 1 | 2 | 5 | 10 |

### 7.8.2 Mercuric nitrate method

*Principle*
Chloride can be titrated with mercuric nitrate, $Hg(NO_3)_2$, because of the formation of soluble, slightly dissociated mercuric chloride. In the pH range 2.3 to 2.8, diphenylcarbazone indicates the titration end-point by formation of a purple complex indicator and an end-point enhancer. Increasing the strength of the titrant and modifying the indicator mixtures extend the range of measurable chloride concentrations.

*Interferences*
Bromide and iodide are titrated with $Hg(NO_3)_2$ in the same manner as chloride. Chromate, ferric and sulphite ions interfere when present in concentrations that exceed 10 mg $l^{-1}$.

*Apparatus*
✓ Erlenmeyer flask, 250 ml.
✓ Microburette, 5 ml with 0.01-ml graduations.

*Reagents*
✓ Sodium chloride standard solution. Dissolve 0.824 g dried sodium chloride, NaCl, in about 200 ml of distilled water in a beaker. Transfer to a 1-litre volumetric flask. Rinse the beaker twice with distilled water and pour the rinsings into the volumetric flask. Make up to the mark with distilled water (1.0 ml ≡ 500 µg $Cl^-$).
✓ Nitric acid, $HNO_3$, 0.1 mol $l^{-1}$.
✓ Sodium hydroxide, NaOH, 0.1 mol $l^{-1}$.

A. *Reagents for chloride concentration below 100 mg $l^{-1}$*
✓ Indicator–acidifier reagent: The $HNO_3$ concentration of this reagent is an important factor in the success of the determination and can be varied as indicated in (i) or (ii) below to suit the alkalinity range of the sample. Reagent (i) contains enough $HNO_3$ to neutralise a total alkalinity of 150 mg $l^{-1}$ as $CaCO_3$ to the proper pH in a 100-ml sample. Adjust the amount of $HNO_3$ to accommodate samples with alkalinity other than 150 mg $l^{-1}$.

Reagent (i). Dissolve, in the order named, 250 mg *s*-diphenylcarbazone, 4.0 ml concentrated $HNO_3$ and 30 mg xylene cyanol FF in 100 ml 95 per cent ethyl alcohol or isopropyl alcohol. Store in a dark bottle in a refrigerator. This reagent is not stable indefinitely. Deterioration causes a slow end-point and high results.
Reagent (ii). Because pH control is critical, adjust pH of highly alkaline or acid samples to 2.5 ± 0.1 with 0.1 mol $l^{-1}$ $HNO_3$ or NaOH, not with sodium carbonate. Use a pH meter with a non-chloride type of reference electrode for pH adjustment. If only the usual chloride-type reference electrode is available for pH adjustment, determine the amount of acid or alkali required to obtain the desired pH and discard this sample portion. Treat a separate sample portion with the determined amount of acid or alkali and continue the analysis. Under these circumstances, omit $HNO_3$ from the indicator reagent.

✓ Standard mercuric nitrate titrant, 0.00705 mol $l^{-1}$. Dissolve 2.3 g $Hg(NO_3)_2.H_2O$ in 100 ml distilled water containing 0.25 ml concentrated $HNO_3$. Dilute to just under 1 litre. Make a preliminary standardisation by titrating against standard sodium chloride solution. Use replicates containing 5.00 ml standard NaCl solution and 10 mg sodium bicarbonate, $NaHCO_3$, diluted to 100 ml with distilled water. Adjust titrant to 0.00705 mol $l^{-1}$ and make a final standardisation: 1.00 ml ≡ 300 µg $Cl^-$. Store away from light in a dark bottle.

B. *Reagent for chloride concentrations greater than 100 mg $l^{-1}$*

✓ Mixed indicator reagent. Dissolve 0.50 g *s*-diphenylcarbazone powder and 0.05 g bromophenol blue powder in 75 ml of 95 per cent ethyl or isopropyl alcohol and dilute to 100 ml with the same alcohol.

✓ Strong standard mercuric nitrate titrant, 0.0705 mol $l^{-1}$. Dissolve 25 g $Hg(NO_3)_2.H_2O$ in 900 ml distilled water containing 5.0 ml concentrated $HNO_3$. Dilute to just under 1 litre and standardise against standard sodium chloride solution. Use replicates containing 25.00 ml standard NaCl solution and 25 ml distilled water. Adjust titrant to 0.0705 mol $l^{-1}$ and make a final standardisation: 1.00 ml = 5.00 mg $Cl^-$.

*Procedure*

A. *Titration of chloride concentrations less than 100 mg $l^{-1}$*

1. Use a 100-ml sample or smaller portion so that the chloride content is less than 10 mg.
2. Add 1.0 ml indicator–acidifier reagent. The colour of the solution should be green-blue at this point. A light green indicates pH less than 2.0, a pure blue indicates pH more than 3.8. For most potable waters, the pH after this addition will be 2.5 ± 0.1. For highly acid or alkaline waters adjust pH to 8 before adding indicator–acidifier reagent.
3. Titrate with 0.00705 mol $l^{-1}$ $Hg(NO_3)_2$ titrant to a definite purple end-point. The solution turns from green-blue to blue a few drops before the end-point.
4. Determine the blank by titrating 100 ml distilled water containing 10 mg $NaHCO_3$.

B. *Titration of chloride concentrations greater than 100 mg $l^{-1}$*

1. Use a sample portion (5 to 50 ml) requiring less than 5 ml titrant to reach the end-point. Measure into a 150-ml beaker.
2. Add approximately 0.5 ml mixed indicator reagent and mix well. The colour should be purple.
3. Add 0.1 mol $l^{-1}$ $HNO_3$ drop by drop until the colour just turns yellow.
4. Titrate with strong $Hg(NO_3)_2$ titrant to first permanent dark purple.
5. Titrate a distilled water blank using the same procedure.

*Calculation*

$$\text{Chloride} = \frac{(A - B) \times M \times 70{,}900}{\text{ml sample}} \text{ mg l}^{-1}$$

where: $A$ = ml titration for sample
$B$ = ml titration for blank
$M$ = molarity of $Hg(NO_3)_2$

*Precision*
A synthetic sample containing $Cl^-$, 241 mg $l^{-1}$; Ca, 108 mg $l^{-1}$; Mg, 82 mg $l^{-1}$; $NO_3^-$-N, 1.1 mg $l^{-1}$; $NO_2^-$-N, 0.25 mg $l^{-1}$; K, 3.1 mg $l^{-1}$; Na, 19.9 mg $l^{-1}$; $SO_4^{2-}$, 259 mg $l^{-1}$ and 42.5 mg total alkalinity (contributed by $NaHCO_3$) in distilled water was analysed in 10 laboratories by the mercuric nitrate method with a relative standard deviation of 3.3 per cent and a relative error of 2.9 per cent.

## 7.9 Chlorophyll *a*

Analysis of the photosynthetic chlorophyll pigment present in aquatic algae is an important biological measurement which is commonly used to assess the total biomass of algae present in water samples (see section 11.3).

*Sampling*
Samples should be taken with an appropriate sampler, such as a depth or grab sampler (see section 5.3.2), a submersible pump or a hose-pipe sampler (see section 5.2.1). For nutrient-poor (high transparency) water up to 6 litres will be required. For eutrophic waters, 1–2 litres are usually adequate.

*Principle*
Three types of chlorophyll (chlorophyll *a*, *b*, and *c*) are found in phytoplankton and may be extracted with acetone. Each type has a characteristic light absorption spectrum with a particular peak absorbance. The acetone extract is analysed in a spectrophotometer at these peaks. The peak height indicates chlorophyll concentration.

When samples are concentrated by filtration for the purposes of analysis, the phytoplankton cells die. Consequently, the chlorophyll immediately starts to degrade and its concentration is thus reduced. The degradation product of chlorophyll *a*, phaeophytin *a*, fluoresces in the same spectral region, and this can lead to errors in results. It is therefore essential to measure the concentration of phaeophytin *a* and to make appropriate corrections to analytical results.

*Apparatus*
✓ Spectrophotometer, with a spectral band width between 0.5 and 2 nm.
✓ Cuvettes, 1 cm; longer path-length cuvettes may be used (usually 4 cm or 10 cm).
✓ Centrifuge.
✓ Tissue-grinder.

✓ Centrifuge tubes, 15 ml, graduated, screw-tops.
✓ Filters, glass fibre GF/C, 4.7 cm diameter.
✓ Filtration cup and pump.
*Note:* As far as possible, all apparatus should be acid- and alkali-free.

*Reagents*
✓ Magnesium carbonate suspension, 1.0 g $MgCO_3$ in 100 ml distilled water. Shake before use.
✓ Acetone solution, 90 per cent acetone.
✓ Hydrochloric acid, 1 mol $l^{-1}$.

*Procedure*

1. After recording the initial water volume, separate the cells from the water by filtration. Filter continuously and do not allow the filter to dry during filtration of a single sample. As filtration ends, add 0.2 ml of $MgCO_3$ suspension to the final few millilitres of water in the filter cup. If extraction is delayed at this point, filters should be placed in individual labelled bags or plastic Petri dishes and stored at –20 °C in darkness. Samples may be transported in this form.
2. Place the filter in the tissue-grinder, add 2–3 ml of 90 per cent acetone, and grind until the filter fibres are separated. Pour the acetone and ground filter into a centrifuge tube; rinse out the grinding tube with another 2 ml of 90 per cent acetone and add this to the centrifuge tube. Make up the total volume in the centrifuge tube to 10 ml with 90 per cent acetone. Place top on tube, label, and store in darkness at 4 °C for 10–12 hours. Samples may also be transported in this form.
3. Centrifuge closed tubes for 15 minutes at 3,000 rev/min to clarify samples. Decant the clear supernatant into a clean centrifuge tube and record the volume.
4. Fill a cuvette with 90 per cent acetone. Record absorbance on the spectrophotometer at 750 nm and 663 nm. Zero on this blank if possible; otherwise record the absorbance and subtract it from sample readings.
5. Place sample in the cuvette and record absorbance at 750 nm and 663 nm (750a and 663a).
6. Add two drops of 1 mol $l^{-1}$ HCl to sample in 1-cm cuvette (increase acid in proportion to volume for larger cuvettes). Agitate gently for 1 minute and record absorbance at 750 nm and 665 nm (750b and 665b).
7. Repeat the procedure for all samples. Some preliminary samples may need to be taken to assess the best sample volume.

*Calculation*

1. Subtract absorbance: 663a – 750a = corrected 663a absorbance
   665b – 750b = corrected 665b absorbance

2. Use these corrected 663a and 665b absorbances to calculate:

$$\text{Chlorophyll } a = \frac{26.73(663\text{a} - 665\text{b}) \times V_e}{V_s \times l} \text{ mg m}^{-3}$$

$$\text{Phaeophytin } a = \frac{26.73[1.7(665_b) - 663_a] \times V_e}{V_s \times l} \text{ mg m}^{-3}$$

where $V_e$ = volume of acetone extract (litres)
$V_s$ = volume of water sample ($m^3$)
$l$ = path length of cuvette (cm)

Chlorophyll *a* concentrations should be recorded. The ratio of chlorophyll *a* to phaeophytin *a* gives an indication of the effectiveness of sample preservation, as well as of the condition of the algal population.

## 7.10 Fluoride

While fluoride is considered to be one of the major ions of seawater, its concentration in seawater, 1.3 mg $kg^{-1}$, is indicative of most natural water concentrations. Rarely, natural waters (mainly groundwaters of arid regions) may contain fluoride concentrations greater than 10 mg $l^{-1}$. Fluoride may be added to drinking water to assist in control of dental caries. Such additions require close control of fluoride concentrations to roughly 1.0 mg $l^{-1}$, as higher levels can cause mottling of the teeth. The guideline value of 1.5 mg $l^{-1}$ in drinking water has been proposed by WHO. The local application of this value must take into account the climatic conditions and levels of water consumption. Mottling of teeth has been classed, in the USA, as a cosmetic effect, but a maximum limit of 4 mg $l^{-1}$ fluoride has been set to prevent skeletal fluorosis (a crippling condition that can result from excessive fluoride intake).

Fluoride is used in certain industrial processes and consequently occurs in the resulting wastewaters. Significant industrial sources of fluoride are the production of coke, glass and ceramics, electronics, steel and aluminium processing, pesticides and fertilisers, and electroplating operations. Waste levels may range from several hundred to several thousand milligrams per litre in untreated wastewaters. It is worthy of note that conventional treatment (lime) seldom reduces fluoride concentrations below 8–15 mg $l^{-1}$ without dilution.

The reference method for the analysis of fluoride ions is potentiometric, using the lanthanum fluoride, solid state, selective ion electrode. The secondary method is a photometric procedure employing the lanthanum alizarin complex.

### *Sample handling*

Generally, clean polyethylene bottles are preferred for collection and storage of samples for fluoride analysis, provided that long-term evaporative loss is not encountered. Glass and borosilicate glass bottles should be avoided; however, they may be used provided that low pH is not maintained, and that the containers have been thoroughly cleaned and have not previously been in contact with solutions of high fluoride concentration. Pretreatment with high levels of sodium thiosulphate (more than 100 mg $l^{-1}$) should be avoided.

### 7.10.1 Selective ion electrode method

*Principle*

The fluoride electrode consists of a single lanthanum fluoride crystal, the internal portion of which is in contact with a constant concentration of fluoride ion and an internal reference electrode. Upon contact of the external electrode surface with the test solution (standard or unknown) a potential difference is set up across the crystal which is related to the fluoride ion concentrations in contact with the crystal surfaces. An external reference electrode in the test solution completes the circuit and allows measurement of the membrane or crystal potential:

| Internal reference electrode | $F^-$ (internal) | $LaF_3$ | Test solution | External reference electrode |
|---|---|---|---|---|

Since the relationship between potential and fluoride ion concentration is described by a form of the Nernst equation ($E = E° - RT \ln a_{F^-}$), it is the log fluoride ion activity that is related to change in measured potential. Consequently, variations in ionic strength between samples and standards and among samples must be prevented. Similarly, since it is the free fluoride ion activity that yields the electrode response, formation of complex species (Al, Fe) or undissociated hydrofluoric acid must be prevented. The procedure is designed to maintain control of these problems.

*Interferences*

Without the addition of a suitable complexing agent, polyvalent cations such as Al(III), Fe(III), Si(IV) will remove free fluoride ion from solution as soluble complexes. Similarly, if pH is not maintained above 5.0, the presence of molecular hydrogen fluoride and $HF_2^-$ reduces the free fluoride ion concentration.

Fluoride ion selectivity is extremely high with respect to most aqueous cations. However, if solution pH is not maintained below 8, hydroxide ion will begin to yield sufficient electrode response to interfere with the fluoride measurement since the relative response, $F^-:OH^-$ is about 10:1. The electrode does not respond to fluoroborate ($BF_4^-$) directly.

*Apparatus*

- ✓ High impedance millivoltmeter. An expanded-scale or digital pH meter or ion-selective meter or any high impedance millivolt potentiometer.
- ✓ Reference electrode. Sleeve type rather than fibre-tip reference electrode should be used, especially for very dilute solutions.
- ✓ Fluoride electrode. Commercial electrode.
- ✓ Magnetic stirrer. Proper insulation should be provided to prevent heat transfer to sample solution during measurement. Stirring bars should be Teflon-coated.
- ✓ Stopwatch or timer.

*Reagents*

✓ Deionised or distilled water. All solutions should be prepared from high-quality distilled or deionised water.

✓ Fluoride standard solution. Dissolve 0.221 g of analytical grade anhydrous sodium fluoride, NaF, in distilled water and dilute to 1 litre (1 ml ≡ 100 µg F).

✓ A working standard of 10 µg $F^-$ per 1.0 ml should be prepared just before starting the analysis.

✓ Total ionic strength adjustment buffer. The total ionic strength adjustment buffer solution (TISAB) may be prepared by the following procedure:
Dissolve 57 ml glacial acetic acid, 58 g sodium chloride and 4.0 g 1,2-cyclohexylene-diaminetetraacetic acid in 500 ml distilled water. Adjust pH to 5.0–5.5 with 6 mol $l^{-1}$ sodium hydroxide. Stir and cool during NaOH addition (about 125 ml will be required). Dilute to final volume of 1 litre with distilled water.

*Procedure*

1. Preparation of calibration graph. Prepare a series of standards over the appropriate concentration range, such that the TISAB constitutes 50 per cent of the solution by volume. Linear millivolt response versus log of concentration should be obtained from roughly 0.2 to 2,000 mg $l^{-1}$. In the linear region, three standards should suffice to determine the standard curve. In non-linear regions, i.e. low levels, more data points are necessary.

2. Instrument operation. Follow manufacturer's recommendation for obtaining millivolt readings. Some instruments read in absolute millivolts while others operate with relative millivolt scales.

3. Sample measurement. Mix sample to be analysed with an equal volume of TISAB in a beaker. If high levels of Al (> 3 mg $l^{-1}$) or Fe (> 200 mg $l^{-1}$) are present, follow the procedure for distillation. Ensure that the sample has achieved room temperature (or that measurement temperature is the same as that of standards). Place beaker on an insulated magnetic stirrer, add a stirring bar, immerse the electrodes and allow 3 minutes for equilibration. Determine fluoride concentration from the calibration curve. The electrodes should be rinsed and dried between samples. Frequent recalibration should be made with an intermediate standard. The detection limit is about 0.02 mg $l^{-1}$ fluoride.

*Calculation*

The concentration of fluoride ion can be determined directly from the calibration curve if the samples and standards are treated alike and the concentration axis is constructed in terms of initial standard concentration before addition of TISAB.

*Precision and accuracy*

Precision is limited by variations in temperature, instrument drift and scale reading errors (if the instrument has a needle read-out). With proper recalibration (roughly once per hour), relative precision of ± 2 per cent has been reported by one manufacturer, independent of concentration.

### 7.10.2 SPADNS method

*Principle*

The SPADNS colorimetric method is based on the reaction between fluoride and a zirconium-dye lake. Fluoride reacts with the dye lake, dissociating a portion of it into a colourless complex anion, $ZrF_6^{2-}$, and the dye. As the amount of fluoride increases, the colour produced becomes progressively lighter.

The reaction rate between fluoride and zirconium ions is greatly influenced by the acidity of the reaction mixture. If the proportion of acid in the reagent is increased, the reaction can be made almost instantaneous. Under such conditions, however, the effect of various ions differs from that in the conventional alizarin methods. The selection of dye for this rapid fluoride method is governed largely by the resulting tolerance to these ions.

*Interferences*

Chlorine, colour and turbidity interfere and must be removed by distillation. Interference caused by alkalinity, chloride, iron, phosphate and sulphate is not linear and so cannot be accounted for mathematically. Whenever any substance is present in quantities large enough to produce an error of 0.1 mg $l^{-1}$, or if the total interfering effect is in doubt, distil the sample. If alkalinity is the only interference it can be neutralised with either hydrochloric or nitric acid.

Volumetric measurements of sample and reagent are extremely important to analytical accuracy. Samples and standards should be at the same temperature or at most have a 2 °C temperature differential. Maintain temperature as nearly constant as possible throughout colour development. Different calibration curves are required for different temperature ranges.

*Apparatus*

✓ Colorimetric equipment. One of the following is required:
 — Spectrophotometer for measurement at 570 nm and capable of providing a light path of 1 cm or longer.
 — Filter photometer providing a light path of at least 1 cm and equipped with a greenish-yellow filter having maximum transmittance at 550 to 580 nm.

*Reagents*

✓ Stock fluoride solution. Dissolve 0.221 g of analytical grade anhydrous sodium fluoride, NaF, in distilled water and dilute to 1 litre (1 ml ≡ 100 µg $F^-$).

✓ Working standard of 10 µg $F^-$ per 1.0 ml, which should be prepared just before the analysis.

✓ SPADNS solution. Dissolve 0.958 g of SPADNS, sodium 2-(parasulphophenylazo)-1,8-dihydroxy-3,6-naphthalene disulphonate, also called 4,5-dihydroxy-3-(parasulphophenylazo)-2,7-naphalene disulphonic acid trisodium salt, in distilled water and dilute to 500 ml. This solution is stable for at least one year if protected from direct sunlight.

✓ Acid-zirconyl reagent. Dissolve 130 mg zirconyl chloride octahydrate, $ZrOCl_2.8H_2O$, in about 25 ml of distilled water. Add 350 ml of concentrated hydrochloric acid and dilute to 500 ml with distilled water.

✓ Acid-zirconyl/SPADNS reagent. Mix equal quantities of SPADNS solution and acid-zirconyl reagent. The combined reagent is stable for at least two years.

✓ Reference solution. Add 10 ml of SPADNS solution to 100 ml distilled water. Dilute 7 ml of concentrated HCl to 10 ml and add to the diluted SPADNS solution. The resulting solution is used for setting the instrument reference point (at zero) and is stable for at least one year. Alternatively, a prepared standard of 0 mg $F^- l^{-1}$ may be used as a reference.

✓ Sodium arsenite solution. Dissolve 5.0 g of sodium arsenite, $NaAsO_2$, in distilled water and dilute to 1 litre.

*Procedure*

1. Preparation of standard curve. Prepare fluoride standards in the range of 0 to 1.4 mg $F^- l^{-1}$ by diluting appropriate quantities of standard fluoride solution to 50 ml with distilled water. Pipette 5.00 ml each of the SPADNS solution and the acid-zirconyl reagent or 10.00 ml of the mixed acid-zirconyl/SPADNS reagent to each standard and mix well. Avoid contamination. Set the photometer to zero absorbance with the reference solution and obtain absorbance readings of the standards. Plot a curve of the relationship between mg fluoride and absorbance. Prepare a new standard curve whenever a fresh reagent is made or a different standard temperature is desired.

   As an alternative to using a reference, set the photometer at a convenient point (0.300 or 0.500 absorbance) with the prepared 0 mg $F^- l^{-1}$ standard.

2. Sample pretreatment. If the sample contains residual chlorine, remove it by adding 1 drop (0.05 ml) $NaAsO_2$ solution per 0.1 mg residual chlorine and mix. (Sodium arsenite concentrations of 1,300 mg $l^{-1}$ produce an error of 0.1 mg $l^{-1}$ when the fluoride concentration is 1.0 mg $F^- l^{-1}$.)

3. Colour development. Use a 50-ml sample or a portion diluted to 50 ml with distilled water. Adjust temperature to that used for the standard curve. Add 5.00 ml each of SPADNS solution and acid-zirconyl reagent (or 10.00 ml of acid-zirconyl/SPADNS reagent), mix well and read absorbance, after setting the reference point of the photometer at zero as described in step 1 above. If the absorbance falls beyond the range of the standard curve, repeat the procedure using a diluted sample.

*Calculation*

$$\text{Fluoride} = \frac{A}{\text{ml sample}} \times \frac{B}{C} \ \text{mg l}^{-1}$$

where $A$ = $F^-$ determined from plotted curve (µg)
$B$ = final volume of diluted sample (ml)
$C$ = volume of diluted sample used for colour development (ml).

### 7.10.3 Fluoride distillation

*Principle*

In the event of interferences that cannot be controlled in the direct fluoride analysis procedures, it is possible to remove the fluoride from solution selectively. Distillation of either fluorosilicic or hydrofluoric acid from an acid solution of higher boiling point separates the fluoride from other constituents in the original sample.

*Apparatus*

Distillation apparatus (shown in Figure 7.1), consisting of a 1-litre round-bottomed, long-necked boiling flask (of borosilicate glass), a connecting tube, an efficient condenser, a thermometer adapter, and a thermometer reading to 200 °C. Any comparable apparatus

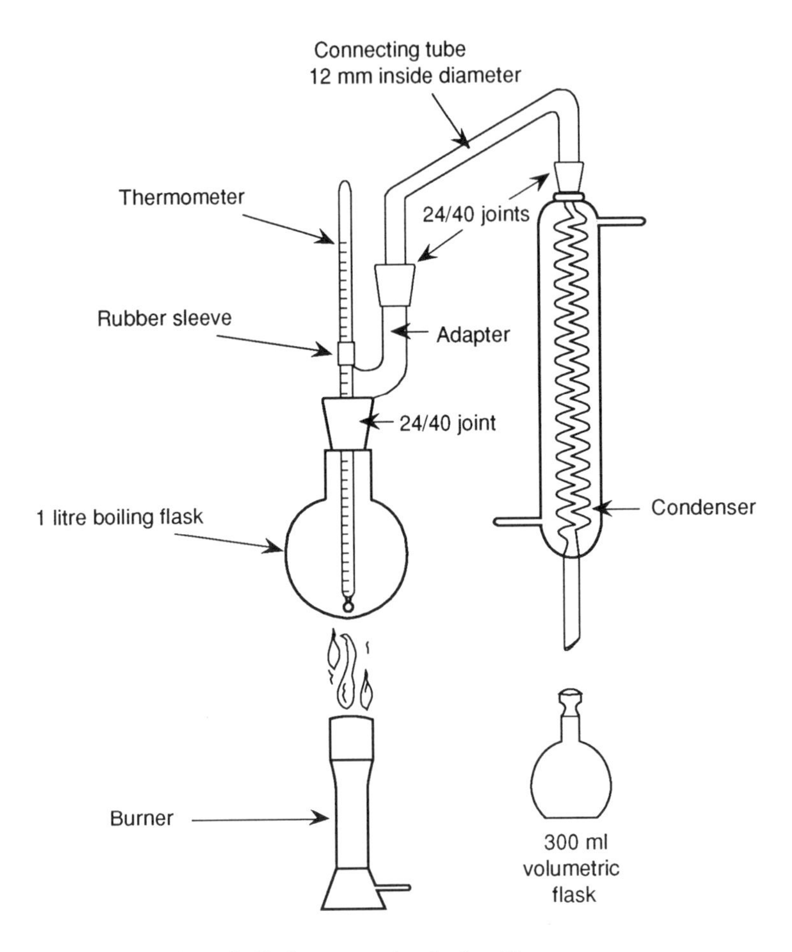


**Figure 7.1** Direct distillation apparatus for fluoride

with the essential design features may be used. The critical points to observe are those that could affect complete fluoride recovery, such as obstruction in the vapour path and trapping of liquid in the adapter and condenser, and conditions that might enhance sulphate carry-over. An asbestos shield or similar device should be used to protect the upper part of the boiling flask from the burner flame. If desired, this apparatus can be modified so that the heat is automatically shut off when distillation is completed.

*Reagents*

✓ Sulphuric acid, $H_2SO_4$, concentrated.
✓ Silver sulphate, $Ag_2SO_4$, crystals.

*Procedure*

1. Place 400 ml distilled water in the boiling flask and carefully add 200 ml concentrated $H_2SO_4$. Swirl until the flask contents are homogeneous. Add 25–35 glass beads and connect the apparatus as shown in Figure 7.1, making sure that all joints are tight.

Begin heating slowly at first, then as rapidly as the efficiency of the condenser will permit (the distillate must be cool) until the temperature of the flask contents reaches exactly 180 °C. Discard the distillate. The process removes fluoride contamination and adjusts the acid:water ratio for subsequent distillations.

2. After cooling the acid mixture remaining after step 1 (to 120 °C or below), add 300 ml of sample, mix thoroughly, and distil as before until the temperature reaches 180 °C. To prevent sulphate carry-over, do not heat above 180 °C.
3. When high-chloride samples are distilled, add $Ag_2SO_4$ to the boiling flask at the rate of 5 mg for each mg of $Cl^-$.
4. Use the sulphuric acid solution in the flask repeatedly until the contaminants from the samples accumulate to such an extent that recovery is affected or interferences appear in the distillate. Check suitability of the acid periodically by distilling standard fluoride samples. After the distillation of high-fluoride samples, flush the still with 300 ml distilled water and combine two fluoride distillates. If necessary, repeat the flushing operation until the fluoride content of the distillates is at a minimum. Include the additional fluoride recovered with that of the first distillation. After periods of inactivity, similarly flush the still and discard the distillate. Use the distillate to conduct fluoride analysis by one of the prescribed methods. Correct the volume relationship if the distillate volume differs from that of the original sample.

## 7.11 Iron

Iron is an abundant element in the earth's crust, but exists generally in minor concentrations in natural water systems. The form and solubility of iron in natural waters are strongly dependent upon the pH and the oxidation–reduction potential of the water. Iron is found in the +2 and +3 oxidation states. In a reducing environment, ferrous (+2) iron is relatively soluble. An increase in the oxidation–reduction potential of the water readily converts ferrous ions to ferric (+3) and allows ferric iron to hydrolyse and precipitate as hydrated ferric oxide. The precipitate is highly insoluble. Consequently, ferric iron is found in solution only at a pH of less than 3. The presence of inorganic or organic complex-forming ions in the natural water system can enhance the solubility of both ferrous and ferric iron.

Surface waters in a normal pH range of 6 to 9 rarely carry more than 1 mg of dissolved iron per litre. However, subsurface water removed from atmospheric oxidative conditions and in contact with iron-bearing minerals may readily contain elevated amounts of ferrous iron. For example, in groundwater systems affected by mining, the quantities of iron routinely measured may be several hundred milligrams per litre.

It is the formation of hydrated ferric oxide that makes iron-laden waters objectionable. This ferric precipitate imparts an orange stain to any settling surfaces, including laundry articles, cooking and eating utensils and plumbing fixtures. Additionally, colloidal suspensions of the ferric precipitate can give the water a uniformly yellow-orange, murky cast. This coloration, along with associated tastes and odour, can make the water undesirable for domestic use when levels exceed 0.3 mg $l^{-1}$.

The reference method of analysis is a photometric method described here in which iron is bound into a colour-forming complex with 1,10-phenanthroline.

*Sample handling*

In the sampling and storage process, iron in solution may undergo changes in oxidation form and it can readily precipitate on the sample container walls or as a partially settleable solid suspension. For total iron measurements, precipitation can be controlled in the sample containers by the addition of 1.5–2.0 ml of concentrated $HNO_3$ per litre of sample immediately after collection. If the pH is not less than 2 after the addition of acid, more $HNO_3$ should be added.

*Principle*

For total iron determinations, precipitated iron is brought into solution by boiling with acid. Ferric iron is reduced to the ferrous state by the addition of hydroxylamine hydrochloride.

Ferrous iron is chelated with 1,10-phenanthroline to form an orange-red complex. Colour intensity is proportional to iron concentration. Absorbance can be measured spectrophotometrically at 510 nm or photometrically using a green filter having maximum transmittance near 510 nm. For cell lengths of 1 cm, Beer's law is obeyed in iron solutions containing 0.1–5 mg $l^{-1}$. The most rapid colour development occurs between pH 2.9 and pH 3.5, while colour intensity is unaffected by a pH between 3 and 9.

If large amounts of organic materials are present, the sample must first be digested with $H_2SO_4$ to destroy organic structures and to bring all the iron into solution. Hydrochloric acid is then added until the HCl concentration is between 7 and 8 mol $l^{-1}$ and the iron is extracted as $FeCl_3$ into diisopropyl ether. Iron is then re-extracted into water and reduced with hydroxylamine.

*Interferences*

Strongly oxidising substances may interfere. Cyanide, nitrite, phosphates, chromium, zinc, cobalt and copper interfere if concentrations exceed 10 times that of iron. Additionally, cobalt or copper present in excess of 5 mg $l^{-1}$ and nickel in excess of 2 mg $l^{-1}$ result in interferences. Bismuth, cadmium, mercury, molybdate and silver cannot be present because they precipitate phenanthroline.

Cyanide and nitrite may be removed by boiling with acid. The same procedure converts polyphosphates into orthophosphates, which cause less interference. Excess hydroxylamine addition will reduce strongly oxidising agents, and excess phenanthroline is required to guarantee complete iron complexation if large concentrations of interfering metal ions are present. The milky solution produced from molybdate interference can be overcome by adjusting the pH to greater than 5.5.

For samples that are highly coloured or that contain large amounts of organic material, ashing procedures should precede analysis. The sample may be wet-ashed with sulphuric acid and nitric acid or dry-ashed at temperatures not exceeding 700 °C.

Preparation for the analysis of highly contaminated water and industrial wastewater must include careful consideration of possible interferences. The general procedures for correcting the interferences described above will aid in dealing with specific interferences. The ultimate choice may be to eliminate interferences by extracting the iron with diisopropyl ether from a hydrochloric acid solution and then back-extracting the iron with water.

*Apparatus*

✓ Colorimetric equipment. One of the following is required:
  - Spectrophotometer, for use at 510 nm, providing a light path of 1 cm or longer.
  - Filter photometer, providing a light path of 1 cm or longer and equipped with a green filter having maximum transmittance near 510 nm.
  - Nessler tubes, matched, 100 ml, tall form.

✓ Acid-washed glassware. Wash all glassware with concentrated hydrochloric acid and rinse with distilled water before use to remove deposits of iron oxide.

✓ Separatory funnels, 125 ml, Squibb form, with ground-glass or TFE stopcocks and stoppers.

*Reagents*

Use reagents low in iron. Use iron-free distilled water in preparing standards and reagent solutions. Store reagents in glass-stoppered bottles. The HCl and ammonium acetate solutions are stable indefinitely if tightly stoppered. The hydroxylamine, phenanthroline and stock iron solutions are stable for several months. The standard iron solutions are not stable; prepare daily as needed by diluting the stock solution. Visual standards in Nessler tubes are stable for several months if sealed and protected from light.

✓ Hydrochloric acid, HCl, concentrated, containing less that 0.00005 per cent iron.

✓ Hydroxylamine hydrochloride solution. Dissolve 10 g $NH_2OH.HCl$ in 100 ml water.

✓ Ammonium acetate buffer solution. Dissolve 250 g $NH_4C_2H_3O_2$ in 150 ml water. Add 700 ml concentrated (glacial) acetic acid. Prepare new reference standards with each buffer preparation because even a good grade of $NH_4C_2H_3O_2$ contains a significant amount of iron.

✓ Phenanthroline solution. Dissolve 100 mg 1,10-phenanthroline monohydrate, $C_{12}H_8N_2.H_2O$, in 100 ml water by stirring and heating to 80 °C. Do not boil. Discard the solution if it darkens. Heating is unnecessary if two drops of concentrated HCl are added to the water. One ml of this reagent is sufficient for no more than 100 µg Fe.

✓ Stock iron solution. Use metal or salt for preparing the stock solution as described below. Method (i). Use electrolytic iron wire or "iron wire for standardising" to prepare the solution. If necessary clean the wire with fine sandpaper to remove any oxide coating and to produce a bright surface. Weigh 200.0 mg wire and place in a 1,000-ml volumetric flask. Dissolve in 20 ml of 3 mol $l^{-1}$ $H_2SO_4$ and dilute to the mark with distilled water (1.00 ml ≡ 200 µg Fe).

**Table 7.6** Selection of light path for various iron concentrations

| Final volume (μg Fe) | | |
|---|---|---|
| 50 ml | 100 ml | Light path (cm) |
| 50–200 | 100–400 | 1 |
| 25–500 | 50–200 | 2 |
| 10–40 | 20–80 | 5 |
| 5–20 | 10–40 | 10 |

Method (ii). If ferrous ammonium sulphate is preferred, slowly add 20 ml concentrated $H_2SO_4$ to 50 ml distilled water and dissolve 1.404 g of $Fe(NH_4)_2(SO_4)_2.6H_2O$. Add 0.1 mol $l^{-1}$ potassium permanganate ($KMnO_4$) drop by drop until a faint pink colour persists. Dilute to 1,000 ml with distilled water and mix (1.00 ml ≡ 200 μg Fe).

✓ Standard iron solutions. Prepare fresh solutions the day they are to be used. Pipette 50.00 ml stock solution into a 1,000-ml volumetric flask and dilute to the mark with distilled water (1.00 ml ≡ 10.0 μg Fe). Pipette 5.00 ml stock solution into a 1,000-ml volumetric flask and dilute to the mark with distilled water (1.00 ml ≡ 1.00 μg Fe).

*Preparation of calibration graph*

1. Prepare a blank and a series of standards to encompass the range of concentrations of iron in the samples.
2. For photometric measurement Table 7.6 can be used as a rough guide for selecting the proper light path at 510 nm. Treat the blank and the standards as indicated in the procedure. Zero the instrument against distilled water and measure the absorbance of the standards and the blank.
3. Subtract the absorbance of the blank from the absorbances of the standards to obtain the net absorbance. Prepare a calibration graph relating net absorbance to the amount of iron.

*Procedure*

1. Mix the sample thoroughly and measure out a volume of solution containing not more than 0.5 mg of iron into a 125-ml Erlenmeyer flask. If necessary, dilute to 50 ml and add 2 ml of concentrated HCl and 1 ml of hydroxylamine hydrochloride solution.
2. Drop in a few glass beads and boil until the volume is reduced to 10–20 ml. Cool to room temperature.
3. Transfer to a 50-ml or 100-ml volumetric flask, add 10 ml of ammonium acetate buffer solution and 2 ml of phenanthroline solution, and dilute to the mark with distilled water. Mix thoroughly and set aside for 10–15 minutes for full colour development.
4. Measure the colour intensity spectrometrically at 510 nm. Subtract the absorbance of the blank from that of the sample to determine the net absorbance.

*Calculation*

Determine the weight of iron (μg Fe) equivalent to the net absorbance from the calibration graph.

$$\text{Concentration of Fe} = \frac{\mu\text{g Fe}}{\text{ml sample}} \text{ mg l}^{-1}$$

*Precision and accuracy*
An overall relative standard deviation of 25.5 per cent and a difference of 13.3 per cent between the overall mean and the true value among 44 laboratories examining iron by the phenanthroline method have been reported. The sample contained 300 µg Fe, 500 µg Al, 50 µg Cd, 110 µg Cr, 470 µg Cu, 70 µg Pb, 120 µg Mn, 150 µg Ag and 650 µg Zn per litre of distilled water.

## 7.12 Magnesium

Magnesium is a relatively abundant element in the earth's crust and hence a common constituent of natural water. Waters associated with granite or siliceous sand may contain less than 5 mg of magnesium per litre. Water in contact with dolomite or magnesium-rich limestone may contain 10–50 mg $l^{-1}$ and several hundred milligrams per litre may be present in water that has been in contact with deposits containing sulphates and chlorides of magnesium.

By a similar action to that of calcium, magnesium imparts hardness to water. This may be reduced by chemical softening or by ion exchange. It should be noted that the difference between total hardness and the calcium concentration can be used to calculate the magnesium concentration.

*Sample handling*
Samples should be collected in plastic or borosilicate glass bottles without the addition of preservative. If any calcium carbonate is formed during sample storage, it must be redissolved before analysis.

### 7.12.1 Titrimetric method

*Principle*
When EDTA is added in a titration to water containing calcium and magnesium ions, it reacts with the calcium before the magnesium. In this procedure calcium is precipitated as its oxalate and is removed by filtration before titration. A small amount of Eriochrome Black T is added to the solution which, when buffered at pH 10.0 ± 0.1, becomes wine-red in colour. When the solution is titrated with EDTA, the magnesium is complexed; at the end-point the solution changes from wine-red to blue. A small amount of the compleximetrically neutral salt of EDTA with magnesium is added to the buffer because magnesium ions must be present to obtain a satisfactory end-point. This introduces sufficient magnesium and eliminates the need for a blank correction.

It should be noted that the procedure subsequent to the precipitation of calcium oxalate (and its removal from solution by filtration) is the same as that for hardness.

*Interferences*

Several metal ions can interfere with the titration by producing fading or indistinct end-points. To minimise these interferences, sodium sulphide solution is added. The sodium sulphide inhibitor precipitates insoluble sulphides and, if these are dark in colour and present in appreciable quantities, they may tend to mask the end-point or alter the colour change.

Not all samples require the use of the inhibitor. To determine whether addition of the inhibitor is necessary, the analyst should compare the results of two titrations, one with and one without inhibitor.

*Apparatus*

- ✓ Porcelain dishes, 100-ml capacity.
- ✓ Burette, 25 or 50 ml.
- ✓ Pipettes.
- ✓ Stirring rods.
- ✓ Graduated cylinder, 50 ml.

*Reagents*

- ✓ Ammonium oxalate solution, 5 per cent. Dissolve 10 g of ammonium oxalate monohydrate, $(NH_4)_2C_2O_4.H_2O$, in 250 ml of distilled water. Filter the solution if it contains a precipitate.
- ✓ Buffer solution. Add 55 ml of concentrated hydrochloric acid to 400 ml of distilled water. Slowly, with stirring, add 310 ml of 2-aminoethanol (also called ethanolamine). Add 5.0 g of magnesium disodium ethylenediaminetetraacetate and dilute to 1 litre with distilled water. This buffer has a normal shelf-life of about 1 month. Discard the buffer when addition of 1 or 2 ml to a sample fails to produce a pH of $10.0 \pm 0.1$ at the end-point of a titration.
- ✓ Sodium sulphide inhibitor. Dissolve 5.0 g of sodium sulphide, $Na_2S.9H_2O$, in 100 ml of distilled water. Store in a glass container with a tightly fitting rubber stopper to reduce deterioration through air oxidation.
- ✓ Eriochrome Black T indicator. Sodium salt of 1-(1-hydroxy-2-naphthylazo)-5-nitro-2-napthol-4-sulphonic acid; No. 203 in the Colour Index. Dissolve 0.5 g of dye in 100 ml of triethanolamine (also known as 2,2',2"-nitrilotriethanol). This may also be prepared by grinding 0.5 g of Eriochrome Black T together with 100 g of NaCl.
- ✓ Standard EDTA titrant, 0.01 mol $l^{-1}$ (see section 7.7, Calcium).
- ✓ Standard calcium solution (see section 7.7, Calcium).

*Sample pretreatment*

Samples containing high concentrations of organic material may give indistinct end-points. If this cannot be overcome by dilution, pretreat samples by nitric acid digestion (see section 7.7, Calcium).

*Procedure*

1. Removal of calcium. Measure 75 ml of the sample into a beaker or similar vessel. Add 2 ml of buffer solution and 1 ml of 5 per cent ammonium oxalate solution. Mix and allow to stand for 5–10 minutes. Filter through Whatman No. 5 filter paper or equivalent. Discard the first 5–10 ml of the filtrate.

2. Titration of sample. Dilute 25.0 ml of the filtered sample to about 50 ml with distilled water in a porcelain dish. Determine whether the pH is $10.0 \pm 0.1$. If it is not, but the deviation is small, additional buffer may be sufficient to adjust it. If the deviation is

large, the pH must be adjusted before the addition of the ammonium oxalate. If the inhibitor is being used, it must be added at this point in the procedure.

3. Add 2 drops of indicator solution or a small amount of dry-powder indicator mixture.
4. Add the standard EDTA titrant slowly, with continuous stirring, until the last reddish tinge disappears from the solution, adding the last few drops at intervals of 3–5 seconds. At the end-point the solution is normally blue. Working in daylight or under a daylight fluorescent lamp is highly recommended. Ordinary incandescent light tends to produce a reddish tinge in the blue at the end-point. If the end-point is indistinct, either the indicator has deteriorated or the inhibitor must be added before titration begins.
5. If sufficient sample is available and there is no interference, improve the accuracy by increasing the sample size. If a large sample volume is taken, add proportionally larger amounts of buffer, inhibitor and indicator.
6. Titrate a distilled water blank of the same volume as the sample. Add identical amounts of buffer, inhibitor and indicator solution.

*Calculation*

$$\text{Concentration of Mg} = \frac{(A-B)\times C\times 243.1}{\text{ml sample}}\ \text{mg l}^{-1}$$

where $A$ = volume of EDTA for titration of sample (ml)
$B$ = volume of EDTA for titration of blank (ml)
$C$ is calculated from the standardisation of the EDTA titrant:

$$C = \frac{\text{ml of standard calcium solution}}{\text{ml of EDTA titrant}}$$

### 7.12.2 Magnesium hardness method

*Principle*

Magnesium may also be determined by calculating the difference between the total hardness and the calcium hardness of the sample. This yields the value of magnesium hardness as mg $l^{-1}$ $CaCO_3$ which, when multiplied by 243.1, will be the concentration of magnesium.

*Apparatus*

- ✓ Porcelain dishes, 100-ml capacity.
- ✓ Burette, 25 or 50 ml.
- ✓ Pipettes.
- ✓ Stirring rods.
- ✓ Graduated cylinder, 50 ml.

*Reagents*

- ✓ Buffer solution. Add 55 ml of concentrated hydrochloric acid to 400 ml of distilled water. Slowly, with stirring, add 310 ml of 2-aminoethanol (also called ethanolamine). Add 5.0 g of magnesium disodium ethylenediaminetetraacetate and dilute to 1 litre with distilled water. This buffer has a normal shelf-life of about 1 month. Discard the buffer when addition of 1 or 2 ml to a sample fails to produce a pH of 10.0 ± 0.1 at the end-point of a titration.

- ✓ Sodium sulphide inhibitor. Dissolve 5.0 g of sodium sulphide, $Na_2S.9H_2O$, in 100 ml of distilled water. Store in a glass container with a tightly fitting rubber stopper to reduce deterioration through air oxidation.
- ✓ Eriochrome Black T indicator. Sodium salt of 1-(1-hydroxy-2-naphthylazo)-5-nitro-2-napthol-4-sulphonic acid; No. 203 in the Colour Index. Dissolve 0.5 g of dye in 100 ml of triethanolamine (also known as 2,2',2"-nitrilotriethanol). This may also be prepared by grinding 0.5 g of Eriochrome Black T together with 100 g of NaCl.
- ✓ Standard EDTA titrant, 0.01 mol $l^{-1}$ (see section 7.7, Calcium).
- ✓ Standard calcium solution (see section 7.7, Calcium).

*Sample pretreatment*

Samples containing high concentrations of organic material may give indistinct end-points. If this cannot be overcome by dilution, follow the nitric acid digestion procedure described under "Sample pretreatment" in section 7.7, Calcium.

*Procedure*

1. Titration of sample. Dilute 25.0 ml of the sample to about 50 ml with distilled water in a porcelain dish. Determine whether the pH is $10.0 \pm 0.1$. If it is not, but the deviation is small, additional buffer may be sufficient to adjust it. If the deviation is large, the pH must be adjusted before the addition of the ammonium oxalate. If the inhibitor is being used, it must be added at this point in the procedure.
2. Add 2 drops of indicator solution or a small amount of dry-powder indicator mixture.
3. Add the standard EDTA titrant slowly, with continuous stirring, until the last reddish tinge disappears from the solution, adding the last few drops at intervals of 3–5 seconds. At the end-point, the solution is normally blue. Working in daylight or under a daylight fluorescent lamp is highly recommended. Ordinary incandescent light tends to produce a reddish tinge in the blue at the end-point. If the end-point is indistinct, either the indicator has deteriorated or the inhibitor must be added before titration begins.
4. If sufficient sample is available and there is no interference, improve the accuracy by increasing the sample size. If a large sample volume is taken, add proportionally larger amounts of buffer, inhibitor and indicator.
5. Titrate a distilled water blank of the same volume as the sample. Add identical amounts of buffer, inhibitor and indicator solution.
6. From the results obtained when following the procedure for the determination of calcium, record as amount "*A*" the number of millilitres of standard EDTA titrant required to reach the orchid purple colour end-point.

*Calculation*

$$\text{Total hardness} = \frac{(T-S)\times C\times 1{,}000}{\text{Volume of sample (ml)}}\ \text{mg l}^{-1}\ \text{as CaCO}_3$$

$$\text{Calcium hardness} = \frac{A\times C\times 1{,}000}{\text{Volume of sample (ml)}}\ \text{mg l}^{-1}\ \text{as CaCO}_3$$

Magnesium hardness = total hardness – calcium hardness

Magnesium hardness (as mg $CaCO_3$ $l^{-1}$) $\times$ 0.2431 = mg Mg $l^{-1}$

where $T$ = volume of EDTA for titration of the total hardness sample (ml)
$S$ = volume of EDTA for titration of the blank (ml)
$A$ = volume of EDTA for titration in the calcium procedure (ml)
$C$ is calculated from the standardisation of the EDTA titrant:

$$C = \frac{\text{ml of standard calcium solution}}{\text{ml of EDTA titrant}}$$

## 7.13 Manganese

Although manganese in groundwater is generally present in the soluble divalent ionic form because of the absence of oxygen, part or all of the manganese in surface waters (or water from other sources) may be in a higher valence state. Determination of total manganese does not differentiate the various valence states. The heptavalent permanganate ion is used to oxidise manganese and/or any organic matter that causes taste. Excess permanganate, complexed trivalent manganese, or a suspension of quadrivalent manganese must be detected with great sensitivity to control treatment processes and to prevent their discharge into a water distribution system. There is evidence that manganese occurs in surface waters both in suspension in the quadrivalent state and in the trivalent state in a relatively stable, soluble complex. Although rarely present at concentrations in excess of 1 mg $l^{-1}$, manganese imparts objectionable and tenacious stains to laundry and plumbing fixtures. The low manganese limits imposed on an acceptable water stem from these, rather than toxicological, considerations. Special means of removal are often necessary, such as chemical precipitation, pH adjustment, aeration and use of special ion exchange materials. Manganese occurs in domestic wastewater, industrial effluents and receiving water bodies.

The analytical method described here is the persulphate method.

*Principle*

Persulphate oxidation of soluble manganese compounds to permanganate is carried out in the presence of silver nitrate. The resulting colour is stable for at least 24 hours if excess persulphate is present and organic matter is absent.

*Interferences*

As much as 0.1 g chloride, $Cl^-$, in a 50-ml sample can be prevented from interfering by adding 1 g mercuric sulphate, $HgSO_4$, to form slightly dissociated complexes. Bromide and iodide will still interfere and only trace amounts may be present. The persulphate procedure can be used for potable water with trace to small amounts of organic matter if the period of heating is increased after more persulphate has been added. For wastewaters containing organic matter, use preliminary digestion with nitric and sulphuric acids, $HNO_3$ and $H_2SO_4$ respectively. If large amounts of $Cl^-$ are also present, boiling with $HNO_3$ helps remove them. Interfering traces of $Cl^-$ are eliminated by $HgSO_4$ in the special reagent.

Coloured solutions from other inorganic ions are compensated for in the final colorimetric step.

Samples that have been exposed to air may give low results because of precipitation of manganese dioxide, $MnO_2$. Add 1 drop of 30 per cent hydrogen peroxide, $H_2O_2$, to the sample (after adding the special reagent) to redissolve precipitated manganese.

*Minimum detectable concentration*

The molar absorptivity of the permanganate ion is about 2,300 l $g^{-1}$ $cm^{-1}$. This corresponds to a minimum detectable concentration (98 per cent transmittance) of 210 µg Mn $l^{-1}$ when a 1-cm cell is used or 42 µg Mn $l^{-1}$ when a 5-cm cell is used.

*Apparatus*

✓ Colorimetric equipment. One of the following is required:
- —Spectrophotometer, for use at 525 nm, providing a light path of 1 cm or longer.
- —Filter photometer, providing a light path of 1 cm or longer and equipped with a green filter having maximum transmittance near 525 nm.
- —Nessler tubes, matched, 100 ml, tall form.

*Reagents*

✓ Special reagent. Dissolve 75 g $HgSO_4$ in 400 ml concentrated $HNO_3$ and 200 ml distilled water. Add 200 ml 85 per cent phosphoric acid, $H_3PO_4$, and 35 mg silver nitrate, $AgNO_3$. Dilute the cooled solution to 1 litre.

✓ Ammonium persulphate, $(NH_4)_2S_2O_8$, solid.

✓ Standard manganese solution. Prepare a 0.1 mol $l^{-1}$ potassium permanganate, $KMnO_4$, solution by dissolving 3.2 g $KMnO_4$ in distilled water and making up to 1 litre. Age for several weeks in sunlight or heat for several hours near the boiling point, then filter through a fine fritted-glass filter crucible and standardise against sodium oxalate, $Na_2C_2O_4$, as follows:

Weigh several 100- to 200-mg samples of $Na_2C_2O_4$ to ± 0.1 mg and transfer to 400-ml beakers. To each beaker add 100 ml distilled water and stir to dissolve. Add 10 ml of 1+1 $H_2SO_4$ and heat rapidly to 90–95 °C. Titrate rapidly with the $KMnO_4$ solution to be standardised, while stirring, to a slight pink end-point colour that persists for at least 1 minute. Do not let temperature fall below 85 °C. If necessary, warm beaker contents during titration; 100 mg $Na_2C_2O_4$ will consume about 15 ml permanganate solution. Run a blank on distilled water and $H_2SO_4$.

$$\text{Molarity of } KMnO_4 = \frac{\text{g } Na_2C_2O_4}{(A-B) \times 0.06701}$$

where A = ml titrant for sample
B = ml titrant for blank

Average the results of several titrations. Calculate the volume of this solution necessary to prepare 1 litre of solution so that 1.00 ml = 50.0 µg Mn as follows:

$$\text{ml KMnO}_4 = \frac{4.55}{\text{molarity KMnO}_4}$$

To this volume add 2–3 ml concentrated $H_2SO_4$ and $NaHSO_3$ solution drop by drop, with stirring, until the permanganate colour disappears. Boil to remove excess $SO_2$, cool, and dilute to 1,000 ml with distilled water. Dilute this solution further to measure small amounts of manganese.

- ✓ Standard manganese solution (alternative). Dissolve 1.000 g manganese metal (99.8 per cent min.) in 10 ml redistilled $HNO_3$. Dilute to 1,000 ml with 1 per cent (*v/v*) HCl (1 ml ≡ 1.000 mg Mn). Dilute 10 ml to 200 ml with distilled water (1 ml ≡ 0.05 mg Mn). Prepare dilute solution daily.
- ✓ Hydrogen peroxide, $H_2O_2$, 30 per cent.
- ✓ Nitric acid, $HNO_3$, concentrated.
- ✓ Sulphuric acid, $H_2SO_4$, concentrated.
- ✓ Sodium nitrite solution. Dissolve 5.0 g $NaNO_2$ in 95 ml distilled water.
- ✓ Sodium oxalate, $Na_2C_2O_4$, primary standard.
- ✓ Sodium bisulphite. Dissolve 10 g $NaHSO_3$ in 100 ml distilled water.

*Procedure*

*Treatment of sample*

1. If a digested sample has been prepared according to directions for reducing organic matter and/or excessive chlorides, pipette a portion containing 0.05–2.0 mg Mn into a 250-ml conical flask. Add distilled water, if necessary, to 90 ml, and proceed as in step 2 below.

2. To a suitable sample portion, add 5 ml special reagent and 1 drop $H_2O_2$. Concentrate to 90 ml by boiling or dilute to 90 ml. Add 1 g $(NH_4)_2S_2O_8$, bring to a boil, and boil for 1 minute. Do not heat on a water-bath. Remove from heat source, let stand for 1 minute, then cool under the tap. (Boiling too long results in decomposition of excess persulphate and subsequent loss of permanganate colour; cooling too slowly has the same effect.) Dilute to 100 ml with distilled water free from reducing substances, and mix. Prepare standards containing 0, 5.00, ... 1,500 µg Mn by treating various amounts of standard Mn solution in the same way.

*Nessler tube comparison*

3. Use standards prepared as in step 2 above and containing 5 to 100 µg Mn per 100 ml final volume. Compare samples and standards visually.

*Photometric determination*

4. Use a series of standards from 0 to 1,500 µg Mn per 100 ml final volume. Make photometric measurements against a distilled water blank. The following table shows the light path length appropriate for various amounts of manganese in 100 ml final volume:

| Mn range (µg) | Light path (cm) |
|---|---|
| 5–200 | 15 |
| 20–400 | 5 |
| 50–1,000 | 2 |
| 100–1,500 | 1 |

5. Prepare a calibration curve of manganese concentration v. absorbance from the standards and determine Mn in the samples from the curve. If turbidity or interfering colour is present, make corrections as in step 6 below.

*Correction for turbidity or interfering colour*

6. Avoid filtration because of possible retention of some permanganate on the filter paper. If visual comparison of colour is used, the effect of turbidity can only be estimated and no correction can be made for interfering colour ions. When photometric measurements are made, use the following "bleaching" method, which also corrects for interfering colour. As soon as the photometer reading has been made, add 0.05 ml $H_2O_2$ solution directly to the sample in the optical cell. Mix and, as soon as the permanganate colour has faded completely and no bubbles remain, read again. Deduct absorbance of bleached solution from initial absorbance to obtain absorbance due to Mn.

*Calculation*

A. When all of the original sample is taken for analysis:

$$\text{Mn} = \frac{\mu\text{g Mn (in 100 ml final volume)}}{\text{ml sample}} \ \text{mg l}^{-1}$$

B. When a portion of the digested sample (100 ml final volume) is taken for analysis:

$$\text{Mn} = \frac{\mu\text{g Mn / 100 ml}}{\text{ml sample}} \times \frac{100}{\text{ml portion}} \ \text{mg l}^{-1}$$

*Precision and accuracy*

A synthetic sample containing Mn, 120 µg $l^{-1}$; Al, 500 µg $l^{-1}$; Cd, 50 µg $l^{-1}$; Cr, 110 µg $l^{-1}$; Cu, 470 µg $l^{-1}$; Fe, 300 µg $l^{-1}$; Pb, 70 µg $l^{-1}$; Ag, 150 µg $l^{-1}$ and Zn, 650 µg $l^{-1}$ in distilled water was analysed in 33 laboratories by the persulphate method, with a relative standard deviation of 26.3 per cent and a relative error of 0 per cent.

A second synthetic sample, similar in all respects except for 50 µg Mn $l^{-1}$ and 1,000 µg Cu $l^{-1}$, was analysed in 17 laboratories by the persulphate method, with a relative standard deviation of 50.3 per cent and a relative error of 7.2 per cent.

## 7.14 Nitrogen, ammonia

When nitrogenous organic matter is destroyed by microbiological activity, ammonia is produced and is therefore found in many surface and groundwaters. Higher concentrations occur in water polluted by sewage, fertilisers, agricultural wastes or industrial wastes containing organic nitrogen, free ammonia or ammonium salts.

Certain aerobic bacteria convert ammonia into nitrites and then into nitrates. Nitrogen compounds, as nutrients for aquatic micro-organisms, may be partially responsible for the eutrophication of lakes and rivers. Ammonia can result from natural reduction processes under anaerobic conditions.

The proportions of the two forms of ammonia nitrogen, i.e. free ammonia and ammonium ions, depend on the pH:

| pH | 6 | 7 | 8 | 9 | 10 | 11 |
|---|---|---|---|---|---|---|
| % $NH_3$ | 0 | 1 | 4 | 25 | 78 | 96 |
| % $NH_4$ | 100 | 99 | 96 | 75 | 22 | |

*Sample handling*

The preferred procedure is to remove ammonia from the sample by distillation. The ammonia may then be determined either by titration or colorimetrically using Nessler's reagent. Direct nesslerisation of the sample is quicker but is subject to considerable interference. The procedure given is the distillation and titration method.

If it is not possible to carry out the determination very soon after sampling, the sample should be refrigerated at 4 °C. Chemical preservation may be achieved by adding either 20–40 mg $HgCl_2$ or 1 ml $H_2SO_4$ to 1 litre of sample.

*Principle*

Ammonia can be quantitatively recovered from a sample by distillation under alkaline conditions into a solution of boric acid followed by titration with standard acid. The method is particularly suitable for the analysis of polluted surface and ground waters that contain sufficient ammonia to neutralise at least 1 ml of 0.00714 mol $l^{-1}$ HCl.

*Interferences*

Volatile amines, if present, interfere with the acid titration. Generally, however, this method is less subject to interferences than other methods.

*Apparatus*

- ✓ Distillation apparatus, consisting of a 1-litre, round-bottomed, heat-resistant glass flask fitted with a splash head, together with a suitable vertical condenser of either the spiral tube or double surface type. The condenser must be arranged so that the outlet tip can be submerged in the liquid in the receiver.
- ✓ Usual laboratory glassware.

*Reagents*

- ✓ Ammonia-free water. This should be prepared fresh for each batch of samples, since it is virtually impossible to store ammonia-free water in the laboratory without contamination from ammonia fumes.

  (i) Distillation. To each litre of tap water add 2 ml of a solution of ferrous sulphate (100 g $l^{-1}$ $FeSO_4.7H_2O$) and sufficient sulphuric acid to give a slight acid reaction to methyl orange. Distil with care, preferably in an all-glass distillation apparatus provided with a splash head. Reject the first 50 ml of distillate and then proceed until three-quarters of the volume of water has distilled over. Test for the absence of ammonia in the distillate with Nessler's reagent in the manner described below.

  (ii) Ion exchange. As an alternative, ammonia may be removed from distilled water by the use of a strongly acidic cation exchange resin (hydrogen form). If only a small quantity of ammonia-free water is needed, add about 3 g of the cation exchange resin to each litre of distilled water and shake for a few minutes. If a regular supply of ammonia-free distilled water is needed, it is convenient to pass distilled water slowly down a column of the resin enclosed in a glass tube (250 mm long and about 25 mm in diameter is suitable). In either case, check that the water is free from ammonia by testing with Nessler's reagent.

✓ Light magnesium oxide.

✓ Indicating boric acid solution. Dissolve 20 g pure boric acid, $H_3BO_3$, in warm water and dilute to approximately 1 litre. Add 20 ml methyl red solution (0.5 g $l^{-1}$) and 0.4 ml methylene blue solution (15 g $l^{-1}$) and mix well. One drop of 0.1 mol $l^{-1}$ NaOH should change the colour of 20 ml of the solution from purple to green. The solution is stable for several months.

✓ Hydrochloric acid, 0.00714 mol $l^{-1}$. Prepare approximately 0.714 ml $l^{-1}$ HCl by diluting 65 ml of hydrochloric acid ($d$ = 1.18) to 1 litre. Standardise by a suitable method and dilute with water to give a solution of the required strength.

*Procedure*

1. Before assembling the apparatus, thoroughly clean the distillation flask, splash head and condenser. In order to free the apparatus from possible contamination by ammonia, add to the flask about 350 ml water (preferably ammonia-free) and distil until the distillate is shown to be ammonia-free by testing with Nessler's reagent. Empty the distillation flask and allow it to cool.
2. Place a suitable volume of the sample in the flask; 100 ml should be sufficient for a purified effluent, while 200 to 400 ml of surface water may be necessary to give a final titration of reasonable magnitude. Use a 400-ml beaker as a receiver and keep the lower end of the delivery tube from the condenser below the surface of the absorbent liquid throughout the distillation.
3. Neutralise the measured volume of sample with NaOH, if necessary.
4. Dilute the measured volume of sample, if necessary, to 400 ml with ammonia-free water in the distillation flask and add about 0.25 g magnesium oxide.
5. Place 50 ml of the indicating $H_3BO_3$ solution in the 400-ml receiving beaker.
6. Distil at a rate of about 10 ml per minute. As the indicating $H_3BO_3$ solution changes colour, titrate with 0.00714 mol $l^{-1}$ HCl, continuing the distillation until the addition of one drop of the standard acid produces a permanent pink colour in the solution.
7. At the completion of the titration, remove the receiver from the apparatus before the source of heat is withdrawn.
8. Carry out a blank determination and correct the final titration values for samples to compensate for any ammonia in the reagent used.

*Calculation*

$$\text{Ammonia nitrogen (as N)} = \frac{100V_2}{V_1}\ \text{mg l}^{-1}$$

where $V_1$ = volume of sample taken (ml)
$V_2$ = volume of 0.00714 mol $l^{-1}$ acid used (ml).

## 7.15 Nitrogen, Kjeldahl

Kjeldahl nitrogen is defined as the sum of ammonia nitrogen and those organic nitrogen compounds converted to ammonium sulphate under the conditions of the digestion procedure described below. The organic Kjeldahl nitrogen is obtained by subtracting the value of ammonia nitrogen from the Kjeldahl nitrogen value.

*Sample handling*
Conversion of organic nitrogen to ammonia may occur in samples between collection and analysis. This effect may be decreased by the addition of 2 ml of sulphuric acid ($d = 1.84$) per litre of sample and storage at 4 °C, but it is prudent to analyse all samples as soon as possible after collection.

Mercuric salts sometimes used for conservation interfere in the colorimetric method.

*Principle*
The sample is heated in the presence of sulphuric acid and a catalyst, alcohol also being added to ensure removal of oxidised nitrogen. After digestion, the solution is diluted. The ammonia is determined by photometry or, if sufficiently large amounts are present, distilled from the solution and titrated.

*Interferences*
Neither this nor any other single method can guarantee that every organic nitrogen compound will be broken down to ammonia. This is unlikely to lead to serious difficulty, but analysts should be watchful for exceptional cases. Nitrate and nitrite are removed in the procedure described below and so cause no important errors.

### 7.15.1 Titrimetric determination following mineralisation and distillation

*Sample handling*
Samples for laboratory analysis must be cooled as soon as possible after being taken and must be kept at a temperature around 4 °C until the time of analysis.

The analysis must be carried out as soon as possible after taking the sample, but if it cannot be carried out within 24 hours the sample may be acidified at the sampling point to pH 2 with concentrated sulphuric acid (in general 2 ml of acid is sufficient for 1 litre of sample).
*Note:* Attention is drawn to the fact that sulphuric acid is capable of entrapping ammonia vapour.

*Principle*
Mineralisation of the organic matter in an acid medium and in the presence of a catalyst. Interference due to oxidised forms of mineral nitrogen may be eliminated by the addition of ethyl alcohol. Steam distillation in an alkaline

medium of the ammonia nitrogen obtained. Titrimetric determination (direct or back titration).

*Apparatus*

- ✓ Usual laboratory equipment. The glassware used for the preparation of the reagents and for determination should not be used for other determinations, and should not come into contact with ammonia or ammonium salts in high concentrations. When using new equipment it is a good idea to carry out the blank test twice in graduated flasks. Between analyses, the flasks or tubes should be filled with water, stoppered and kept in the dark.
- ✓ Kjeldahl flask, 500-ml capacity, and flask heater (or mineralisation burners).
- ✓ Distillation or steam distillation apparatus. It is recommended that the base of the refrigerant be fitted with a glass cap to prevent possible condensation running into the flask in which the distillate is collected. When distillation equipment is exposed to the atmosphere of the laboratory, its walls may pick up traces of ammonia nitrogen that cannot be removed by rinsing in water. For the determination of low nitrogen contents it is essential to clean the apparatus by carrying out one or two "blank" distillations of a sodium hydroxide solution before every series of measurements. Between tests the equipment should be kept protected from the laboratory atmosphere.
- ✓ Burette, 20 ml, 0.1-ml graduations.

*Reagents*

- ✓ Only water recently demineralised on strong cationic resin, or water of equivalent purity with an insignificant nitrogen content, should be used in the analysis and for preparation of the reagents. The reagents themselves must be of analytical purity.
- ✓ Concentrated sulphuric acid ($d$ = 1.84 g $ml^{-1}$).
- ✓ Sulphuric acid, solution at roughly 5 g $l^{-1}$.
- ✓ Sodium hydroxide, solution at 400 g $l^{-1}$ (10 mol $l^{-1}$).
- ✓ Boric acid, solution at 10 g $l^{-1}$.
- ✓ Sulphuric acid, standard solution, 0.05 mol $l^{-1}$.
- ✓ Sulphuric acid, standard solution, 0.01 mol $l^{-1}$.
- ✓ Mineralisation catalyst: Prepare a homogeneous mixture of 995 g of potassium sulphate, $K_2SO_4$, and 5 g of powdered selenium.
- ✓ Indicator: solution of methyl red and bromocresol green in 95 per cent ethyl alcohol (by volume).
- ✓ Solution of methyl red in water, 10 g $l^{-1}$.
- ✓ Ethyl alcohol.

*Procedure*

1. Take at least 50 ml of sample so as to have between 0.2 and 20 mg of nitrogen expressed as N.

*Mineralisation*

2. Place the sample for analysis in a Kjeldahl flask containing a boiling regulator (glass beads, ceramic chips or pumice). Add 1 g of catalyst and, if appropriate, an antifoam agent. Add 10 ml of ethyl alcohol. Add 10 ml of concentrated sulphuric acid. Fit a funnel to the flask, inserting the stem into the neck of the flask.

3. Bring slowly to the boil and evaporate until a white vapour is given off. Thereafter increase the heat and continue to heat for 2 hours. Leave to cool to room temperature.

   *Note:* If mineralisation is incomplete (turbid or highly coloured liquid), the operation must be repeated, reducing the size of sample for analysis or carrying out preliminary dilution (in the case of samples with a heavy organic load).

*Distillation and determination*

4. If necessary, decant the contents of the flask and the washing water (about 200 ml to 250 ml) into the distillation apparatus.

5. Add 50 ml of 10 mol $l^{-1}$ sodium hydroxide. Admit steam for distillation of the ammonia nitrogen.

6. Collect about 200 ml of distillate in a receptacle containing 10 ml of boric acid solution and 3 or 4 drops of indicator. Add sufficient water, if required, to ensure that the end of the extension tube of the refrigerant is submerged in the solution.

7. Carry out titration, using:

   (i) the 0.05 mol $l^{-1}$ standard solution of sulphuric acid if the sample for analysis contains between 2 and 20 mg of nitrogen, expressed as N, or

   (ii) the 0.01 mol $l^{-1}$ standard solution of sulphuric acid if the sample for analysis contains between 0.2 and 2 mg of nitrogen, expressed as N.

8. Note the volume $V_1$ required.

*Blank test*

9. Carry out a blank test under the same conditions of mineralisation and determination as for the sample being analysed. Let $V_0$ be the required volume of 0.05 or 0.01 mol $l^{-1}$ sulphuric acid in millilitres.

*Calculation*

The Kjeldahl nitrogen content, expressed as mg N $l^{-1}$, is given by:

$$N = \frac{(V_1 - V_0)c \times 1{,}000 \times 28}{V} \text{ mg l}^{-1}$$

where:

$V_1$ = the volume of 0.05 or 0.01 mol $l^{-1}$ sulphuric acid (ml) used for the determination

$V_0$ = the volume of 0.05 or 0.01 mol $l^{-1}$ sulphuric acid (ml) used for the blank test

$c$ = the molarity (0.05 or 0.01 mol $l^{-1}$) of the standard solution of sulphuric acid used for the determination

$V$ = the volume of the sample taken for analysis (ml)

### 7.15.2 Spectrophotometric method after mineralisation

*Sample handling*

Samples for the analytical laboratory must be cooled to a temperature of around 4 °C as soon as possible after being taken. It is recommended that the analysis be carried out as soon as possible and that acidification of the sample be avoided for the determination of low concentrations because sulphuric acid is capable of entrapping ammonia vapour.

*Principle*

Mineralisation of the organic matter in an acid medium in the presence of a catalyst. Determination by indophenol blue spectrophotometry at a wavelength around 630 nm.

*Apparatus*

- ✓ Usual laboratory equipment. The glassware used to prepare the reagents and carry out the determination must not be used for other determinations, nor must it come into contact with ammonia or ammonium salts in high concentration. When using new equipment it is a good idea to carry out the blank test twice in the tubes in which the colour will be developed. Between analyses the flasks or tubes should be filled with water, stoppered, and kept in the dark.
- ✓ Kjeldahl flasks, 500-ml capacity, and flask heater (or mineralisation burners).
- ✓ Spectrophotometer, 630 nm, fitted with measuring tubes 10 mm deep.

*Reagents*

- ✓ Concentrated sulphuric acid ($d$ = 1.84 g $ml^{-1}$).
- ✓ Sodium hydroxide, 400 g $l^{-1}$ solution.
- ✓ Alkaline solution, complexing and chlorinated: Dissolve 20 g of sodium hydroxide tablets and 380 g of dihydrated trisodium citrate ($C_6H_5Na_3O_7.2H_2O$) in about 800 ml of water. Raise to boiling point and keep simmering gently for 20 minutes. Cool, add 4 g of dihydrated dichloro-1,3-isocyanuric acid sodium salt ($C_3NaCl_2N_3O_3.2H_2O$) and make up the volume to 1,000 ml with water. Store this solution at around 4 °C.
- ✓ Phenol and nitroprusside solution. Dissolve 35 g of phenol and 0.4 g of sodium nitroprusside ($Na_2[Fe(CN)_5NO].2H_2O$) in water and make up the volume to 1,000 ml with water. Store this solution in a brown glass flask at around 4 °C. Never place it in direct sunlight. Handle in subdued light.
- ✓ Mineralisation catalyst. Prepare a homogeneous mixture of 995 g of potassium sulphate ($K_2SO_4$) and 5 g of powdered selenium.
- ✓ Ethyl alcohol.
- ✓ Stock nitrogen solution corresponding to 100 mg N per litre. Dissolve 381.9 mg of anhydrous ammonium chloride in water and make up to 1,000 ml in a volumetric flask.
- ✓ Standard nitrogen solution corresponding to 10 mg N per litre. At the time of use, dilute the stock solution to one-tenth.
- ✓ 0.1 paranitrophenol solution in water (*m/v*).

*Procedure*

1. Sample for analysis. Take at least 50 ml from the sample so that the sample for analysis contains between 20 and 200 µg of nitrogen expressed as N.

2. Preparation of calibration solutions. For spectrophotometric measurements carried out in tubes 10 mm deep, place increasing volumes (e.g. 0, 1, 2, 5, 10, 15, 20 ml) of the standard 10 mg N $l^{-1}$ solution in a series of 100-ml volumetric flasks. Make up to the volume and homogenise.

3. Mineralisation. Place the sample for analysis and each of the standard solutions in a series of Kjeldahl flasks containing a boiling regulator (glass beads, ceramic chips or pumice). Add 0.2–0.5 g of catalyst followed respectively by 2 to 5 ml of concentrated $H_2SO_4$ and 5 ml of ethyl alcohol. Fit a funnel to the flask, inserting the stem into the neck of the flask. Bring slowly to the boil and evaporate until white vapour appears. Thereafter increase the heat and continue to heat for 1 hour. Cool to room temperature.

   *Note:* If mineralisation is incomplete (turbid or highly coloured liquid), recommence the operation, reducing the size of the sample for analysis or carrying out preliminary dilution.

4. Determination. Collect the mineralisates obtained in step 3 and the wash water in a series of 200-ml volumetric flasks to obtain a volume of about 150 ml. Neutralise with sodium hydroxide solution in the presence of a drop of paranitrophenol. Make up to

the volume with water. Homogenise and allow to settle if necessary. Take 20 ml of each solution and place in a series of test-tubes. Add in sequence, without waiting between each addition:
1.0 ml of the phenol and nitroprusside solution, stir;
1.0 ml of the complexing and chlorinated alkaline solution.

5. Stir and place in the dark for at least 6 hours.
6. Carry out the spectrophotometric measurements at the maximum of the absorption curve (wavelength generally around 630 nm) after having adjusted the apparatus to zero absorbance relative to the control.
7. Control. Carry out a control test alongside the determination, replacing the sample by water.

*Calculation*
Taking the concentrations of the calibration solutions as a starting point, establish a calibration curve and use it to deduce the Kjeldahl nitrogen content of the sample being analysed, having regard to any possible dilutions. Express the result in milligrams of N per litre.

## 7.16 Nitrogen, nitrate

Nitrate, the most highly oxidised form of nitrogen compounds, is commonly present in surface and ground waters, because it is the end product of the aerobic decomposition of organic nitrogenous matter. Significant sources of nitrate are chemical fertilisers from cultivated land and drainage from livestock feedlots, as well as domestic and some industrial waters.

The determination of nitrate helps the assessment of the character and degree of oxidation in surface waters, in groundwater penetrating through soil layers, in biological processes and in the advanced treatment of wastewater.

Unpolluted natural waters usually contain only minute amounts of nitrate. In surface water, nitrate is a nutrient taken up by plants and assimilated into cell protein. Stimulation of plant growth, especially of algae, may cause water quality problems associated with eutrophication. The subsequent death and decay of algae produces secondary effects on water quality, which may also be undesirable. High concentrations of nitrate in drinking water may present a risk to bottle-fed babies under three months of age because the low acidity of their stomachs favours the reduction of nitrates to nitrites by microbial action. Nitrite is readily absorbed into the blood where it combines irreversibly with haemoglobin to form methaemoglobin, which is ineffective as an oxygen carrier in the blood. In severe cases a condition known as infantile methaemoglobinaemia may occur which can be fatal for young babies.

The determination of nitrate in water is difficult because of interferences, and much more difficult in wastewaters because of higher concentrations of numerous interfering substances.

The first method given here, Devarda's alloy method, involves oxidation, distillation and titration. One of its attractive features is that it can be performed on the residue remaining in the flask after the distillation process required in the determination of ammonia nitrogen. The distillation step in the method also eliminates many interferences and is often used for the analysis of wastewater samples. In the second method given, nitrate is reduced to nitrite in a cadmium column and nitrite is then determined by the method given in section 7.17.

*Sample handling*

To prevent any change in the nitrogen balance through biological activity, the nitrate determination should be started as soon as possible after sampling. If storage is necessary, samples should be kept at a temperature just above the freezing point, with or without preservatives, such as 0.8 ml of concentrated sulphuric acid ($d = 1.84$) or 40 mg of mercury (as mercuric chloride) per litre of sample. If acid preservation is employed, the sample should be neutralised to about pH 7 immediately before the analysis is begun.

### 7.16.1 Devarda's alloy method (reduction to ammonia)

*Principle*

This method is suitable for nitrate concentrations exceeding 1 mg $l^{-1}$, especially for wastewater and polluted surface water. The analysis may be carried out either on the original sample or on the residue from the determination of ammonia.

Nitrate is reduced to ammonia by nascent hydrogen, by the use of Devarda's alloy (59 per cent Al, 39 per cent Cu, 2 per cent Zn). The resulting ammonia is distilled and its concentration determined by titration.

Nitrites are also reduced by Devarda's alloy and their separate determination can be carried out rapidly and readily. The nitrate concentration can therefore be satisfactorily determined by subtracting the nitrite fraction from the total oxidised nitrogen.

Sometimes, especially when the proportion of nitrogen present as nitrite is small, the report of the analysis is confined to total oxidised nitrogen.

*Interferences*

Ammonia must be removed from the sample before the main procedure is started. This is achieved either by pretreatment or by using the distillation residue from the determination of ammonia. The air in the laboratory and the distilled water used for solutions and during the procedure should be free of ammonia. If the sample or any reagent needs to be filtered, only nitrogen-free filters should be used. Nitrite is determined separately and subtracted from the result.

*Apparatus*

✓ The same glassware and distilling apparatus as for the ammonia nitrogen determination (see section 7.14).

*Reagents*

The following reagents are required in addition to those used for the ammonia nitrogen determination:

✓ Devarda's alloy, powdered. If fine powder is not available, the material should be ground to pass through a 0.07–0.1 mm sieve (200–140 mesh). It should be as free as possible from nitrogen. It can be purchased as such but a blank determination of its nitrogen content should always be carried out under the conditions of the test.

✓ Sodium hydroxide, 10 mol $l^{-1}$ (needed only if an original sample is used — see procedure).

*Procedure*

*Determination on an original sample*

1. Before assembling the apparatus, thoroughly clean the distillation flask, splash head and condenser. In order to free the apparatus from possible contamination by ammonia, add to the flask about 350 ml water, preferably ammonia-free, and distil until the distillate is shown to be free from ammonia by testing with Nessler's reagent. Empty the distillation flask and allow it to cool.
2. Measure 200 ml of the sample into the flask.
3. Add 10 ml of 10 mol $l^{-1}$ sodium hydroxide. Evaporate in the distillation flask to 100 ml. Allow the residue to cool.
4. Continue as indicated in step 6, below.

*Determination on the residue from the analysis for ammonia*

5. At the end of the distillation procedure for the analysis for ammonia, allow the residue to cool.
6. To the cooled residue add sufficient ammonia-free water to bring the volume in the distillation flask to about 350 ml. Add 1 g Devarda's alloy, and immediately connect the flask to the condenser.
7. After some minutes, start the distillation, keeping the lower end of the delivery tube from the condenser below the surface of the liquid in the receiver throughout the distillation.
8. Place 50 ml of the indicating boric acid solution in the receiver and distil at a rate of about 10 ml per minute.
9. As the absorbent solution changes colour, titrate with 0.00714 mol $l^{-1}$ hydrochloric acid, continuing the distillation until the addition of one drop of the standard acid produces a permanent pink colour in the solution.
10. At the completion of the titration, remove the receiver from the apparatus before the source of heat is withdrawn.
11. Carry out blank determinations as appropriate. For each sample, correct the final titration figure for any ammonia in the reagent used.

*Calculation*

$$\text{Nitrate nitrogen (as N)} = \frac{(a-b)\times 100}{V} - n \quad \text{mg l}^{-1}$$

where $a$ = volume of 0.00714 mol $l^{-1}$ acid used for titration of the distillate of the sample (ml)
$b$ = volume of 0.00714 mol $l^{-1}$ acid used for titration of the distillate of the blank (ml)
$V$ = volume of the undiluted sample (ml)
$n$ = concentration of nitrite nitrogen in mg $l^{-1}$ N, determined separately

*Note:* The same calculation without subtracting nitrite gives the result for total oxidised nitrogen.

The result is reported as nitrate nitrogen (as N) mg $l^{-1}$ and should be rounded to two significant figures.

### 7.16.2 Cadmium reduction method

*Principle*

Nitrate is reduced to nitrite when a sample is passed through a column containing amalgamated cadmium filings. Nitrite, that originally present plus that reduced from nitrate, is then determined. A separate determination for the concentration of nitrite alone is necessary. The applicable range of this method is 0.01 to 1.0 mg $l^{-1}$ of nitrate plus nitrite nitrogen.

*Interferences*

Build-up of suspended material in the reduction column will restrict sample flow. Since nitrate nitrogen occurs in a soluble state, the sample may be filtered through a glass-fibre filter or a 0.45 µm pore diameter membrane filter. Highly turbid samples may be treated with zinc sulphate before filtration to remove the bulk of the particulate matter present in the sample.

Low results may be obtained for samples that contain high concentrations of iron, copper or other metals. Addition of EDTA to the samples will eliminate this interference. Samples containing oil and/or grease will coat the surface of the cadmium and inhibit the reduction process. This interference may be eliminated by pre-extracting the sample with an organic solvent.

*Apparatus*

- ✓ Reduction column. This can be constructed from three pieces of tubing joined together as shown in Figure 7.2: 100 mm of 50 mm diameter tubing is joined to 300 mm of 10 mm diameter tubing which is, in turn, joined to 350 mm of 2 mm diameter tubing (all diameters are internal). The 2 mm diameter tubing is bent at its lower end so that it will be parallel to the 10 mm diameter tube and also bent at its upper end to form a siphon. This last bend should be just level with the top of the 10 mm diameter tube as shown in Figure 7.2. This arrangement allows liquid placed in the top reservoir to flow out of the system but to stop when the liquid level reaches the top of the cadmium filings. Place a mark on the reservoir to indicate the level that will be reached when 80 ml of liquid is added to the system.
- ✓ Nessler tubes, 50 ml, or volumetric flasks, 50 ml.
- ✓ Erlenmeyer flasks, 125 ml.

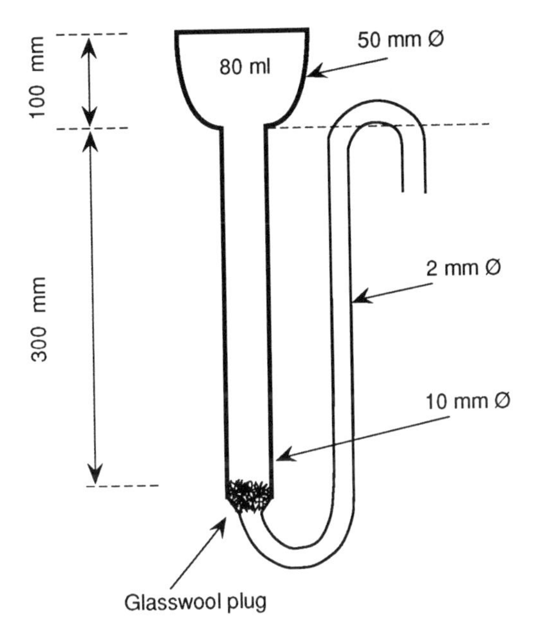


**Figure 7.2** Cadmium reduction column

✓ Graduated cylinders, 50 ml.
✓ Colorimetric equipment, either a spectrophotometer equipped with 10 mm or larger cells for use at 540 nm or a filter photometer with light path 1 cm or longer and equipped with a filter having maximum transmittance near 540 nm.

*Reagents*
✓ Ammonium chloride solution, concentrated. Dissolve 100 g of $NH_4Cl$ in 500 ml of distilled water and store in a glass bottle.
✓ Dilute ammonium chloride solution. Dilute 50 ml of concentrated ammonium chloride solution to 2 litres with distilled water. Store in a glass or plastic bottle.
✓ Amalgamated cadmium filings. File sticks of pure cadmium metal (reagent grade, 99.9 per cent Cd) with a coarse metal hand file and collect the fraction that passes a sieve with 5 mm openings but is retained on a sieve with 0.2 mm openings. Stir about 300 g of filings with 300 ml of 10 g $l^{-1}$ mercuric chloride solution for 3 minutes. This amount should be sufficient to charge six reduction columns. Allow the metal to settle and decant off the liquid. Wash the amalgamated filings several times with distilled water then briefly with dilute (1+99) nitric acid followed by several washings with dilute (1+99) hydrochloric acid. Wash thoroughly with distilled water until no nitrite can be detected in the supernatant fluid. Store the filings in the dark under dilute ammonium chloride solution.
✓ Stock nitrate solution. Dissolve 7.218 g of potassium nitrate, $KNO_3$, in distilled water and dilute to 1 litre. Preserve with 2 ml of chloroform per litre. This solution is stable for about 6 months (1.0 ml ≡ 1.0 mg nitrate N).
✓ Working nitrate solution. Dilute 10.0 ml of stock nitrate solution to 1 litre with distilled water (1.0 ml ≡ 0.01 mg nitrate N).

The reagents required for the nitrite nitrogen procedure are described in section 7.17.

*Procedure*

1. Preparation of reduction columns. Pack a plug of glass wool in the bottom of a reduction column and fill the column with distilled water. Pour in sufficient amalgamated cadmium filings to produce a column 300 mm in length. Use the column size specified because columns of smaller length and diameter give erratic results and show rapid deterioration. Wash the column thoroughly with dilute ammonium chloride solution. Use a flow rate no greater than 8 ml per minute. If the flow rate is too fast, slow it by constricting the end of the outlet siphon or by packing more glass wool at the base of the column. On the other hand, flow rates less than 5 ml per minute unnecessarily increase the time required for an analysis and may give low results. When not in use, cover the metal in the column with dilute ammonium chloride solution. Prepare several reduction columns similarly.

2. Sample pretreatment:

   (i) If turbidity or suspended solids are present, filter the sample through a glass fibre filter or a membrane filter.

   (ii) If the sample contains oil or grease, adjust the pH to 2 by the addition of hydrochloric acid to 100 ml of filtered sample. Extract the oil and grease from the filtrate with two 25-ml portions of a non-polar solvent (chloroform or equivalent).

   (iii) Adjust the pH of the sample to between 7 and 9 using a pH meter and by adding either ammonia or hydrochloric acid as appropriate.

3. Place 80–90 ml of sample in a 125-ml Erlenmeyer flask and add 2.0 ml of concentrated ammonium chloride solution. Mix and pour the sample from the flask onto the column until the marked level in the top of the column is reached. Place a 50-ml graduated cylinder under the outlet to collect the effluent and discard any of the sample remaining in the Erlenmeyer flask. Shake the flask as dry as possible and retain it for collecting the main portion of the effluent.

4. Allow between 25 and 30 ml of effluent to collect in the cylinder and then replace the cylinder with the Erlenmeyer flask that contained the sample. Discard the contents of the cylinder. The passage of 25 to 30 ml of sample through the column removes the ammonium chloride solution or traces of a previous sample from the voids in the column. The volume flushed through is not critical provided that it exceeds 25 ml, but sufficient sample should be left in the column so that 50 ml will be available for subsequent analysis. The volume of flushing effluent should therefore not exceed 30 ml. A maximum of about eight columns can be handled conveniently at one time by one technician. The technician should experiment to find a suitable time delay between adding samples to successive columns so that there will be time to reject the flushing liquid from one column and replace the cylinder by the Erlenmeyer flask before too much flushing liquid has escaped from the next column in line.

5. Allow the remainder of the reduced sample to collect in the flask. When the flow of the sample from the column has ceased, pour exactly 50 ml from the flask into the 50-ml measuring cylinder. Drain and discard the remaining reduced sample from the Erlenmeyer flask, shake the flask as dry as possible and then pour the 50 ml of reduced sample back into the flask.

   *Note:* There is no need to wash columns between samples, but if columns are not to be re-used for several hours or longer, pour 50 ml of dilute ammonium chloride solution into the top and allow it to pass through the system. Store the cadmium filings in this ammonium chloride solution and never allow them to get dry. The cadmium filings can be regenerated by washing them briefly with $HNO_3$ (1+99), rinsing them thoroughly with distilled water, drying, re-sieving and re-amalgamating.

6. Prepare a series of standards in 50-ml Nessler tubes as follows:

| Volume of working nitrate solution (ml) | Concentration when diluted to 100 ml (mg $l^{-1}$ $NO_2$-N) |
|---|---|
| 0.0 (blank) | 0.0 |
| 0.5 | 0.05 |
| 1.0 | 0.10 |
| 2.0 | 0.20 |
| 5.0 | 0.50 |
| 10.0 | 1.00 |

7. Colour development. Proceed as described in steps 4 and 5 of section 7.17.

*Calculation*
Obtain a standard curve by plotting absorbance of standards against $NO_3$-N concentration. Compute sample concentrations directly from the standard curve. Report the results as milligrams of oxidised N per litre (the sum $NO_2$-N plus $NO_3$-N). If the concentration of $NO_2$-N has been determined separately, subtract it from the value obtained for oxidised N to obtain mg $l^{-1}$ $NO_3$-N.

## 7.17 Nitrogen, nitrite

Nitrite is an unstable, intermediate stage in the nitrogen cycle and is formed in water either by the oxidation of ammonia or by the reduction of nitrate. Thus, biochemical processes can cause a rapid change in the nitrite concentration in a water sample. In natural waters nitrite is normally present only in low concentrations (a few tenths of a milligram per litre). Higher concentrations may be present in sewage and industrial wastes, in treated sewage effluents and in polluted waters.

*Sample handling*
The determination should be made promptly on fresh samples to prevent bacterial conversion of the nitrite to nitrate or ammonia. In no case should acid preservation be used for samples to be analysed for nitrite. Short-term preservation for 1 to 2 days is possible by the addition of 40 mg mercuric ion as $HgCl_2$ per litre of sample, with storage at 4 °C.

*Principle*
Nitrite reacts, in strongly acid medium, with sulphanilamide. The resulting diazo compound is coupled with *N*-(1-naphthyl)-ethylenediamine dihydrochloride to form an intensely red-coloured azo-compound. The absorbance of the dye is proportional to the concentration of nitrite present.

The method is applicable in the range of 0.01–1.0 mg $l^{-1}$ nitrite nitrogen. Samples containing higher concentrations must be diluted.

*Interferences*

There are very few known interferences at concentrations less than 1,000 times that of the nitrite. However, the presence of strong oxidants or reductants in the samples will readily affect the nitrite concentrations. High alkalinity (> 600 mg $l^{-1}$ as $CaCO_3$) will give low results owing to a shift in pH.

*Apparatus*

✓ Spectrophotometer equipped with 10-mm or larger cells for use at 540 nm.
✓ Nessler tubes, 50 ml, or volumetric flasks, 50 ml.

*Reagents*

✓ Hydrochloric acid, 3 mol $l^{-1}$. Combine 1 part concentrated HCl with 3 parts distilled water.
✓ Nitrite- and nitrate-free distilled water. Add 1 ml of concentrated sulphuric acid and 0.2 ml manganous sulphate ($MnSO_4$) solution to 1 litre of distilled water and make it pink with 1–3 ml potassium permanganate solution (440 mg $KMnO_4$ in 100 ml distilled water). Redistil in an all borosilicate glass still. Discard the first 50 ml of distillate. Test each subsequent 100 ml fraction of distillate by the addition of DPD indicator, discarding those with a reddish colour that indicates the presence of permanganate.
✓ DPD indicator solution. Dissolve 1 g of DPD oxalate or 1.5 g of *p*-amino-*N*,*N*-diethylaniline sulphate in chlorine-free distilled water containing 8 ml of $H_2SO_4$, 18 mol $l^{-1}$, and 200 mg of disodium ethylenediaminetetraacetate dihydrate (EDTA). Make up to 1 litre and store in a brown bottle and discard the solution when it becomes discoloured.
*Note:* DPD indicator solution is commercially available.
✓ Buffer–colour reagent. Distilled water free of nitrite and nitrate must be used in the preparation of this reagent. To 250 ml of distilled water add 105 ml concentrated hydrochloric acid, 5.0 g sulphanilamide and 0.5 g *N*-(1-naphthyl)-ethylenediamine dihydrochloride. Stir until dissolved. Add 136 g of sodium acetate ($CH_3COONa.3H_2O$) and again stir until dissolved. Dilute to 500 ml with distilled water. This solution is stable for several weeks if stored in the dark.
✓ Stock nitrite solution. Distilled water free of nitrite and nitrate must be used in the preparation of this solution. Dissolve 0.4926 g of dried anhydrous sodium nitrite (24 hours in desiccator) in distilled water and dilute to 1 litre. Preserve with 2 ml chloroform per litre (1.0 ml ≡ 100 µg nitrite N).
✓ Working nitrite solution. Distilled water free of nitrite and nitrate must be used in the preparation of this solution. Dilute 10.0 ml of the stock solution to 1 litre (1.0 ml ≡ 1.0 µg nitrite N).

*Procedure*

1. If the sample has a pH greater than 10 or a total alkalinity in excess of 600 mg $l^{-1}$ (expressed as $CaCO_3$), adjust to approximately pH 6 with 3 mol $l^{-1}$ HCl.
2. If necessary, pass the sample through a filter of pore size 0.45 µm, using the first portion of filtrate to rinse the filter flask.
3. Place 50 ml of sample, or an aliquot diluted to 50 ml, in a 50-ml Nessler tube and set aside until preparation of standards is completed. At the same time, prepare a series of standards in 50-ml Nessler tubes as follows:

| Volume of working nitrate solution (ml) | Concentration when diluted to 50 ml (mg $l^{-1}$ of $NO_2$-N) |
|---|---|
| 0.0 (blank) | 0.0 |
| 0.5 | 0.01 |
| 1.0 | 0.02 |
| 1.5 | 0.03 |
| 2.0 | 0.04 |
| 3.0 | 0.06 |
| 4.0 | 0.08 |
| 5.0 | 0.10 |
| 10.0 | 0.20 |

4. Add 2 ml of buffer–colour reagent to each standard and sample, mix, and allow colour to develop for at least 15 minutes. The pH values of the solutions at this stage should be between 1.5 and 2.0.
5. Measure the absorbance of the standards and samples at 540 nm. Prepare a standard curve by plotting the absorbance of the standards against the concentration of $NO_2$-N.

*Calculation*
Read the concentration of $NO_2$-N in samples directly from the calibration curve. If less than 50 ml of sample is taken, calculate the concentrations as follows:

$$\text{Nitrite nitrogen (as N)} = \frac{\text{mg l}^{-1}\text{ from standard curve} \times 50}{\text{ml sample}} \text{ mg l}^{-1}$$

## 7.18 Phosphorus

Groundwaters rarely contain more than 0.1 mg $l^{-1}$ phosphorus unless they have passed through soil containing phosphate or have been polluted by organic matter.

Phosphorus compounds are present in fertilisers and in many detergents. Consequently, they are carried into both ground and surface waters with sewage, industrial wastes and storm run-off. High concentrations of phosphorus compounds may produce a secondary problem in water bodies where algal growth is normally limited by phosphorus. In such situations the presence of additional phosphorus compounds can stimulate algal productivity and enhance eutrophication processes.

*Principle*
Organically combined phosphorus and all phosphates are first converted to orthophosphate. To release phosphorus from combination with organic matter, a digestion or wet oxidation technique is necessary. The least tedious method, wet oxidation with potassium peroxydisulphate, is recommended.

Orthophosphate reacts with ammonium molybdate to form molybdophosphoric acid. This is transformed by reductants to the intensely coloured complex known as molybdenum blue. The method based on reduction with ascorbic acid is preferable. Addition of potassium antimonyl tartrate increases the coloration and the reaction velocity at room temperature.

For concentrations of phosphate below 20 μg $l^{-1}$, the recommended procedure involves extraction of the molybdenum blue complex from up to 200 ml of water into a relatively small volume of hexanol, so that a considerable increase in sensitivity is obtained.

*Interferences*

The method is relatively free from interferences. Changes in temperature of ± 10 °C do not affect the result.

*Apparatus*

✓ Heating equipment. One of the following is required:
- —Hotplate, a 30 cm x 50 cm heating surface is adequate.
- —Autoclave. An autoclave or pressure-cooker capable of developing 1.1–1.4 kPa $cm^{-2}$ may be used in place of a hotplate.

✓ Colorimetric equipment. One of the following is required:
- —Spectrophotometer, with infrared phototube, for use at 880 nm, providing a light path of 40 mm or longer. If this wavelength is not obtainable on the spectrophotometer, a wavelength of 710 nm may be used but there will be some loss of sensitivity.
- —Absorptiometer, equipped with a red colour filter and a light path of 40 mm or longer.

✓ Acid-washed glassware. The glassware used, including sample bottles, should be reserved for the determination of phosphate, should not be used for any other purpose and should be left full of sulphuric acid (4.5 mol $l^{-1}$) until required for use. If necessary, glassware may be cleaned with chromic acid, equal mixtures of nitric and hydrochloric acids, or pure sulphuric acid. Detergents containing phosphate compounds must not be used.

*Reagents*

✓ Phenolphthalein indicator solution. Dissolve 0.5 g of phenolphthalein in 50 ml of 95 per cent ethyl alcohol, and add 50 ml of distilled water. Add a dilute (e.g. 0.01 or 0.05 mol $l^{-1}$) carbon dioxide-free solution of sodium hydroxide, a drop at a time, until the indicator turns faintly pink.

✓ Potassium peroxydisulphate solution. Dissolve 5 g $K_2S_2O_8$ in 100 ml distilled water. Prepare daily.

✓ Phosphate stock solution. Dissolve 4.390 g potassium dihydrogen phosphate, $KH_2PO_4$, in 1 litre of water. Add one or two drops of toluene as a preservative (1.0 ml stock solution is equivalent to 1.0 mg P).

✓ Phosphate working solution. Dilute 10 ml of stock solution to 100 ml with distilled water and mix well. Then dilute 10 ml of this solution to 1 litre and mix. The dilute solution thus obtained will not keep very long and should be freshly prepared when required (1.0 ml working solution is equivalent to 1.0 μg P).

✓ Ammonium molybdate tetrahydrate solution, 40 g $l^{-1}$.

✓ Sodium hydroxide, approximately 5 mol $l^{-1}$.

✓ Sulphuric acid, approximately 2.5 mol $l^{-1}$. Add carefully, with mixing, 140 ml sulphuric acid ($d$ = 1.84) to water, cool and make up to 1 litre.

✓ Potassium antimonyl tartrate solution. Dissolve 2.7 g potassium antimonyl tartrate in water and make up to 1 litre.

✓ Reducing agent. Mix together 250 ml of 2.5 mol $l^{-1}$ sulphuric acid, 75 ml ammonium molybdate solution and 150 ml distilled water. Add 25 ml potassium antimonyl tartrate solution and again mix well. The solution should be kept in a refrigerator and is stable for several weeks. Immediately before use of the reagent in step 9 of the procedure, pour an aliquot into an Erlenmeyer flask and add 1.73 g ascorbic acid for each 100 ml of reagent. After the addition of ascorbic acid the solution is unstable and cannot be stored. For each standard and each sample, 8 ml of solution is required.

*Procedure*

1. Take 100 ml of thoroughly mixed sample.
2. Add 1 drop (0.05 ml) of phenolphthalein indicator solution. If a red colour develops, add sulphuric acid solution drop by drop to just discharge the colour.
3. Add 2 ml sulphuric acid solution and 15 ml potassium peroxydisulphate solution.
4. Boil gently for at least 90 minutes, adding distilled water to keep the volume between 25 and 50 ml. Alternatively, heat for 30 minutes in an autoclave or pressure-cooker at 1.1–1.4 kPa $cm^{-2}$.
5. Cool, add 1 drop (0.05 ml) phenolphthalein indicator solution, and neutralise to a faint pink colour with sodium hydroxide solution.
6. Restore the volume to 100 ml with distilled water and set aside.
7. Prepare a series of standards in 50-ml volumetric flasks as follows:

| Volume of working phosphate solution (ml) | Concentration, when diluted to 40 ml (μg $l^{-1}$ of phosphorus) |
|---|---|
| 0.0 (blank) | 0 |
| 1.0 | 25 |
| 2.0 | 50 |
| 3.0 | 75 |
| 4.0 | 100 |
| 8.0 | 200 |
| 12.0 | 300 |
| 16.0 | 400 |

8. Place 40 ml of sample in a stoppered 50-ml volumetric flask.
9. Add 8 ml of the mixed reducing agent to the standards and samples, make up to 50 ml with distilled water and mix. Allow to stand for 10 minutes.
10. Measure the absorbance of the blank and each of the standards. Prepare a calibration graph by plotting absorbance against the concentration of phosphorus in μg $l^{-1}$.
11. Measure the absorbance of the samples and, from the calibration graph, read the number of μg $l^{-1}$ of phosphorus in the samples.

## 7.19 Potassium

Although potassium is a relatively abundant element, its concentration in natural fresh waters is usually less than 20 mg $l^{-1}$. Brines and seawater, however, may contain as much as 400 mg $l^{-1}$ of potassium or more. The colorimetric method of analysis given here avoids the need for sophisticated laboratory equipment but does require the use of a centrifuge. Potassium concentrations may also be determined by flame photometry (see section 8.3).

*Principle*

Potassium is determined by precipitating it with sodium cobaltinitrite, oxidising the dipotassium sodium cobaltinitrite with standard potassium dichromate solution in the presence of sulphuric acid, and measuring the excess dichromate colorimetrically. A series of standards with known concentrations of potassium must be carried through the procedure with each set of samples because the temperature and the time of preparation can significantly affect the results.

*Interferences*

Ammonium ions interfere and should not be present. Silica may interfere if a turbid silica gel is formed as a result of either evaporation or the addition of the reagent, and this must be removed by filtration of the coloured sample. The procedure permits measurement of potassium with an accuracy of ± 0.5 mg $l^{-1}$.

*Apparatus*

- ✓ Colorimetric equipment. One of the following is required:
  - —Spectrophotometer, for use at 425 nm and providing a light path of 1 cm or longer.
  - —Filter photometer, equipped with a violet filter, with a maximum transmittance of 425 nm and providing a light path of 1 cm or longer (for concentrations below 20 mg $l^{-1}$, a longer light path is desirable).
  - —Nessler tubes, matched, 100 ml tall form.
- ✓ Centrifuge with 25-ml centrifuge tubes.
- ✓ Small-diameter glass stirring rods to stir the precipitate in the centrifuge tubes.

*Reagents*

- ✓ Nitric acid, $HNO_3$, 1 mol $l^{-1}$.
- ✓ Nitric acid, 0.01 mol $l^{-1}$. Dilute 10 ml of the 1 mol $l^{-1}$ nitric acid to 1 litre with distilled water.
- ✓ Trisodium cobaltinitrite solution. Dissolve 10 mg $Na_3Co(NO_2)_6$ in 50 ml distilled water. Prepare fresh daily and filter before use.
- ✓ Standard potassium dichromate solution. Dissolve 4.904 g anhydrous $K_2Cr_2O_7$ in distilled water and make up to 1 litre in a volumetric flask. This solution is stable for at least one year.
- ✓ Sulphuric acid, $H_2SO_4$, concentrated.
- ✓ Standard potassium solution. Dissolve 1.907 g KCl, dried at 110 °C, and dilute to 1 litre with deionised distilled water (1 ml ≡ 1.00 mg K).

*Procedure*

1. For normal surface and groundwaters, take a 100-ml sample and concentrate its potassium content by evaporation until about only 5 ml remain. Transfer this concentrated sample to a 25-ml centrifuge tube and make up to 10.0 ml with deionised distilled water.

2. The reaction is dependent on time and temperature, so both of these should be kept reasonably constant for all samples and standards in a series of tests: ± 15 minutes and ± 5 °C.

3. At room temperature add, with mixing, 1 ml of the 1 mol $l^{-1}$ nitric acid and 5 ml of the trisodium cobaltinitrite solution. Let stand for 2 hours.

4. Centrifuge for 10 minutes. Carefully pour off the liquid and wash the precipitate with 15 ml of the 0.01 mol $l^{-1}$ nitric acid. Mix with a small glass stirring rod to ensure contact between the precipitate and the wash solution.
5. Centrifuge again for 10 minutes. Pour off the liquid and add, with mixing, 10.00 ml of standard potassium dichromate solution and 5 ml concentrated sulphuric acid.
6. Cool to room temperature. Make up to 100 ml with deionised distilled water. If the solution is turbid, filter it into a Nessler tube and make up to 100 ml.
7. Preparation of standards. Pipette portions of 1, 2, 3, 4, 5, 6 and 7 ml of the standard potassium solution into a series of 25-ml centrifuge tubes, and make up to 10 ml with deionised distilled water. Treat all tubes in the manner described for the sample in steps 3 to 6 above to obtain colour standards containing 1.00 to 7.00 mg K.
8. If a spectrophotometer is being used, prepare a calibration curve with absorbance plotted against mg K. Measure the absorbance of the sample and determine the concentration of potassium from the calibration curve.
9. If visual colour comparison is being made, compare the colour of the sample with the colour of the standards and select the standard that is the closest.

*Calculation*

$$\text{Potassium} = \frac{\text{mg K} \times 1{,}000}{\text{ml sample}} \text{ mg l}^{-1}$$

## 7.20 Selenium

The chemistry of selenium is similar in many respects to that of sulphur, but selenium is a much less common element. The selenium concentrations usually found in water are of the order of a few micrograms per litre, but may reach 50–300 µg $l^{-1}$ in seleniferous areas and have been reported to reach 1 mg $l^{-1}$ in drainage water from irrigated seleniferous soil. Well water containing 9 mg of selenium per litre has been reported.

Little is known about the oxidation state of selenium in water. Selenium appears in the soil as basic ferric selenite, as calcium selenate and as elemental selenium. Although the solubility of elemental selenium is limited, selenium may be present in water in the elemental form as well as the selenate ($SeO_4^{2-}$), selenite ($SeO_3^{2-}$) and selenide ($Se^{2-}$) anions. In addition, many organic compounds of selenium are known. The geochemical control of selenium concentrations in water is not understood, but adsorption by sediments and suspended materials appears to be of importance.

Selenium is an essential, beneficial element required by animals in trace amounts but toxic when ingested at higher levels. A guideline value of 0.01 mg $l^{-1}$ for selenium in drinking water has been recommended by WHO on the basis of long-term health effects. In humans, the symptoms of selenium toxicity are similar to those of arsenic and the toxic effects of long-term exposure are manifested in nails, hair and liver. Selenium poisoning has occurred in animals grazing exclusively in areas where the vegetation contains toxic levels of selenium because of highly seleniferous soils. In general, such soils are found in arid or semi-arid areas of limited agricultural activity. Selenium deficiency in animals occurs in many areas of the world and causes large losses in animal production.

Two methods of selenium analysis are given and both are secondary methods of analysis. The photometric diaminobenzidine method is less sensitive than the fluorometric, but with sample preconcentration the limit of detection is generally acceptable.

*Sample handling*

Selenium in concentrations of around 1 μg $l^{-1}$ has been found to be adsorbed on Pyrex glass and on polyethylene containers. Collect the sample in a polyethylene bottle and acidify by the addition of 1.5 ml of concentrated $HNO_3$ per litre if the sample is to be stored.

*Principle*

Oxidation by acid permanganate converts all selenium compounds to selenate. Many carbon compounds are not completely oxidised by acid permanganate, but it is unlikely that the selenium–carbon bond will remain intact through this treatment. Experiments demonstrate that inorganic forms of selenium are oxidised by acid permanganate in the presence of much greater concentrations of organic matter than would be expected in water supplies. There is substantial loss of selenium when solutions of sodium selenate are evaporated to complete dryness, but in the presence of calcium all the selenium is recovered. An excess of calcium over the selenate is not necessary.

Selenate is reduced to selenite in warm 4 mol $l^{-1}$ HCl. Temperature, time and acid concentrations are specified to obtain quantitative reduction without loss of selenium. The optimum pH for the formation of piazselenol is approximately 1.5. Above pH 2, the rate of formation of the coloured compound is critically dependent on the pH. When indicators are used to adjust the pH, the results are frequently erratic. Extraction of piazselenol is not quantitative, but equilibrium is attained rapidly. Above pH 6, the partition ratio of piazselenol between water and toluene is almost independent of the hydrogen ion concentration.

*Interferences*

No inorganic compounds give a positive interference. It is possible that coloured organic compounds exist that are extracted by cyclohexane but it seems unlikely that interference of this nature will resist the initial acid oxidation or other treatment to remove dissolved organics. Negative interference

**Table 7.7** Percentage selenium recovery in the presence of iodide and bromide interference

| Iodide (mg) | $Br^-$ 0 mg | $Br^-$ 1.25 mg | $Br^-$ 2.50 mg |
|---|---|---|---|
| 0 | 100 | 100 | 96 |
| 0.5 | 95 | 94 | 95 |
| 1.25 | 84 | 80 | – |
| 2.50 | 75 | – | 70 |

results from compounds that lower the concentration of diaminonaphthalene by oxidising this reagent. The addition of EDTA eliminates negative interference from at least 2.5 mg ferric iron. Manganese has no effect in any reasonable concentration, probably because it is reduced along with the selenate. Iodide and, to a lesser extent, bromide cause low results. The recovery of selenium from a standard containing 25 µg Se in the presence of varying amounts of iodide and bromide is shown in Table 7.7. The percentage recovery improves slightly as the amount of selenium is decreased. The minimum detectable concentration is 10 µg Se $l^{-1}$.

*Apparatus*

✓ Colorimetric equipment. A spectrophotometer, for use at 480 nm, providing a light path of 1 cm or longer.
✓ Separatory funnel 250 ml, preferably with a fluorocarbon stopcock.
✓ Thermostatically controlled water-bath (50 °C), with cover.
✓ pH meter.
✓ Centrifuge, for 50-ml tubes (optional).
✓ Centrifuge bottles, 60 ml, screw-capped, fluorocarbon.

*Reagents*

✓ Stock selenium solution. Dissolve 1.633 g selenious acid, $H_2SeO_3$, in distilled water and dilute to 1 litre (1.00 ml ≡ 1.00 mg Se(IV)).
✓ Standard selenium solution. Dilute an appropriate volume of stock selenium solution with distilled water to produce a series of working standards spanning the concentration range of interest.
✓ Methyl orange indicator solution.
✓ Calcium chloride solution. Dissolve 30 g $CaCl_2.2H_2O$ in distilled water and dilute to 1 litre.
✓ Potassium permanganate, 0.1 mol $l^{-1}$. Dissolve 3.2 g $KMnO_4$ in 1,000 ml distilled water.
✓ Sodium hydroxide, 0.1 mol $l^{-1}$.
✓ Hydrochloric acid, concentrated and 0.1 mol $l^{-1}$.
✓ Ammonium chloride solution. Dissolve 250 g $NH_4Cl$ in 1 litre distilled water.
✓ EDTA–sulphate reagent. Dissolve 100 g disodium ethylenediaminetetraacetate dihydrate (also called ethylenedinitrilotetraacetic acid disodium salt) and 200 g sodium sulphate in 1 litre distilled water. Add concentrated ammonium hydroxide drop by drop while stirring until the dissolution is complete.

- ✓ 2,3-diaminonaphthalene (DAN) solution. Dissolve 200 mg DAN in 200 ml 0.1 mol $l^{-1}$ HCl. Shake for 5 minutes. Extract three times with 25-ml portions of cyclohexane, retaining the aqueous phase and discarding organic portions. Filter through a Whatman No. 42 or equivalent filter paper and store in a cool dark place for no longer than 8 hours. CAUTION: This reagent is TOXIC. Handle it with extreme care.
- ✓ Cyclohexane, $C_6H_{12}$.
- ✓ Toluene.
- ✓ Sodium sulphate, anhydrous (required if no centrifuge is available).
- ✓ Ammonium hydroxide, $NH_4OH$, 50 per cent strength by volume.

*Procedure*

*Oxidation to selenate*

1. Prepare standards containing 0, 10.0, 25.0 and 50.0 µg Se in 500-ml Erlenmeyer flasks. Dilute to approximately 250 ml, add 10 drops methyl orange indicator solution, 2 ml 0.1 mol $l^{-1}$ HCl, 5 ml $CaCl_2$ solution, 3 drops 0.1 mol $l^{-1}$ $KMnO_4$, and a 5-ml measure of glass beads to prevent bumping. Boil vigorously for approximately 5 minutes.
2. To a 1,000-ml sample in a 2-litre beaker add 10 drops methyl orange indicator solution. Titrate to the methyl orange end-point with 0.1 mol $l^{-1}$ HCl and add 2 ml excess.
3. Add 3 drops $KMnO_4$, 5 ml $CaCl_2$ solution and a 5 ml measure of glass beads to prevent bumping. Heat to boiling, adding $KMnO_4$ as required to maintain a purple tint. Ignore a precipitate of $MnO_2$ because it will have no adverse effect.
4. After the volume has been reduced to approximately 250 ml, transfer the solution to a 500-ml Erlenmeyer flask.

*Evaporation*

5. Add 5 ml 0.1 mol $l^{-1}$ NaOH to each flask and evaporate to dryness. Avoid prolonged heating of the residue.

*Reduction to selenite*

6. Cool the flask, add 5 ml concentrated HCl and 10 ml $NH_4Cl$ solution. Heat in a boiling water-bath for 10 ± 0.5 minutes.

*Formation of piazselenol*

7. Transfer the warm solution and ammonium chloride precipitate, if present, from the flask to a beaker suitable for pH adjustment, washing the flask with 5 ml EDTA–sulphate reagent and 3 ml 50 per cent $NH_4OH$. Adjust the pH to 1.5 ± 0.3 with $NH_4OH$, using a pH meter. The precipitate of EDTA will not interfere. Add 1 ml diaminobenzidine solution and heat in a thermostatically controlled water-bath at 50 °C for approximately 30 min.

*Extraction of piazselenol*

8. Cool and then add $NH_4OH$ to adjust the pH to 8 ± 1; the precipitate of EDTA will dissolve. Pour the sample into a 50-ml graduated cylinder and adjust the volume to 50 ± 1 ml with washings from the beaker.
9. Pour the contents of the graduated cylinder into a 250-ml separatory funnel. Add 10 ml toluene and shake for 30 ± 5 seconds. Discard the aqueous layer and transfer the organic phase to a centrifuge tube. Centrifuge briefly to clear the toluene from water droplets. If a centrifuge is not available, filter the organic phase through a dry filter paper to which approximately 0.1 g anhydrous $Na_2SO_4$ has been added.

*Determination of absorbance*

10. Read the absorbance at approximately 420 nm, using toluene to establish zero absorbance. The piazselenol colour is stable, but evaporation of toluene concentrates the colour to a marked degree in a few hours. Beer's law is obeyed up to 50 µg.

*Calculation*

$$\text{Concentration of selenium} = \frac{\mu g\ Se}{ml\ sample}\ mg\ l^{-1}$$

*Precision and accuracy*

A synthetic unknown sample containing 20 µg $l^{-1}$ Se, 40 µg $l^{-1}$ As, 250 µg $l^{-1}$ Be, 240 µg $l^{-1}$ B and 6 µg $l^{-1}$ V in distilled water was determined by the diaminobenzidine method, with a relative standard deviation of 21.2 per cent and a relative error of 5.0 per cent in 35 laboratories.

## 7.21 Reactive silica

After oxygen, silicon is the most abundant element in the earth's crust. It is a major constituent of igneous and metamorphic rocks, of clay minerals such as kaolin, and of feldspars and quartz. Although crystalline silica is a major constituent of many igneous rocks and sandstones, it has low solubility and is therefore of limited importance as a source of silica in water. It is likely that most of the dissolved silica in water originates from the chemical breakdown of silicates in the processes of metamorphism or weathering.

The concentration of silica in most natural waters is in the range 1–30 mg $l^{-1}$. Up to 100 mg $l^{-1}$ is not uncommon. Over 100 mg $l^{-1}$ is relatively rare, although more than 1,000 mg $l^{-1}$ is occasionally found in some brackish waters and brines.

*Sample handling*

Samples should be stored in plastic bottles to prevent leaching of silica from glass. Samples should be passed through a membrane filter of 0.45 µm pore size as soon as possible after sample collection and should be stored at 4 °C without preservatives. Analysis should be performed within 1 week of sample collection.

*Principle*

Ammonium molybdate at a pH of approximately 1.2 reacts with silica and phosphate to form heteropoly acids. Addition of oxalic acid destroys any molybdophosphoric acid but not the molybdosilicic acid. The yellow molybdosilicic acid is reduced by aminonaphtholsulphonic acid to heteropoly blue. The blue colour of the heteropoly blue is more intense than the yellow colour of the molybdosilicic acid, so that this reaction increases the sensitivity of the method.

Silica can be measured at either 815 nm or 650 nm. The sensitivity at 650 nm is approximately half that at 815 nm.

*Interferences*

Both the apparatus and the reagents may contribute silica. The use of glassware should be avoided as far as possible and only reagents low in silica should be used. A blank determination should be carried out to correct for silica introduced from these sources. Tannin, large amounts of iron, colour, turbidity, sulphide and phosphate are potential sources of interference. The treatment with oxalic acid eliminates the interference from phosphate and decreases the interference from tannin. Photometric compensation may be used to cancel interference from colour or turbidity in the sample.

*Apparatus*

✓ Colorimetric equipment. One of the following is required:
—Spectrophotometer for measurement at 815 nm, providing a light path of at least 1 cm.
—Absorptiometer for measurement at a wavelength of 815 nm.
*Note:* If no instrument is available for measurement at this wavelength, an instrument capable of measuring at 650 nm can be used, but this leads to a decrease in sensitivity.

*Reagents*

✓ Hydrochloric acid, 6 mol $l^{-1}$. Combine 1 volume of concentrated HCl with an equal volume of water.

✓ Ammonium molybdate reagent. Dissolve 10 g of ammonium molybdate tetrahydrate, $(NH_4)_6Mo_7O_{24}.4H_2O$, in distilled water, with stirring and gentle warming, and dilute to 100 ml. Filter if necessary. Adjust to pH 7–8 with silica-free $NH_4OH$ or NaOH and store in a polyethylene bottle. If the pH is not adjusted, a precipitate gradually forms. If the solution is stored in glass, silica may leach out, causing high blanks. If necessary, prepare silica-free $NH_4OH$ by passing gaseous $NH_3$ into distilled water contained in a plastic bottle.

✓ Oxalic acid solution. Dissolve 10 g of oxalic acid dihydrate, $H_2C_2O_4.2H_2O$, in distilled water and dilute to 100 ml.

✓ Reducing agent. Dissolve 0.5 g of 1-amino-2-naphthol-4-sulphonic acid and 1 g of $Na_2SO_3$ in 50 ml of distilled water, with gentle warming if necessary. Add this solution to a solution of 30 g of $NaHSO_3$ in 150 ml of distilled water. Filter into a plastic bottle and store in a refrigerator away from light. Discard the solution when it becomes dark. Do not use aminonaphtholsulphonic acid that is incompletely soluble or that produces reagents that are dark even when freshly prepared; such material is not suitable for silica determinations.

✓ Silica standard stock solution. Dissolve 313.0 mg of sodium hexafluorosilicate, $Na_2SiF_6$, in 1,000 ml of distilled water (1 ml ≡ 0.1 mg $SiO_2$).

✓ Silica standard working solution. Dilute 100 ml of standard stock solution to 1,000 ml (1 ml ≡ 10 µg $SiO_2$).

*Procedure*

1. Prepare standards and a blank by pipetting portions of the silica working standard into tall-form 50-ml Nessler tubes and make up to 50 ml with distilled water. If measurement is to be made at 650 nm, use portions of 0–30 ml (giving 0–300 µg $l^{-1}$ concentrations in the Nessler tubes); if measurement is to be made at 815 nm use portions of 0–10 ml (0–100 µg $l^{-1}$). If cells with a light path longer than 1 cm are to be used, the silica concentrations should be proportionally reduced.
*Note:* Reduction by aminonaphtholsulphonic acid is temperature-dependent. The best results will be obtained if tubes are immersed in a water-bath, thermostatically controlled at a temperature near 25 or 30 °C.

2. To each standard add, in rapid succession, 1.0 ml of HCl, 6 mol $l^{-1}$, and 2.0 ml of ammonium molybdate reagent. Mix by inverting the tubes at least six times.
3. Add 1.5 ml of oxalic acid solution and mix thoroughly. At least 2 minutes, but less than 15 minutes, after addition of the oxalic acid, add 2.0 ml of reducing agent and mix thoroughly.
4. After 5 minutes, measure the absorbance at 650 or 815 nm.
5. Subtract the absorbance of the blank from the absorbances of the standards to obtain the net absorbances.
6. Prepare a calibration graph relating the net absorbance to the amount of silica ($SiO_2$).
7. Measure 50 ml of sample into a tall-form Nessler tube and add, in rapid succession, 1.0 ml of HCl, 6 mol $l^{-1}$, and 2.0 ml of ammonium molybdate reagent. Mix by inverting the tube at least six times.
8. Add 1.5 ml of oxalic acid solution and mix thoroughly. At least 2 minutes, but less than 15 minutes, after addition of the oxalic acid, add 2.0 ml of reducing agent and mix thoroughly.
9. After 5 minutes, measure the absorbance at 650 or 815 nm.
10. Subtract the absorbance of the blank from the absorbance of the sample to obtain the net absorbance.

*Calculation*

Determine the amount of silica ($SiO_2$) equivalent to the net absorbance from the calibration graph and divide by the volume of the sample (ml) to obtain the concentration of $SiO_2$ in mg $l^{-1}$.

$$\text{Concentration of silica} = \frac{SiO_2 \text{ (from graph)}}{\text{ml of sample}} \text{ mg } l^{-1}$$

## 7.22 Sodium

Sodium is a common element, the sixth most abundant, and present to some extent in most natural waters. Concentrations vary from negligible in freshwater to considerable in seawater and brackish water. The permeability of agricultural soil is harmed by a high ratio of sodium ions to total cations. Sodium concentrations higher than a few milligrams per litre are undesirable in feed water for high-pressure boilers. When compounded with certain anions (e.g. chloride), sodium imparts a salty taste to drinking water and, if the concentration is sufficiently high, consumers may not be willing to drink it.

Sodium concentrations can be determined by flame photometry (see section 8.3) but the method presented below does not require any sophisticated laboratory equipment.

*Principle*

Sodium is precipitated as sodium zinc uranyl acetate hexahydrate, $NaC_2H_3O_2.Zn(C_2H_3O_2)_2.3UO_2(C_2H_3O_2)_2.6H_2O$, by adding a large volume of zinc uranyl acetate reagent, previously saturated with the sodium salt, to a small volume of concentrated sample. At least 10 ml of reagent must be used for each millilitre of sample and the mixture must be allowed to stand for 60 minutes or more. After collection in a filter crucible, the precipitate is washed with successive small portions of the salt-saturated reagent, then with 95 per cent ethyl alcohol, also saturated with the triple salt. It is then washed with diethyl ether to remove the alcohol, and the ether is evaporated by drawing a stream of air through the sample. The air-dried sample is then weighed.

*Apparatus*

- ✓ Beakers, 20 or 50 ml, borosilicate glass.
- ✓ Fritted glass crucibles, 30 ml, borosilicate glass or porcelain of medium porosity.
- ✓ Vacuum pump or aspirator, with manifold and individual petcocks.

*Reagents*

- ✓ Zinc uranyl acetate reagent. Mix 2 ml concentrated acetic acid with 100 ml distilled water. Add 10 g uranyl acetate dihydrate, $UO_2(C_2H_3O_2)_2.2H_2O$, and 30 g zinc acetate dihydrate, $Zn(C_2H_3O_2)_2.2H_2O$, and warm to dissolve. Cool, add 2–3 mg NaCl, let stand for 24 hours and filter off the precipitate of zinc uranyl acetate. The filtrate is the triple salt saturated reagent and should be stored in a borosilicate glass bottle.
- ✓ Pure sodium zinc uranyl acetate. Add 25 ml of zinc uranyl acetate solution to 2 ml saturated sodium chloride solution. Stir. Collect the precipitate by filtering through a fritted glass crucible. Wash the precipitate three times with concentrated acetic acid, then three times with diethyl ether.
- ✓ Ethyl alcohol wash solution. Saturate 95 per cent ethyl alcohol with pure sodium zinc uranyl acetate. Decant or filter the solution just before use.
- ✓ Diethyl ether.

*Procedure*

1. If necessary, remove any suspended matter from 50–100 ml of the sample by filtration.
2. Select a portion of the clear sample containing less than 8 mg sodium. This could be as little as 2 ml of a very salty water such as seawater or as much as 30 ml of a relatively salt-free "fresh" water. Pipette the measured portion into a borosilicate glass beaker of appropriate size.
3. Evaporate to dryness over a steam bath or hot water-bath.
4. Cool the residue to room temperature. Add 1.0 ml distilled water and stir with a glass rod to dissolve the residue. If all residue does not dissolve, add distilled water in 1.0 ml increments until all of the residue is dissolved.
5. Add zinc uranyl acetate reagent in the ratio of 10 ml of reagent for each 1.0 ml of distilled water used to dissolve the residue. Mix, cover the beaker and let stand for 1 hour, stirring periodically to prevent the formation of a supersaturated solution.
6. Arrange an apparatus for suction filtration using a medium porosity fritted glass (or porous-bottomed porcelain) crucible that has been previously weighed.
7. Pour the contents of the beaker into the crucible and filter under suction. Rinse the beaker at least five times with 2-ml portions of the zinc uranyl acetate reagent, pouring the rinsings into the crucible. After the last rinse, maintain suction for several minutes to remove all possible traces of the reagent.

8. Maintain suction while washing the contents of the crucible five times with 2-ml portions of the ethyl alcohol wash solution. Then conclude the washing with three small portions of diethyl ether.
9. Continue suction for a few minutes until the ether has evaporated and the precipitate in the crucible is dry.
10. If salts have crystallised on the outside of the crucible, wipe it clean with a soft cloth or tissue.
11. Weigh the crucible and contents. Repeat the weighing after a 15-minute interval and a third time after a further 15 minutes to check on the constancy of the weight.
12. Return the crucible to the suction apparatus, apply suction and add warmed distilled water in portions of approximately 10 ml, using a total of about 100 ml to dissolve all traces of the sodium zinc uranyl acetate. Dry the crucible with ethyl alcohol wash solution and diethyl ether as in step 8 above, and reweigh the crucible as in step 11.
13. Subtract the weight obtained in step 12 from that obtained in step 11. This represents the weight of the sodium zinc uranyl acetate.

*Calculation*

$$\text{Concentration of sodium} = \frac{A \times 14.95}{\text{ml sample}} \text{ mg l}^{-1}$$

where $A$ = the weight of the sodium zinc uranyl acetate (mg)

## 7.23 Sulphate

Sulphate is an abundant ion in the earth's crust and its concentration in water can range from a few milligrams to several thousand milligrams per litre. Industrial wastes and mine drainage may contain high concentrations of sulphate. Sulphate also results from the breakdown of sulphur-containing organic compounds.

Sulphate is one of the least toxic anions and WHO does not recommend any guideline value for it in drinking water. However, catharsis, dehydration and gastrointestinal irritation have been observed at high concentrations in drinking water and WHO therefore suggests that health authorities should be notified when concentrations of sulphate in drinking water exceed 500 mg $l^{-1}$.

*Sample handling*

Samples may be stored, refrigerated, in either plastic or glass containers for not more than seven days.

### 7.23.1 Titrimetric method

*Principle*

Barium ions in water are titrated as hardness in the EDTA hardness determination. Sulphate can therefore be indirectly determined on the basis of the difference in hardness of the sample before and after the addition of barium ions in excess of that required for sulphate precipitation.

*Apparatus*

✓ Porcelain dishes, 100-ml capacity.
✓ Burette, 25 or 50 ml.

*Reagents*

✓ Hydrochloric acid, approximately 1 mol $l^{-1}$. Carefully add 83 ml HCl ($d$ = 1.18) to about 500 ml distilled water in a large beaker. Stir and cool to room temperature before transferring, with rinsings, to a 1-litre volumetric flask. Make up to the mark with distilled water. Store in a tightly stoppered glass bottle.
✓ Standard EDTA titrant, 0.01 mol $l^{-1}$ (see section 7.7, Calcium).
✓ Eriochrome Black T indicator (see section 7.7, Calcium).
✓ Buffer solution (see section 7.12, Magnesium).
✓ Barium chloride solution, approximately 0.01 mol $l^{-1}$. Dissolve 2.443 g of barium chloride dihydrate, $BaCl.2H_2O$, in 500 ml of distilled water. Transfer, with rinsings, to a 1-litre volumetric flask. Add 2.0 ml concentrated HCl and make up to the mark with distilled water. Store in a polyethylene bottle. The solution is stable for six months.

*Procedure*

1. Place 50 ml of sample in a porcelain dish. Add 1 to 2 ml of buffer solution. The pH should be 10.0 ± 0.1. Adjust as necessary.
2. Add 2 drops of indicator solution (or a small portion of dry powder indicator mixture).
3. Titrate slowly with EDTA standard titrant, stirring continuously, until the last reddish tinge disappears and the sample becomes blue. The titration should be completed within 5 minutes of the addition of buffer. Record the amount of EDTA titrant used.
4. If more than 15 ml of titrant are used, repeat steps 1 to 3 using 25 ml of sample diluted to 50 ml with distilled water. If less than 1 ml is used, repeat using 100 ml of sample, or more if necessary.
5. Calculate hardness as $CaCO_3$ mg $l^{-1}$ (see calculation, below).
6. Measure 100 ml of sample and pour into a beaker. Neutralise the alkalinity to pH 4.5 with 1 mol $l^{-1}$ HCl or $HNO_3$. Add 1 ml more of the acid.
7. Bring the sample to the boil to expel carbon dioxide. Add 10 ml of barium chloride standard solution to the boiling sample. After the volume has been reduced to less than 100 ml, remove from heat and allow to cool.
8. Transfer with rinsings to a 100-ml graduated cylinder and make up to the 100-ml mark with distilled water. Allow any precipitate to settle.
9. Pour 50 ml of the clear supernatant into a porcelain dish. Add 2 ml of buffer solution; the pH should be 10.0 ± 0.1. Adjust as necessary.
10. Add 2 drops of the indicator solution (or a small portion of the dry powder indicator mixture).
11. Titrate slowly with EDTA standard titrant, stirring continuously, until the last reddish tinge disappears and the sample becomes blue. The titration should be completed within 5 minutes of the addition of buffer. Record the amount of EDTA titrant used.
12. If more than 15 ml of titrant are used, repeat steps 9 to 11 using 25 ml of sample diluted to 50 ml with distilled water. If less than 1 ml is used, repeat using 100 ml of sample, or more if necessary.

*Calculation*
The calculation is in three parts. First, the hardness is calculated from the results of the titration in step 3 of the procedure.

$$\text{Hardness} = \frac{1{,}000 \times H}{V_1} \text{ mg l}^{-1} \text{ as CaCO}_3$$

where $H$ = volume of titrant used in step 3 of procedure (ml)
$V_1$ = volume of sample used in the titration at step 3 (ml)

The second part of the calculation yields a value that is the hardness plus the amount of barium chloride that did not combine with sulphate in the sample. This is determined from the results of the titration in step 11 of the procedure.

$$\text{Hardness + BaCl}_u = \frac{1{,}000 \times T}{V_2} \text{ mg l}^{-1} \text{ as CaCO}_3$$

where $T$ = volume of titrant used in step 11 of the procedure (ml)
$V_2$ = volume of sample used in the titration in step 11 (ml)
$BaCl_u$ = volume of barium chloride that did not combine with $SO_4$ (ml)

The final part of the calculation is to add the result of the first titration to the concentration of the barium chloride solution (1,000 mg $l^{-1}$), to subtract the result of the second titration and to convert the result to mg $l^{-1}$ $SO_4$.

$$\{\text{Hardness (mg l}^{-1}) + 1{,}000 - [\text{Hardness + BaCl}_u]\} \times 0.96 = \text{mg l}^{-1}\ SO_4$$

### 7.23.2 Gravimetric method with drying of residue

*Principle*
Sulphate is precipitated in a hydrochloric acid solution as barium sulphate ($BaSO_4$) by the addition of barium chloride ($BaCl_2$). The precipitation is carried out near the boiling temperature and after a period of digestion the precipitate is filtered, washed with water until free of $Cl^-$, dried and weighed as $BaSO_4$.

The gravimetric determination of $SO_4^{2-}$ is subject to many errors, both positive and negative. Interferences that lead to high results are suspended matter, silica, $BaCl_2$ precipitant, $NO_3^-$, $SO_3^{2-}$ and water occluded in the precipitant. Interferences leading to low results are alkali metal sulphates, heavy metals (especially chromium and iron) and the solubility of the $BaSO_4$ precipitant, especially in acid solution.

*Apparatus*
- ✓ Steam bath.
- ✓ Drying oven equipped with thermostatic control.
- ✓ Desiccator.
- ✓ Analytical balance capable of weighing to 0.1 mg.
- ✓ Filter: either fritted glass filter, fine porosity with a maximum pore size of 5 µm, or membrane filter with a pore size of about 0.45 µm.
- ✓ Vacuum oven.

*Reagents*

✓ Methyl red indicator solution. Dissolve 100 mg methyl red sodium salt in distilled water and dilute to 100 ml.

✓ Hydrochloric acid, HCl 1+1.

✓ Barium chloride solution. Dissolve 100 g $BaCl_2.2H_2O$ in 1 litre distilled water. Filter through a membrane filter or hard-finish filter paper before use. One millilitre is capable of precipitating approximately 40 mg $SO_4^{2-}$.

✓ Silver nitrate–nitric acid reagent. Dissolve 8.5 g $AgNO_3$ and 0.5 ml concentrated $HNO_3$ in 500 ml distilled water.

✓ Silicone fluid.

*Procedure*

*Removal of silica*

1. If the silica concentration exceeds 25 g $l^{-1}$, evaporate sample nearly to dryness in a platinum dish on a steam bath. Add 1 ml of HCl, tilt and rotate dish until acid comes in complete contact with the residue. Continue evaporation to dryness. Complete drying in an oven at 180 °C and, if organic matter is present, char over the flame of a burner. Moisten residue with 2 ml of distilled water and 1 ml of HCl and evaporate to dryness on a steam bath. Add 2 ml of HCl, take up soluble residue in hot water and filter. Wash insoluble silica with several small portions of hot distilled water. Combine filtration and washing. Discard residue.

*Precipitation of barium sulphate*

2. Adjust volume of clarified sample to contain approximately 50 mg $SO_4^{2-}$ in a 250-ml volume. Lower concentrations of $SO_4^{2-}$ may be tolerated if it is impracticable to concentrate the sample to the optimum level, but in such cases limit the total volume to 150 ml. Adjust pH with HCl to pH 4.5 to 5.0 using a pH meter or the orange colour of methyl red indicator. Add 1 to 2 ml of HCl. Heat to boiling and, while stirring gently, slowly add warm $BaCl_2$ solution until precipitation appears to be complete, then add 2 ml excess. If amount of precipitate is small, add a total of 5 ml $BaCl_2$ solution. Digest precipitate at 80–90 °C, preferably overnight but for not less than 2 hours.

*Preparation of filters*

3. (i) Fritted glass filter: dry to constant weight in an oven maintained at 105 °C or higher, cool in a desiccator and weigh.

   (ii) Membrane filter: place filter on a piece of filter paper or a watch glass and dry to constant weight in a vacuum oven at 80 °C while maintaining a vacuum of at least 85 kPa or in a conventional oven at 103–105 °C. Cool in a desiccator and weigh membrane only.

*Filtration and weighing*

4. Filter $BaSO_4$ at room temperature. Wash precipitate with several small portions of warm distilled water until washings are free of $Cl^-$ as indicated by testing with $AgNO_3.HNO_3$ reagent. If a membrane filter is used, add a few drops of silicone fluid to the suspension before filtering to prevent adherence of precipitate to holder. Dry filter and precipitate by the same procedure used in preparing the filter. Cool in a desiccator and weigh.

*Calculation*

$$\text{Concentration of sulphate} = \frac{\text{mg } BaSO_4 \times 411.6}{\text{ml sample}} \text{ mg l}^{-1}$$

## 7.24 Total dissolved solids

The substances remaining after evaporation and drying of a water sample are termed the "residue". The residue is approximately equivalent to the total content of the dissolved and suspended matter in the water sample. Non-filterable residue corresponds to the total suspended solids (see section 7.25) and the filterable residue is the total dissolved solids (TDS).

The results of a determination of total dissolved solids (TDS) can be used to check the accuracy of analyses when relatively complete analyses have been made on a water sample. This is accomplished by comparing the value of calculated TDS with the measured value. Ion concentrations in mg $l^{-1}$ of constituents required to calculate the TDS are as follows:

Calculated TDS = 0.6 (alkalinity) + Na + K + Ca + Mg + Cl + $SO_4$ + $SiO_3$ + ($NO_3$-N) + F

The measured TDS concentration should be higher than the calculated value, because a significant contributor may not be included in the calculation. If the measured value is less than the calculated value, all values are suspect. If the measured value is higher than the calculated value, the low ion sum is suspect and selected constituents should be reanalysed. The acceptable ratio is:

$$1.0 < \frac{\text{measured TDS}}{\text{calculated TDS}} < 1.2$$

*Principle*

A well mixed, measured portion of a sample is filtered through a standard glass-fibre filter and the filtrate is evaporated to dryness in a weighed dish and dried to constant weight at 180 °C. The increase in dish weight represents the total dissolved solids.

*Apparatus*

- ✓ Evaporating dishes made of porcelain (90 mm), platinum or high-silica glass.
- ✓ Steam bath.
- ✓ Desiccator provided with a desiccant containing a colour indicator for moisture concentration.
- ✓ Analytical balance capable of weighing to 0.1 mg.
- ✓ Glass-fibre filter discs without organic binder.
- ✓ Filtration apparatus. All of the following are suitable:
  - —Membrane filter funnel.
  - —Gooch crucible, capacity 25–40 ml, with Gooch crucible adapter.
  - —Filtration apparatus with reservoir and coarse (40–60 µm) fritted disc as filter support.
- ✓ Suction flask, size large enough for chosen sample size.
- ✓ Drying oven, for operation at 180 ± 2 °C.

*Procedure*

1. Prepare the glass-fibre filter disc by placing it, wrinkled side up, in the filtration apparatus. Apply vacuum and wash the disc with three successive 20 ml washings of distilled water. Continue suction to remove all traces of water. Discard washings.

2. Prepare the evaporating dish by heating a clean dish to 180 ± 2 °C in an oven for 1 hour. Cool and store in desiccator until needed. Weigh immediately before use.
3. Choose a sample volume to yield between 2.5 and 200 mg dried residue. If more than 10 minutes are required to complete filtration, increase filter size or decrease sample volume but do not produce less than 2.5 mg of residue.
4. Filter measured volume of well mixed sample through the glass-fibre filter, wash with three successive 10-ml volumes of distilled water, allowing complete draining between washings. Continue suction for about 3 minutes after filtration is complete.
5. Transfer filtrate to a weighed evaporating dish and evaporate to dryness on a steam bath. If filtrate volume exceeds dish capacity, add successive portions to the same dish after evaporation.
6. Dry for at least 1 hour in an oven at 180 ± 2 °C, cool in a desiccator to balance temperature and weigh.
7. Repeat cycle of drying, cooling, desiccating and weighing until a constant weight is obtained or until weight loss between successive weighings is less than 4 per cent or 0.5 mg, whichever is less.

*Calculation*

$$\text{Total dissolved solids} = \frac{(A - B) \times 1{,}000}{\text{sample volume in ml}} \text{ mg l}^{-1}$$

where $A$ = weight of dish + solids (mg)
$B$ = weight of dish before use (mg)

## 7.25 Total suspended solids

The term "total suspended solids" (TSS) applies to the dry weight of the material that is removed from a measured volume of water sample by filtration through a standard filter. The test is basically empirical and is not subject to the usual criteria of accuracy. To achieve reproducibility and comparability of results requires close attention to procedural details, especially filter characteristics and time and temperature of drying. The method described is based on the following conditions: filtering by glass fibre filter (Whatman GF/C grade or equivalent) and drying at a temperature of 103–105 °C for 2 hours to a constant weight, i.e. a variability of not more than 0.5 mg. If other filters (paper, membrane, etc.) or other temperatures are used, it is necessary to report the specifications followed (e.g. total suspended solids at .... °C , type of filter and pore size or number).

Results may be of questionable value if the total suspended solids include oils, grease or other volatile material. It is obvious that the result of a test cannot include materials that are volatile under the conditions of the procedure. Possible sources of error should be recognised when results are interpreted.

*Sample handling*

Non-homogeneous particulates such as leaves, sticks, fish and lumps of faecal matter should be excluded from the sample. Too much residue on the filter will retain water and may require prolonged drying.

Long-term preservation of a sample is not practical; analysis should therefore begin as soon as possible. Transportation and short-term storage of a sample will not normally affect the results of the test.

*Interferences*

Volatile material in a sample will distort results.

*Apparatus*

✓ Filter holder. All of the following types of filter holder are suitable:
  —Membrane filter funnel.
  —Buchner funnel, with adapter to fit suction flask.
  —Gooch crucible, capacity 25–40 ml size, with Gooch crucible adapter.

✓ Glass-fibre filter discs, Whatman GF/C or equivalent, of a size compatible with the filter holder.

✓ Suction flask, 500-ml capacity.

✓ Drying oven, 103–105 °C.

✓ Desiccator.

✓ Analytical balance, capacity 200 g (or more), accuracy 0.1 mg.

✓ Vacuum pump or aspirator.

*Procedure*

*Preparation of glass-fibre filter discs*

1. Place a filter disc on the filter holder. Assemble filter holder in suction flask apparatus, connect to vacuum source and apply vacuum.
2. Wash the filter disc with three successive 20-ml portions of distilled water. Continue to apply vacuum for 2–3 minutes after the water has passed through the filter. Discard the filtrate.
3. Remove the filter paper from the membrane filter funnel or the Buchner funnel and place it on a supporting surface in a drying oven. If the Gooch crucible/filter combination is being used, place it in the drying oven. The oven should be maintained at 103–105 °C and drying should be continued for at least 1 hour.
4. Cool the filter or Gooch combination in a desiccator and weigh it on an analytical balance.
5. Repeat the cycle of drying, desiccating and weighing until the weight loss between two successive series of operations is less than 0.5 mg.
6. Store filter(s) or Gooch crucible(s) in the desiccator until required.

*Sample analysis*

1. Remove the filter disc or Gooch crucible from the desiccator, weigh it and record its weight.
2. Place the filter in the filter holder and assemble the filter holder in the suction flask apparatus. Connect to the vacuum source and apply vacuum.

3. Wet the filter with a few drops of distilled water to seat the filter.
4. Shake the sample vigorously and measure out 100 ml in a 100-ml graduated cylinder or volumetric flask. Pour this portion of the sample into the filter funnel, being careful not to disturb the seating of the filter disc.
5. Rinse out the measuring flask or cylinder with a small quantity of distilled water. If the sample is very low in suspended material, a larger volume of sample may be used.
6. When filtration is complete, carefully remove the filter disc from the filter holder with tweezers (or remove the Gooch crucible from its supporting socket with a pair of tongs) and place it in the drying oven. Dry for at least 1 hour at 103–105 °C. Cool in a desiccator and weigh.
7. Repeat the drying, desiccating and weighing cycle until the weight loss between two successive weighings is less than 0.5 mg.
8. Record the final weight obtained.

*Calculation*

$$\text{Total suspended solids} = \frac{A - B}{C} \times 10^6 \quad \text{mg l}^{-1}$$

where $A$ = weight of filter + solids (g)
$B$ = weight of filter (g)
$C$ = volume of sample filtered (ml)

Report the results as:
Total suspended solids dried at ....°C, .... mg $l^{-1}$

*Precision and accuracy*
Precision data are not available. Accuracy data on actual samples cannot be obtained.

## 7.26 Source literature and further reading

APHA 1992 *Standard Methods for the Examination of Water and Wastewater*, 18th edition. American Public Health Association (APHA), American Water Works Association (AWWA), Water Pollution Control Federation (WPCF), Washington, DC.

Hach Company, 1989 *Hach Water Analysis Handbook*. Hach Company, Loveland, CO.

HMSO 1979 *Methods for the Examination of Waters and Associated Materials: Sulphate in Waters, Effluents and Solids*. Her Majesty's Stationery Office, London.

ISO 1990 *Water Quality — Determination of Kjeldahl Nitrogen, Method after Mineralization with Selenium*. International Standard ISO 5663, International Organization for Standardization, Geneva.

ISO 1990 *Water Quality — Determination of Dissolved Oxygen, Iodometric Method*. International Standard ISO 5813, International Organization for Standardization, Geneva.

ISO 1990 *Water Quality — Determination of Dissolved Oxygen, Electrochemical Probe Method.* International Standard ISO 5814, International Organization for Standardization, Geneva.

ISO 1990 *Water Quality — Determination of Biochemical Oxygen Demand after 5 Days ($BOD_5$), Dilution and Seeding Method.* International Standard ISO 5815, International Organization for Standardization, Geneva.

ISO 1990 *Water Quality — Determination of Calcium Content, EDTA Titrimetric Method.* International Standard ISO 6058, International Organization for Standardization, Geneva.

ISO 1990 *Water Quality — Determination of the Sum of Calcium and Magnesium, EDTA Titrimetric Method.* International Standard ISO 6059, International Organization for Standardization, Geneva.

ISO 1990 *Water Quality — Determination of Chemical Oxygen Demand.* International Standard ISO 6060, International Organization for Standardization, Geneva.

ISO 1990 *Water Quality — Determination of Iron, Spectrometric Method Using 1,10-Phenanthroline.* International Standard ISO 6332, International Organization for Standardization, Geneva.

ISO 1990 *Water Quality — Determination of Manganese, Formaldoxine Spectrometric Method.* International Standard ISO 6333, International Organization for Standardization, Geneva.

ISO 1990 *Water Quality — Determination of Nitrite, Molecular Absorption Spectrometric Method.* International Standard ISO 6777, International Organization for Standardization, Geneva.

ISO 1990 *Water Quality — Determination of Phosphorus Part 1: Ammonium Molybdate Spectrometric Method.* International Standard ISO 6878-1, International Organization for Standardization, Geneva.

ISO 1990 *Water Quality — Determination of Electrical Conductivity.* International Standard ISO 7888, International Organization for Standardization, Geneva.

ISO 1990 *Water Quality — Determination of Nitrate Part 1: 2,6 - Dimethyl Phenol Spectrometric Method.* International Standard ISO 7980-1, International Organization for Standardization, Geneva.

ISO 1990 *Water Quality — Determination of Nitrate Part 2: 4 - Fluorophenol Spectrometric Method after Distillation.* International Standard ISO 7980-2, International Organization for Standardization, Geneva.

ISO 1990 *Water Quality — Determination of Nitrate Part 3: Spectrometric Method Using Sulfasalicylic Acid.* International Standard ISO 7980-3, International Organization for Standardization, Geneva.

ISO 1990 *Water Quality — Guidelines for the Determination of Total Organic Carbon (TOC)*. International Standard ISO 8245, International Organization for Standardization, Geneva.

ISO 1990 *Water Quality — Determination of Sulfate, Gravimetric Method Using Barium Chloride*. International Standard ISO 9280, International Organization for Standardization, Geneva.

ISO 1990 *Water Quality — Determination of Chloride, Silver Nitrate Titration with Chromate Indicator (Mohr's Method)*. International Standard ISO 9297, International Organization for Standardization, Geneva.

Suess, M.J. [Ed.] 1982 *Examination of Water for Pollution Control. Volume 2, Physical, Chemical and Radiological Examination*. Pergamon Press, London.

Thomas, L.C. and Chamberlin, G.J. 1980 *Colorimetric Chemical Analysis Methods, 9th edition*. The Tintometer Ltd, Salisbury.

WHO 1992 *GEMS/WATER Operational Guide*. Third edition, Unpublished WHO document GEMS/W.92.1, World Health Organization, Geneva.

WHO 1993 *Guidelines for Drinking-water Quality. Volume 1 Recommendations*. Second edition, World Health Organization, Geneva.

# Chapter 8

# ADVANCED INSTRUMENTAL ANALYSIS

This chapter describes some of the more advanced instrumental methods which may be used for the determination of nutrients, major ions and trace elements, together with the analytical techniques for total, inorganic and organic carbon. Some of these techniques are particularly useful for the detailed analysis of sediments, particularly suspended sediments (see Chapter 13) and for the chemical analysis of biota (see Chapter 11). Although these techniques have been classified here as "advanced instrumental analysis" some, such as flame photometry, do not require expensive equipment and are often reliable and appropriate methods for water quality monitoring.

## 8.1 Atomic absorption spectrophotometry (AAS)

Atomic absorption spectrophotometry is commonly used in many analytical laboratories for determination of trace elements in water samples and in acid digests of sediment or biological tissues.

*Principle*

While a sample is being aspirated into a flame, a light-beam is directed through the flame into a monochromator and onto a detector that measures the amount of light absorbed by the atomised element in the flame. A source lamp composed of the element of interest is used because each element has its own characteristic wavelength. This makes the method relatively free from spectral or radiation interferences. The amount of energy at the characteristic wavelength absorbed in the flame is proportional to the concentration of the element in the sample over a limited concentration range. Most atomic absorption instruments are also equipped for operation in an emission mode.

*Interferences*

Many metals can be determined by direct aspiration of sample into an air–acetylene flame. So called "chemical" interference occurs when the flame is not hot enough to dissociate the molecules or when the dissociated atoms are oxidised to a compound that will not dissociate further at the flame temperature. Such interferences can sometimes be overcome by adding specific elements or compounds to the sample solution. Dissociation of the

*This chapter was prepared by R. Ballance*

molecules of silicon, aluminium, barium, beryllium and vanadium requires a hotter flame, and nitrous oxide–acetylene is used. Molecular absorption and light scattering caused by solid particles in the flame can cause high absorption values and consequently positive errors. Background correction techniques can be used to obtain correct values.

*Apparatus*

✓ Atomic absorption spectrophotometer consisting of a light source emitting the line spectrum of an element, a device for vaporising the sample, a means of isolating an absorption line and a photoelectric detector with its associated electronic amplifying and measuring equipment.
✓ Burner. The most common type of burner is a premix, but the type of burner head recommended by the manufacturer of the spectrophotometer should be used.
✓ Readout. Most instruments are equipped with either a digital or a null meter readout mechanism, and modern instruments have microprocessors capable of integrating absorption signals over time and linearising the calibration curve at high concentrations.
✓ Lamps. Either a hollow cathode lamp or an electrodeless discharge lamp may be used. A separate lamp is needed for each element being measured.
✓ Pressure-reducing valves are needed to reduce the high pressure of fuel and oxidant gases in the storage tanks to the controlled operating pressure of the instrument.
✓ Vent. A vent located about 30 cm above the burner will remove fumes and vapours from the flame, thus protecting laboratory staff from toxic vapours and the instrument from corrosive fumes. An air flow rate is usually recommended by the manufacturer.

*Reagents*

✓ Air, cleaned and dried through a suitable filter to remove oil, water and other foreign substances.
✓ Acetylene, standard commercial grade.
✓ Metal-free water is essential for the preparation of all reagents and calibration standards.
✓ Calcium solution. Dissolve 630 mg calcium carbonate, $CaCO_3$, in 50 ml of 1+5 HCl. If necessary, boil gently to obtain complete solution. Cool and dilute to 1,000 ml with water.
✓ Hydrochloric acid, HCl, 1 per cent, 10 per cent, 20 per cent, 1+5, 1+1 and concentrated.
✓ Lanthanum solution. Dissolve 58.65 g lanthanum oxide, $La_2O_3$, in 250 ml concentrated HCl. Add acid slowly until the material is dissolved and dilute to 1,000 ml with water.
✓ Hydrogen peroxide, 30 per cent.
✓ Nitric acid, $HNO_3$, 2 per cent, 1+1 and concentrated.

*Preparation of standards*

Prepare standard solutions of known metal concentrations in water with a matrix similar to the sample. Standards should bracket the expected sample concentration and must be within the method's working range. Very dilute standards should be prepared daily from standard stock solutions having concentrations of at least 100 mg $l^{-1}$ (which can be obtained from commercial suppliers). If sample digestion is used, standards should be carried through the same digestion procedures. The standard stock solutions described below have a concentration of 100 mg $l^{-1}$ (1.00 ml ≡ 100 µg).

*Cadmium:* dissolve 0.100 g Cd metal in 4 ml concentrated $HNO_3$, add 8 ml concentrated $HNO_3$ and make up to 1,000 ml with water.

*Calcium:* suspend 0.2497 g $CaCO_3$ in $H_2O$, dissolve with a minimum of 1+1 $HNO_3$, add 10 ml concentrated $HNO_3$ and make up to 1,000 ml with water.

*Chromium:* dissolve 0.1923 g $CrO_3$ in water, add 10 ml concentrated $HNO_3$, make up to 1,000 ml with water.

*Copper:* dissolve 0.100 g Cu metal in 2 ml concentrated $HNO_3$, add 10 ml concentrated $HNO_3$ and make up to 1,000 ml with water.
*Iron:* dissolve 0.100 g Fe wire in a mixture of 10 ml 1+1 HCl and 3 ml concentrated $HNO_3$, add 5 ml concentrated $HNO_3$, make up to 1,000 ml with water.
*Lead:* dissolve 0.1598 g $Pb(NO_3)_2$ in a minimum of $HNO_3$ and make up to 1,000 ml with water.
*Magnesium:* dissolve 0.1658 g MgO in a minimum of 1+1 $HNO_3$, add 10 ml concentrated $HNO_3$ and make up to 1,000 ml with water.
*Manganese:* dissolve 0.1000 g Mn metal in 10 ml concentrated HCl mixed with 1 ml concentrated $HNO_3$, make up to 1,000 ml with water.
*Nickel:* dissolve 0.1000 g Ni metal in 10 ml hot concentrated $HNO_3$, cool and make up to 1,000 ml with water.
*Potassium:* dissolve 0.1907 g KCl in water and make up to 1,000 ml with water.
*Sodium:* dissolve 0.2542 g NaCl in water, add 10 ml concentrated $HNO_3$ and make up to 1,000 ml with water.
*Tin:* dissolve 1.000 g Sn metal in 100 ml concentrated HCl and make up to 1,000 ml with water.
*Zinc:* dissolve 1.000 g Zn metal in 20 ml 1+1 HCl and make up to 1,000 ml with water.

*Procedure*
It is not possible to provide an operating procedure that would be correct for all atomic absorption spectrophotometers because of differences between makes and models of instrument. The manufacturer's operating manual should be followed. A general procedure contains three components as described below.

*Zero the instrument*

1. Install a hollow cathode lamp for the desired element in the instrument and set the wavelength dial to the setting appropriate for the element.
2. Set the slit width according to the manufacturer's suggested value for the element being measured.
3. Turn on the instrument and adjust the lamp current to the level suggested by the manufacturer.
4. Let the instrument warm up, 10–20 minutes, and readjust current as necessary.
5. Adjust wavelength dial until optimum energy gain is obtained.
6. Align lamp in accordance with the directions in the operating manual.
7. Install suitable burner head and adjust its position.
8. Turn on air. Adjust flow to the rate recommended to give maximum sensitivity for the metal being measured.
9. Turn on acetylene. Adjust flow to recommended rate, ignite and allow a few minutes for the flame to stabilise.
10. Aspirate a blank of deionised water that has been given the same treatment and acid concentration as the standards and samples.
11. Zero the instrument.
12. Aspirate a standard solution. Adjust the aspiration rate to obtain maximum sensitivity. Adjust the burner horizontally and vertically to obtain maximum response.

*Prepare calibration curve*

13. Select at least three concentrations of each standard metal solution. There should be one concentration greater and one less than that expected in the sample(s).
14. Aspirate a blank and zero the instrument.

15. Aspirate each standard in turn into the flame and record the absorbance.
16. Prepare a calibration curve by plotting the absorbance of the standards against their concentrations. This step is not necessary for instruments with direct concentration readout.

*Analysis of samples*

17. Rinse nebuliser by aspirating with water containing 1.5 ml $HNO_3$ per litre. Atomise blank and zero the instrument.
18. Atomise a sample and determine its absorbance.
19. Change lamps and repeat the procedure for each element.

*Calculation*

Refer to the appropriate calibration curves and determine the concentration of each metal ion, in µg $l^{-1}$ for trace elements and in mg $l^{-1}$ for the more common metals. Concentrations may be read directly from instruments with a direct readout capability. If a sample has been diluted, apply the appropriate dilution factor.

## 8.2 Gas chromatography

Gas chromatography is a highly sophisticated analytical procedure. It should be used only by analysts experienced in the techniques required and competent to evaluate and interpret the data.

*Principle*

In gas chromatography (GC) a carrier phase (a carrier gas) and a stationary phase (column packing or capillary column coating) are used to separate individual compounds. The carrier gas is nitrogen, argon–methane, helium or hydrogen. For packed columns, the stationary phase is a liquid that has been coated on an inert granular solid (the column packing) that is held in a length of borosilicate glass tubing. The column is installed in an oven with the inlet attached to a heated injector block and the outlet attached to a detector. Precise and constant temperature control of the injector block, oven and detector is maintained. Stationary phase material and concentration, column length and diameter, oven temperature, carrier gas flow and detector type are the controlled variables.

When the sample solution is introduced into the column the organic compounds are vaporised and moved through the column by the carrier gas. They travel through the column at different rates depending on differences in partition coefficients between the mobile and stationary phases.

*Interferences*

Some interferences occur because of sample, solvent or carrier gas contamination or because large amounts of a compound were injected into the GC and some of it lingered in the detector. Methylene chloride, chloroform and other halocarbon and hydrocarbon solvents are frequent contaminants. These solvents should not be used anywhere in the vicinity of the equipment. Another contaminant is sulphur; interference may be eliminated by adding a small amount of copper filings or mercury to samples to precipitate sulphur as metallic sulphide.

There may also be sources of interference within the equipment itself. Septum bleed occurs when silicon compounds used in the construction of the septum on the injection port "bleed" from the heated septum. This can be prevented by septum sweep — passing clean carrier gas over the septum to flush out bleed compounds. Column bleed can occur when column temperatures are high and water or oxygen is introduced into the system. Solvent injection can damage the stationary phase and some organic compounds can degrade the column coating. Injection of certain surface-active agents can completely destroy GC columns.

Ghost peaks can occur because a sample that has been passed through the system contained either a large quantity of a given compound or a compound that adsorbed to the column coating. Measurements on subsequent sample(s) will show a peak resulting from the residue of the previous sample. This can be avoided by selecting a column coating that precludes such interactions or by flushing out the system with a solvent blank between samples.

*Detectors*

Various types of detector are available for use with GC systems. Brief descriptions of some of these are provided here for general information.

The *electrolytic conductivity detector* contains reference and analytical electrodes, a gas–liquid contactor and a gas–liquid separator. The conductivity solvent enters the cell and flows by the reference electrode. It combines with the gaseous reaction products in the gas–liquid contactor. This mixture is separated into gas and liquid phases in the gas–liquid separator with the liquid phase flowing past the analytical electrode. The electrometer monitors the difference in conductivity between the reference electrode and the analytical electrode. Only organic compounds containing halogen, nitrogen, sulphur or nitrosamine can be detected in this way.

The *electron capture detector* (ECD) is operated by passing the effluent from the gas chromatographic column over a radioactive beta-particle emitter, usually nickel-63 or tritium, adsorbed on platinum or titanium foil. An electron from the emitter ionises the carrier gas and produces a burst of electrons. About 100 secondary electrons are produced for each initial beta-particle. After further collisions the energy of these electrons is reduced to the thermal level and they can be captured by electrophilic sample molecules. The electron population is collected by applying a voltage pulse to the cell electrodes, and the pulse interval is automatically adjusted to maintain a constant current. The change in the pulse rate when a sample enters the detector is related to the concentration of contaminant in the sample. The detector is highly sensitive to molecules containing halogens, peroxides, quinones and nitro groups but is insensitive to functional groups such as amines, alcohols and hydrocarbons.

The *flame ionisation detector* (FID) consists of a small hydrogen/air diffusion flame burning at the end of a jet. When organic compounds enter the flame from the column, electrically charged intermediates are formed. These are collected by applying a voltage across the flame; the resulting current is measured after amplification by an electrometer. The FID is sensitive to nearly all organic carbon-containing compounds but does not respond to carrier gas impurities such as water and carbon dioxide. It has a wide linear response range, is relatively insensitive to small changes in flow rate, and is reliable, rugged and easy to use. It is, however, a destructive detector that changes irreversibly the physical and chemical characteristics of the sample. Nevertheless, the FID is probably the most widely used detector for gas chromatography.

The *photoionisation detector* (PID) detects organic and some inorganic species in the effluent of a gas chromatograph, with detection limits in the picogram range. The detector is equipped with an ultraviolet light source that emits photons that pass through an optically transparent window into an ionisation chamber where they are absorbed by the eluted species. Compounds with ionisation potential less than the UV source are ionised. A positively biased high-voltage electrode accelerates the resulting ions to a collecting electrode. An electrometer measures the resulting current, which is proportional to the concentration of the eluted species. The PID has high sensitivity, low noise and excellent linearity, is non-destructive and can be used in series with a second detector for a more selective detection.

The *mass spectrophotometer* (MS) has the ability to detect a wide variety of compounds, coupled with the capacity to deduce compound structures from fragmentation patterns. It detects compounds by ionising molecules into charged species with a 7-eV beam. The ions are accelerated towards a quadrupole mass filter through a series of lenses and the differently sized charged fragments are separated according to mass-to-charge ratio. A computer control permits fragments of only one mass-to-charge ratio to pass at any one time and be detected by an electron multiplier. Most chemicals have a unique fragmentation pattern (mass spectrum), and the computer searches an internal library of known mass spectra to identify an unknown compound exhibiting a particular spectrum.

*Procedure*

It is not possible to provide a procedure for gas chromatography that would be applicable to all of the commercially available instruments. The instruction manual supplied by the manufacturer will deal with the specific characteristics of the particular model to which it refers. It must be emphasised that the technique is sophisticated and technicians should undergo a period of training in the use of the equipment. Appropriate training can normally be provided by the manufacturer of the equipment.

## 8.3 Flame photometry

Flame photometry makes possible the determination of trace amounts of lithium, potassium, sodium and strontium, although other methods of analysis for lithium and strontium are preferred.

*Principle*

The sample, after dilution if necessary, is sprayed into a butane–air or propane–air flame. The alkali metals absorb energy from the flame and become raised to an excited energy state in their atomic form. As these individual atoms "cool" they fall back into their original unexcited or ground state and re-emit their absorbed energy by radiation of specific wavelengths, some of which are within the visible region of the electromagnetic spectrum. This discrete emission is isolated by an optical filter and, for low concentrations, is proportional to the number of atoms returning to the ground state. This, in turn, is proportional to the number of atoms excited and, hence, to the concentration of the element in the solution.

*Minimum detectable concentration and range*

The minimum detection level for both potassium and sodium is approximately 100 $\mu g\ l^{-1}$. The upper limit is approximately 10.0 $mg\ l^{-1}$, but this may be extended by diluting the samples.

*Apparatus*

- ✓ Flame photometer, either direct reading or internal standard type, or an atomic absorption spectrophotometer in the flame emission mode.
- ✓ Normal laboratory glassware. Rinse glassware with 1+15 $HNO_3$ followed by several flushes with deionised distilled water.

*Reagents*

- ✓ Deionised distilled water. Use deionised distilled water to prepare all calibration standards and reagents and as dilution water.
- ✓ Stock sodium solution. Dissolve 2.542 g NaCl dried at 140 °C and dilute to 1,000 ml with water (1.00 ml ≡ 1.00 mg Na).
- ✓ Intermediate sodium solution. Dilute 10.00 ml stock sodium solution to 100 ml with water (1.00 ml ≡ 100 µg Na). This solution is used to prepare the calibration curve in the range 1 to 10 $mg\ l^{-1}$ sodium.
- ✓ Standard sodium solution. Dilute 10.00 ml intermediate sodium solution to 100 ml with water (1.00 ml ≡ 10.0 µg Na). This solution is used to prepare the calibration curve in the range 0.1 to 1.0 $mg\ l^{-1}$ sodium.

- ✓ Stock potassium solution. Dissolve 1.907 g KCl dried at 110 °C and dilute to 1,000 ml with water (1.00 ml ≡ 1.00 mg K).
- ✓ Intermediate potassium solution. Dilute 10.00 ml stock potassium solution to 100 ml with water (1.00 ml ≡ 100 µg K). This solution is used to prepare the calibration curve in the range 1 to 10 mg $l^{-1}$ potassium.
- ✓ Standard potassium solution. Dilute 10.00 ml intermediate potassium solution to 100 ml with water (1.00 ml ≡ 10.0 µg K). This solution is used to prepare the calibration curve in the range 0.1 to 1.0 mg $l^{-1}$ potassium.

*Procedure*

It is impossible to provide detailed operating instructions that would be universally correct because of differences between the various models of instruments made by different manufacturers. Follow the recommendations contained in the operating manual supplied with the instrument. Select the correct photocell and wavelength (589 nm for sodium, 766.5 nm for potassium), slit width and sensitivity, and fuel and air (or oxygen) pressures. Follow recommended steps for warm-up, making corrections for interferences and flame background, rinsing of burner, igniting of sample and measuring emission intensity.

*Calibration*

1. Prepare a blank and a stepped set of sodium calibration standards in the range 0 to 1.0 mg $l^{-1}$ or 0 to 10.0 mg $l^{-1}$ as appropriate.
2. Starting with the highest calibration standard and working towards the blank, measure emission at 589 nm. Repeat this step enough times to obtain a reliable average reading for each calibration standard.
3. Construct a calibration curve from the sodium standards.
4. Prepare a blank and a stepped set of potassium calibration standards in the range 0 to 1.0 mg $l^{-1}$ or 0 to 10.0 mg $l^{-1}$ as appropriate.
5. Starting with the highest calibration standard and working towards the blank, measure emission at 766.5 nm. Repeat this step enough times to obtain a reliable average reading for each calibration standard.
6. Construct a calibration curve from the potassium standards.

*Measurement*

7. Spray a sample into the flame and measure emission at 589 nm. Repeat enough times to obtain a reliable average value. Determine the concentration of sodium from the sodium calibration curve.
8. Spray a sample into the flame and measure emission at 766.5 nm. Repeat enough times to obtain a reliable average value. Determine the concentration of potassium from the potassium calibration curve.

*Precision and bias*

A synthetic sample containing sodium 19.9 mg $l^{-1}$, potassium 3.1 mg $l^{-1}$, calcium 108 mg $l^{-1}$, magnesium 82 mg $l^{-1}$, chloride 241 mg $l^{-1}$, nitrite nitrogen 0.25 mg $l^{-1}$, nitrate nitrogen 1.1 mg $l^{-1}$, sulphate 259 mg $l^{-1}$ and total alkalinity 42.5 mg $l^{-1}$ was analysed in 35 laboratories by the flame photometric method with the following results:

Sodium: relative standard deviation 17.3 per cent, relative error 4.0 per cent.

Potassium: relative standard deviation 15.5 per cent, relative error 2.3 per cent.

## 8.4 Total, organic and inorganic carbon

The following definitions apply to this procedure:

- Total carbon (TC) is all of the carbon present as dissolved matter and/or in suspension in the water.
- Total inorganic carbon (TIC) is all of the carbon present as inorganic matter, dissolved and/or in suspension in the water.
- Total organic carbon (TOC) is all of the carbon present as organic matter, dissolved and/or in suspension in the water.

Measurement of TOC is a much more rapid means of determining the organic content of water and wastewater than is the measurement of biochemical oxygen demand (BOD). In addition, two of the methods also provide more rapid measurement than the chemical oxygen demand (COD) test. Because of the presence of non-biodegradable organic compounds, BOD is not directly related to total organic carbon, and COD analyses may include reduced inorganic compounds. However, if the relative concentrations of organic compounds in the samples do not change greatly, empirical relationships can be established between TOC and BOD or COD to permit speedy and convenient estimations of the latter. Measurement of TOC can be used to monitor processes for the treatment or removal of organic contaminants without undue dependence on the oxidation states, and is valid at low concentrations.

The concentration of organic carbon present in surface water is generally less than 10 mg $l^{-1}$, except where a high concentration of municipal or industrial waste is present. Higher levels of organic carbon may be encountered in highly coloured water, and water collected from swamps may have organic carbon concentrations exceeding 100 mg $l^{-1}$. For municipal wastewater treatment plants, influent TOC concentrations may reach several hundred milligrams per litre, but effluent concentrations from a secondary treatment facility are typically less than 50 mg of organic carbon per litre.

*Sample handling*

Collect and store samples in bottles made of glass, preferably brown. Plastic containers are acceptable after tests have demonstrated the absence of extractable carbonaceous substances. Samples that cannot be examined promptly should be protected from decomposition or oxidation by preservation at 0–4 °C, minimal exposure to light and atmosphere, or acidification with sulphuric acid to a pH not greater than 2. Storage time should be kept to a minimum under any conditions. It should not exceed seven days and, depending on the type of sample, even shorter storage may be indicated.

*Methods*

So many techniques and types of equipment exist for the analysis of organic carbon that it is impossible to recommend a procedure applicable in all situations. The advice that follows is that given by the French Standardisation Association (Association Française de Normalisation (AFNOR), Paris) for selection of an appropriate analytical procedure. Sections 8.4.1, 8.4.2 and 8.4.3 describe how some of these recommendations are related to the type of water to be analysed.

*Principle*

The principle of all methods for the determination of total carbon (TC) in water is oxidation of the carbon to carbon dioxide ($CO_2$). Oxidation may be carried out by combustion, chemical reaction by the wet method using appropriate oxidising agents, UV irradiation or any other appropriate procedure. The carbon dioxide formed may be determined directly, or indirectly following reduction to another component (methane, for example).

Various analytical methods have been suggested, some of which are:

- IR spectrometry,
- volumetric determination,
- thermal conductivity,
- conductimetric measurement,
- coulometric measurement,
- specific $CO_2$ electrode,
- flame ionisation following methanisation.

*Principal forms of carbon in water*

A water sample may contain variable amounts of:

- dissolved and particulate organic carbon,
- organic carbon originating from more or less volatile substances, and
- dissolved mineral carbon (carbonates, carbon dioxide) and particulate carbon (active charcoal).

The different matrices of the specimens that result from the presence of these forms of carbon in variable proportions must be taken into consideration before the analysis, because they largely determine what apparatus and procedure to select.

*Selection of procedure in relation to the matrix of the sample*

A. Presence of dissolved carbonates and carbon dioxide

When the carbon derived from dissolved carbonates and $CO_2$ is considerably in excess of the TOC, estimation by the separate measurement of total carbon and mineral carbon to arrive at the arithmetic difference may be somewhat inaccurate because of the errors connected with each measurement. Carbonates and $CO_2$ may be eliminated by acidification at pH 1 with $H_3PO_4$ followed by degassing in a stream of gas which is free of $CO_2$ and organic components. Thereafter, TOC is determined in the degassed sample. In this case it is important to check the efficiency of the degassing apparatus of each user (gas flow, duration of degassing and geometry of the degasser).

B. Presence of particulate mineral carbon and suspended matter

Using the proposed techniques, it is impossible to ensure total oxidation of particles by low-temperature wet methods. In addition, the difficulties inherent in the sampling of heterogeneous media still persist.

C. Presence of volatile substances

Pretreatment by degassing modifies the initial composition of the sample by almost completely eliminating the volatile components. The presence of surfactants may interfere with degassing.

D. Presence of dissolved mineral matter

Depending on the method used, the dissolved mineral load may interfere with the operation of the apparatus (for example, the sooting of UV lamps in the photochemical technique; clogging of the catalytic mass of the furnaces in high-temperature and low-temperature oxidation procedures; precipitation of calcium salts; release of chlorine resulting from the oxidation of chlorides).

*Note:* Depending on the method, the carbon present in some inorganic compounds (cyanides, cyanates, thiocyanates, carbon disulphide, etc.) will be regarded as organic carbon.

E. Effect of the nature of the inorganic matter

The oxidation of organic matter is related to its structure, and the degree to which complete oxidation is achieved will depend on the analytical method used.

*Apparatus*

✓ Ordinary laboratory glassware. All the glassware used must be kept perfectly clean in order to limit contamination and must be reserved exclusively for this determination. Regular verification of the quality of the glassware using water of satisfactory purity is needed to restrict contamination by "dirty" glassware. Several cleaning techniques have proved to be effective, including the following:

(i) Leave glassware to soak for several (4 12) hours in a chromic acid mixture (concentrated $H_2SO_4$ saturated in $K_2Cr_2O_7$) or an effective laboratory detergent solution.

(ii) Rinse copiously with the water used to prepare the reagents.

(iii) Allow to dry away from dust, and keep the containers closed.

Should this treatment be inadequate, non-graduated glassware may be raised to a temperature of 350–400 °C for 2 hours in an oxidising atmosphere.
Apparatus for determination of TOC must be set up in a way that will limit outside contamination by carbonaceous compounds.

*Reagents*

All the reagents must be of recognised analytical quality. The total concentration of carbon in the water used for dilution and to prepare the standards must be negligible by comparison with the total carbon of the sample and the standards.

✓ Stock solution of potassium hydrogen phthalate with a carbon content of 1,000 mg $l^{-1}$. In about 700 ml of fresh distilled water in a 1-litre volumetric flask, dissolve 2.125 g of potassium hydrogen phthalate, $C_8H_5KO_4$, that has first been dried for 2 hours at a temperature not exceeding 120 °C. Make up to the 1-litre mark with distilled water. This solution is stable for about a month if stored at 4 °C in a stoppered, opaque glass bottle.

✓ Calibration solutions. These solutions, which must be made up daily, are prepared by diluting the stock solution of potassium hydrogen phthalate.

✓ Control solution with a carbon content of 1,000 mg $l^{-1}$ for determination of total inorganic carbon. In a 1,000-ml graduated flask containing 500 ml of water, dissolve 4.42 g of sodium carbonate, $Na_2CO_3$, that has first been dried at 120 °C for 1 hour. Add 3.497 g of sodium bicarbonate, $NaHCO_3$, that has been stored in a desiccator, and make up to 1,000 ml with water. This solution remains stable for about 1 month at 4 °C.

✓ Concentrated phosphoric acid, $H_3PO_4$. Used to acidify the sample before degassing (in general 2 to 3 drops in 20 ml of water).

✓ Gases. The gases used must be free of $CO_2$ and organic impurities.

*Procedure*

1. Follow the manufacturer's instructions for the use of the equipment in the analysis of TOC. Before TOC is measured, care must be taken to ensure that the control tests are carried out at least as frequently as recommended by the manufacturer.

2. The equipment must be calibrated daily. A calibration curve is produced by the use of potassium hydrogen phthalate solutions at appropriate concentrations (4 or 5 points for each measurement range when the equipment permits).

3. As with all measurements made by instruments, there are many variables that affect the total response of the system. The two main variables to be verified for all technologies are oxidation chemistry and detection. In the first of these, because of the disparity of the physicochemical principles involved, no general rules for verificatory procedures can be given. For the second, the course of the process may be assessed by computer visualisation of the output signal (background noise, shape of peak, return to base-line and replicability).

*Calculation of the results*

Determine the concentrations of TOC in the specimens using the calibration curve or the data processing system for the instrument concerned.

### 8.4.1 Surface waters and water intended for human consumption or domestic use

The characteristics of surface waters and water intended for human consumption or domestic use are:

- low TOC concentrations (of the order of 0.1 to 10 mg $l^{-1}$), with the proportion of the carbon derived from volatile organic compounds generally being slight as compared with TOC,
- a content of dissolved salts that does not appreciably affect the operation of the apparatus,

- a concentration of mineral carbon (carbonates and bicarbonates) that is generally greatly in excess of the TOC, and
- small amounts of suspended matter.

*Choice of technique*

For TOC concentrations greater than 2 mg $l^{-1}$, any of the various analytical procedures is acceptable provided that normal care is used in their performance. Precise measurement of TOC is more difficult for concentrations below 2 mg $l^{-1}$. Laboratory experience has shown that, of the apparatus currently available, photochemical systems are preferable for analysis of waters of the type under consideration. The contribution of carbon derived from volatile organic compounds (carbon usually less than 1,000 μg $l^{-1}$) may be regarded as insignificant compared with measurement errors.

In methods that depend on the difference between TC and TIC, the measurement errors for TC and TIC are at least 5 per cent for waters containing about 50 mg TC per litre or about 200 mg of carbonates per litre. Consequently the calculated value of TOC (TC – TIC) will not agree with the measured value of TOC.

### 8.4.2 Seawater or water with a high content of dissolved salts

These recommendations are applicable to seawater or water with a high content of dissolved salts. The characteristics of these waters are:

- a high chloride content,
- a low TOC concentration, and
- a high TIC concentration.

*Choice of technique*

The main difficulty arises from the presence of chlorides which interfere with the operation of the apparatus, either by the destruction of catalyst masses and corrosion of high-temperature furnaces, or by considerable modification of oxidation kinetics in photochemical oxidation techniques that necessitate the use of a special oxidising medium.

In general, the addition of a soluble mercury salt is recommended to complex the chlorides and thus prevent their oxidation.

### 8.4.3 Wastewater and surface water with a high TOC content

These recommendations apply to wastewater and surface waters with a high TOC content. The characteristics of these waters are:

- a high content of organic matter,
- a large amount of suspended matter, and
- more frequent presence of volatile compounds.

*Choice of technique*

The main difficulty arises from suspended matter because of the heterogeneity of the medium (which makes representative sampling a difficult operation when the volume of the analytical sample is small) and mechanical incompatibility of solid particles with some parts of the apparatus (valves, etc.).

Recognising the diverse origins of organic matter, some substances, especially in suspension, may prove difficult to oxidise by photochemical oxidation techniques (incomplete oxidation is often revealed by protracted peaks). It is sometimes possible to overcome this problem by modifying the oxidising medium and increasing the length of the reaction time. No clear choice of technique is possible. However, techniques employing a high-temperature furnace system are usually best.

## 8.5 Source literature and further reading

AFNOR 1987 *Guide pour la Détermination du Carbone Organique Total.* Association Française de Normalisation, Paris (T90-102).

APHA 1992 *Standard Methods for the Examination of Water and Wastewater*, 18th edition. American Public Health Association (APHA), American Water Works Association (AWWA) and Water Pollution Control Federation (WPCF), Washington, DC.

ISO 1990 *Water Quality — Guidelines for the Determination of Total Organic Carbon (TOC)*. International Standard ISO 8245, International Organization for Standardization, Geneva.

WHO 1992 *GEMS/WATER Operational Guide*. Third edition, Unpublished WHO document GEMS/W.92.1, World Health Organization, Geneva.

# Chapter 9

# ANALYTICAL QUALITY ASSURANCE

This chapter outlines the various techniques that may be employed in analytical quality assurance (AQA), and the reasons why they should be used. Reliability of data for a water quality monitoring programme depends on strict adherence to a wide range of operating procedures for water sampling and analysis. It is the consistent application and monitoring of these procedures that is referred to as quality assurance. The subject can be confusing, especially if more than one reference work is used as an information source. Different authors may use different terms to describe the same thing or the same term to describe different things. A number of the terms used in analytical quality assurance are defined in Table 9.1. The definitions are based on the vocabulary approved by the International Organization for Standardization (ISO) but may not have been universally adopted by those involved in AQA.

In order to demonstrate that a laboratory is producing data of adequate precision, accuracy and sensitivity it is necessary to assess all laboratory procedures at all stages from sampling to reporting. This is a time consuming and costly process and, for this reason, it is important to ensure that the necessary standards of performance are clearly defined and adhered to. In most laboratories, AQA will start with the examination and documentation of all aspects of laboratory management. This will include clearly identifying lines of communication and responsibility, the description and documentation of all procedures which are carried out, and the documentation of instrumental and analytical checks. Within this there should be specific control and assessment procedures designed to monitor quantitatively the accuracy and precision of specific assays.

Analytical quality assurance procedures should be based on a system of traceability and feedback. Traceability, in this context, requires that all steps in a procedure can be checked, wherever possible, by reference to documented results, calibrations, standards, calculations, etc. For example, where a balance is used in a laboratory, the accuracy of measurement must be regularly checked. The weights used for this purpose should either have a certificate demonstrating that they conform to a standard, or the balance must

*This chapter was prepared by R. Briggs*

**Table 9.1** Definitions associated with analytical quality assurance

| Term | Definition |
|---|---|
| Quality | The totality of characteristics of an entity that bear on its ability to satisfy stated and implied needs. |
| Quality policy | The overall intentions and direction of an organisation with regard to quality, as formally expressed by top management. The quality policy forms one element of corporate policy and is authorised by top management. |
| Quality assurance | All the planned and systematic activities implemented within the quality system and demonstrated as needed, to provide adequate confidence that an entity will fulfil requirements for quality. |
| Quality system | Organisational structure, procedures, processes, and resources needed to implement quality management. |
| Organisational structure | The responsibilities, authorities and relationships through which an organisation performs its functions. |
| Procedure | A specified way to perform an activity. When a procedure is documented, the terms "Standard Operating Procedure", "written procedure" or "documented procedure" are frequently used. A documented procedure usually contains the purposes and scope of an activity; what shall be done and by whom; when, where and how it shall be done; what materials, equipment and documents shall be used; and how it shall be controlled and recorded. |
| Process | A set of inter-related resources and activities that transform inputs into outputs. Resources may include personnel, finance, facilities, equipment, techniques and methods. |
| Quality management | All activities of the overall management function that determine the quality policy, objectives and responsibilities, and implement them by means such as quality planning, quality control, quality assurance, and quality improvement within the quality system. |
| Quality control | Operational techniques and activities that are used to fulfil requirements for quality. The terms "internal quality control" and "external quality control" are commonly used. The former refers to activities conducted within a laboratory to monitor performance and the latter refers to activities leading to comparison with other reference laboratories or consensus results amongst several laboratories. |
| Quality audit | Systematic and independent examination to determine whether quality activities and related results comply with planned arrangements and whether these arrangements are implemented effectively and are suitable to achieve objectives. |
| Traceability | Ability to trace the history, application or location of an entity by means of recorded identifications. In the context of calibration, it relates measuring equipment to national or international standards, primary standards, basic physical constants or properties, or reference materials. In the context of data collection, it relates calculations and data generated back to the quality requirements for an entity. |

be regularly checked against such standards by the regular use of check weights which are well documented and thus can be linked within the laboratory to the calibration standard. This principle also applies to the calibration of other equipment.

Feedback is the principle that problems or omissions in the AQA system should be brought to the attention of management. Where standards in the laboratory fall below acceptable limits, procedures should ensure that this is easily recognised and corrected. Criteria for recognition and correction of poor performance, as well as responsibilities for corrective action, must be identified. The procedures for achieving this recognition and correction must be clearly established.

Statistically based assay control systems, as used in internal and external quality control programmes, should also conform to the principles of traceability and feedback to ensure that correct criteria for adequate quality are adopted, and that any problems are quickly recognised and corrected.

Properly implemented AQA should demonstrate the standard to which a laboratory is working, ensure that this standard is monitored effectively and provide the means to correct any deviations from that standard. It is sometimes argued that the value of quality assurance does not justify its cost but, without it, the reliability of data is doubtful and money spent on producing unreliable data is wasted. If 10 per cent of a laboratory's budget is spent on quality assurance, the number of samples that can be analysed will be about 90 per cent of that possible if there were no quality assurance programme. However, the results obtained for that 90 per cent will be accurate, reliable and of consistent value to the monitoring programme.

## 9.1 Quality assurance

Quality assurance (QA) refers to the full range of practices employed to ensure that laboratory results are reliable. The term encompasses internal and external quality control, but these specific aspects of AQA will be covered later. Quality assurance may be defined as the system of documenting and cross referencing the management procedures of the laboratory. Its objective is to have clear and concise records of all procedures which may have a bearing on the quality of data, so that those procedures may be monitored with a view to ensuring that quality is maintained.

Quality assurance achieves these objectives by establishing protocols and quality criteria for all aspects of laboratory work, and provides a framework within which internal quality control (IQC) and external quality control (EQC) programmes can be effective. It is primarily a management system, and as such is analyte-independent, because it deals with the overall running of the laboratory rather than focusing on individual analyses.

Aspects of QA which are specifically applicable to microbiological analyses of water samples are described in Chapter 10.

### 9.1.1 Components of quality assurance

Given the wide scope of quality assurance, the definition given above provides only the haziest picture of what is required to implement a QA programme. In order to provide a fuller explanation, it is more pertinent to study the components of a quality assurance programme and to examine the procedures that are required for each one.

*Management*

One of the most important components of the quality assurance programme in a laboratory are the comprehensive management documents which should describe, in detail, the management structure of the laboratory. Such documentation should provide clearly defined communication channels and a clear reporting structure. Within that structure each member of staff should be able to locate his or her own job description and responsibilities and their relationship with other staff members who are subordinate or superior. From a stable management base all the other components of quality assurance can be put in place. Without this the level of control necessary to ensure that all other components are effective is impossible.

Management documents should specify the role of quality assurance within the laboratory and clearly define who is responsible for each area and activity. The documents should also identify the records that should be kept of routine operations, such as equipment calibration and maintenance, thus ensuring that a logical, coherent system of record keeping is adopted. Such documentation should be brought together as a single Quality Manual which will act a reference text for the whole quality assurance programme.

In larger laboratories, proper management of QA will require the appointment of a quality assurance officer to liaise with management, to manage data archives, to conduct regular audits and reviews of the QA system and to report on any QA issues to the programme or institution manager (see also section 4.3.4). The officer is responsible for regularly inspecting all aspects of the system to ensure staff compliance, for reporting on such inspections and audits to management, and for recommending improvements. In practice, this will involve regularly checking facilities and procedures as they are performed and conducting regular audits, by tracing an analytical sample back through the system from report to sample receipt, and ensuring that all appropriate records have been kept.

The QA officer's duties must be clearly defined within the management documents in order for the role to be effective. Appointment of a quality assurance officer may be difficult in a small laboratory for financial or organisational reasons. In such cases the responsibilities for QA should be delegated to a member of staff. This may create conflicts of interest if the member of staff has to monitor the work conducted in his or her section. Senior management, who are always ultimately responsible for QA, should ensure that such conflicts are minimised.

The QA officer's role should be to monitor the system, to report on any deviations from the system, and to recommend to management any changes that might be required. In order to be able to do this effectively the QA officer should be free from management interference, while remaining responsible to management for undertaking the required duties. As a consequence it is better if a QA officer is in a middle management position, thus allowing effective communication with Laboratory Section Heads. In larger organisations QA is the responsibility of a separate section. In such a situation many of the management difficulties are minimised because the QA section is structured in a similar way to other sections of the organisation. Whichever approach is used, it is necessary that management provide adequate resources for this activity and ensure that all staff are clearly informed of their responsibilities within the QA system.

*Training*

It is important that all staff are adequately trained for the task they have to perform (see also section 4.4). Training must be documented in order that management and other personnel can verify that staff are competent to conduct the duties required of them. The level of training required for each procedure should also be clearly defined to ensure that staff ability and training are matched to procedural requirements. Criteria for the correct levels of training or competence for particular procedures, and job roles, are often specified by national and international agencies and, in some cases, by professional associations. In line with the principle of traceability outlined above, laboratory criteria for training standards should reflect the external criteria which apply. This should be clearly demonstrated in the documentation.

*Standard Operating Procedures*

Standard Operating Procedures (SOPs) provide the core of most of the day to day running of any quality assurance programme. They are the documents describing in detail every procedure conducted by the laboratory. This includes sampling, transportation, analysis, use of equipment, quality control, calibration, production of reports, etc. They are the laboratory's internal reference manual for the particular procedure to which they are dedicated and, for that reason, SOPs must document every relevant step in the procedure. Thus, anyone of the appropriate training grade should be able to apply the procedure when following the SOP. In addition, the SOP must cross reference and, where necessary, expand any other SOPs which are related to it.

Standard operating procedures often cause confusion when first introduced into a laboratory because many people feel that they are not required by virtue of either experience, availability of manuals or the use of papers from the literature or other published references. In practice, an SOP should present the procedure in a way that avoids all potential differences in interpretation, thereby avoiding subtle changes in the way methods are performed or equipment is used. Such differences can, and do, have a marked effect on accuracy and precision. An SOP should be clear, concise and contain all the relevant information to perform the procedure it describes. In addition, it should include the methods and the frequency of calibration, maintenance and quality control, as well as the remedial action to be taken in the event of malfunction or loss of control.

The SOP is the laboratory's reference to a given procedure and, therefore, it must be regularly reviewed and, if necessary, updated. Issue and availability of SOPs should be carefully controlled to ensure that they are used only by appropriately trained staff and to ensure that out of date copies of SOPs do not remain in circulation (thereby defeating their original objective). When a new or amended SOP is published in a laboratory all copies of the old SOP must be taken out of circulation. Consequently, it is necessary to have an issue log for all SOPs in the system, so that all copies of each SOP can be located.

While all procedures require SOPs, it is not necessary to generate new documents where appropriate ones exist. For example, standard analytical methods published by recognised authorities (such as the United States Environmental Protection Agency), or manufacturers manuals for specific pieces of equipment, may be adopted as SOPs if they meet the need of the laboratory and if this is properly documented and sanctioned by the management and, or, the QA officer.

*Laboratory facilities*

Resources are required for regular laboratory work as well as for the additional workload associated with quality assurance (see also section 4.1).

It is essential that these resources, i.e. space, staff, equipment and supplies, are sufficient for the volume of work to be done. Space should be adequate and sufficient equipment should be available to allow the procedures performed in the laboratory to be conducted efficiently. The environment in which the work is conducted must be well controlled. It should be clean and tidy, have adequate space in which to work without risk to personnel or to the analytical sample, and there should be sufficient storage space for glassware, chemicals, samples and consumables. It is also essential that there are adequate numbers of appropriately trained staff available to undertake all the required tasks. Management policy should ensure that these facilities are available before any laboratory work is commenced. In practice, anything that restricts the efficient running of the laboratory would be a cause for concern, and should lead to noncompliance with the quality assurance system.

*Equipment maintenance and calibration*

All equipment must be maintained on a regular basis, consistent with the documented criteria of the laboratory and normally accepted codes of practice. The laboratory must apply standards which are well within the limits normally established and recommended for the care of the particular piece of equipment. This should be checked by the quality assurance officer, and be corrected if inappropriate. These principles apply to general laboratory equipment such as glassware as well as to sophisticated analytical instruments. The care and cleaning of this type of equipment is extremely important to ensure quality and should not be overlooked. Frequent checks on the reliability of equipment must also be performed. This includes calibration checks on all relevant equipment, such as balances, pipettes, etc. The frequency of these checks will depend on the stability of the equipment in question. In some instances calibration checks may be done as a part of normal maintenance. Again, the criteria for checking should be based on established acceptable practice.

Equipment calibration and maintenance records should be kept for all equipment, thus allowing the repair status of each piece of apparatus to be monitored. This reduces the likelihood that malfunctioning equipment will be used for analysis (thereby leading to poor analytical data), and allows any problems with equipment to be more quickly diagnosed and corrected.

*Sampling*

Procedures for sampling operations should be carefully documented. In particular, clear details should be given for precautions to be taken while sampling and the sampling strategies to be employed. Careful documentation during sampling is required so that all relevant information on the nature of the sample (when it was taken, where it was taken and under what conditions it was taken) are clearly recorded on site at the time of sampling by the person conducting the sampling. This is necessary because variations in sampling procedures can have a marked effect on the results of analysis. It is very difficult to quantify these effects and, therefore, the most practical way to control this stage of the analytical process is to document sampling conditions as fully as possible. It is very important to ensure that all relevant information is made available to the analyst. Quality assurance of sampling can be achieved in the following ways:

- Strictly adhere to standard operating procedures for sampling.
- Ensure all equipment is clean and in working order.
- Record all conditions which applied during sampling.
- Take strict precautions to avoid contamination.

Following those simple procedures should help to ensure that the quality of samples matches the quality of analysis.

*Sample receipt, storage and disposal*

Almost as important as proper sampling, is the proper storage of samples prior to analysis. It is important to ensure that the passage of a sample through the laboratory's analytical systems is fully documented, and corresponds to the practices laid down in the relevant SOPs. Equally important are the arrangements for disposal of samples. This should be done when the sample exceeds its stable storage time. With some forms of analysis which are required for legal or for regulatory reasons there may be a requirement to store a suitable aliquot of a sample safely, for a given time, to allow for re-examination should this be considered necessary. The systems in place should take these factors into account.

Procedures for sample handling should ensure that the sample is not compromised. The sample should be logged in and stored in such a way as to minimise its deterioration. The condition of each sample and its storage location should be recorded and, where appropriate, the analyses to which it is to be subjected should also be recorded. Sub-sampling, splitting of the sample to allow for different storage conditions, or sample pretreatment to increase stability must be recorded and the samples clearly and uniquely marked to ensure that no confusion exists about the source and identity of any sample.

*Reporting of results*

The final products of the laboratory are the data that it reports. It, therefore, follows that the efforts of quality assurance are directed towards seeing that these data are suitable for use in an assessment. This includes the final stage of reporting and interpreting the results which have been generated.

The first stage in this process is examination of the data to determine whether the results are fit to report. Data should be examined at many stages in the quality assurance system and no data should be reported from assays that are out of control (see sections 9.2 and 9.3 below). However, once data are ready to report it is important to ensure that they are reported accurately and in a manner that facilitates interpretation. Consequently, it is often necessary to include information which may have a bearing on interpretation, such as that related to the nature of the sample or the analytical procedure which was applied. All such information must be available to the reporting analyst. Reports must be prepared according to an agreed procedure and they must accurately reflect the findings of the study. They should include reference to all calibration and quality control data and to any problems that were encountered during the study (e.g. rejected analytical batches, loss of sample, etc.). All data included should have been comprehensively checked by an experienced analyst.

Many laboratories have a system which requires checking of data records and countersigning of analytical reports to act as a safeguard against erroneous or misleading data leaving the laboratory. This type of system is only effective when conscientiously applied. Automatic signing of reports with minimal checking is all too common and should be avoided.

### 9.1.2 Implementation of quality assurance

The ultimate objective of a QA programme is to ensure that the laboratory functions efficiently and effectively. The benefits in terms of increased reliability of results has already been mentioned. A number of other benefits are also evident. The clear assignment of duties and adherence to written and agreed protocols ensure that staff clearly understand their responsibilities. This allows lines of communication to be clearly identified, making staff management easier. Calibration, maintenance and record keeping, in general, assist laboratory staff to identify developing problems with equipment earlier than would otherwise be the case. In addition, the sources of analytical problems can be more rapidly identified leading to their rapid solution.

The implementation of a QA programme is, in principle, very simple and involves putting in place the components listed above. In practice, this requires a considerable amount of effort and commitment from all staff and takes a long time to set up. A clear plan of action must be formulated and clear objectives and time scales identified for each stage. Staff who are involved with the implementation should be well briefed on what is required of them. It is also wise to ask for the opinion of staff on the proposed QA system, as this ensures that impractical procedures are avoided.

One logical way to tackle the task is first to write the Quality Manual, then to put in place documentation such as SOPs and laboratory records, then to test run the system for a limited period (i.e. three to six months) and then, finally, to conduct a detailed review which identifies successes and failures within the system. This is best done by inspection of key areas such as laboratory records and by conducting audits. An efficient auditing system is to pick data at random and then trace the documentation pertaining to those data back to sampling and sample receipt. Any breaks in the traceability of the data will become apparent as a gap in the linking documentation. Deficiencies that become apparent should be corrected at this stage. The review should also seek to identify and to remove any inefficient or bureaucratic systems which serve no useful purpose.

A common method of implementing a QA programme is to apply for accreditation. Accreditation is the implementation of a QA programme in conformity with a recognised QA system, such as Good Laboratory Practice (GLP) or the National Measurement Accreditation System (NAMAS). Quality Assurance is often linked with accreditation although this does not have to be the case. While implementing QA in this way allows the programme to be independently assessed against an agreed standard, it can be costly. Alternatively, QA can be implemented by reference to international standards such as ISO 9000 without necessarily going to the expense of accreditation. However, commercial, legal or political considerations may require that formal accreditation is adopted by the laboratory because this formally records compliance with a recognised QA system (i.e. compliance that has been validated by an official third party).

### 9.1.3 Checking compliance

In order to maintain the quality assurance system it is necessary to check periodically each area of the laboratory for compliance with the QA system. This involves auditing of the component parts to assess whether they continue to meet the original criteria. As with any other aspect of quality assurance the procedures to be adopted for checking compliance should be formally documented. Reports on all audits should be made available to management, and to the individuals responsible for the work concerned. Deviations from required standards must be corrected immediately. As with any check of such a complicated structure, the audit must be extensive and systematic in order to test every part of the system. Such audits are also better done in a way that makes them hard to predict, thereby minimising abuse of the system.

The audit must also be independent, hence the need for a quality assurance officer who reports directly to the highest level of management. Regular comprehensive audit often requires a large input of resources in order to be effective.

## 9.2 Internal quality control

Internal quality control (IQC) consists of the operational techniques used by the laboratory staff for continuous assessment of the quality of the results of individual analytical procedures. The focus is principally on monitoring precision but also, to a lesser degree, on accuracy. It is necessarily part of the wider quality assurance programme, but differs from it by virtue of the emphasis placed on quantifying precision and accuracy. Whereas quality assurance strives to achieve quality by regulating procedures using what are essentially management techniques, IQC focuses on the individual method and tests its performance against mathematically derived quality criteria.

### 9.2.1 Choice of analytical method

A variety of different analytical methods are usually available for determining the concentration of any variable in a water sample. The choice of method is critical for ensuring that the results of the analysis meet the laboratory's requirements, because different methods have different precisions and sensitivities and are subject to different potential interferences. Consideration must be given to these parameters before a method is chosen, although the technical literature does not always provide adequate information. Nevertheless, a number of standard methods which have procedures described in sufficient detail are available for most of the analytical determinations involved in water quality monitoring. These standard methods frequently include extensive validation data that allows them to be easily evaluated and many are sanctioned by appropriate international or national organisations. It is not recommended, however, that a laboratory purchases equipment and reagents and starts to follow the procedure of a standard method without considering whether the method meets the programme requirements. The performance of a method can be unpredictably affected by many factors which can lead to serious problems. Before any analytical method is put into routine use it is essential that it is properly validated. The following experiments should be performed as a minimum programme of validation.

- *Linearity:* The calibration point should be determined and a linear response curve demonstrated if possible. If the calibrants do not show a linear response, linear transformation of the data should be investigated.

- *Limit of Detection:* The lowest concentration of the variable that can be distinguished from zero with 95 per cent confidence should be determined.
- *Precision:* Within day and between day coefficients of variation should be performed at three concentration levels.
- *Accuracy:* Analysis of reference materials with known concentrations of the variable (i.e. CRMs, see below) or comparison analyses with existing methods in other laboratories where possible.

### 9.2.2 Validity checking

After a method has been validated, found to be suitable and introduced into routine use in the laboratory, it is necessary to ensure that it continues to produce satisfactory results. Validity checks should be made on every batch of samples or at frequent, regular intervals if batches are large or if testing is continuous. Validity checking is an extension of the checks carried out before the method was selected and is intended to confirm regularly the conclusions reached at that time.

*Calibration check*

If a calibration curve is being used, standard solutions should be analysed from time to time within the required range of concentration. The ideal calibration curve is linear within its most useful range, with a regression coefficient of 0.99 or greater. The response of the measuring equipment to the concentration of the variable in a standard solution (in terms of absorbance or some other physical parameter) should be recorded when it is expected that this parameter will be comparable from assay to assay. In addition, the deviation of individual calibration points from the line of best fit can be used to assess the precision of the calibration, which should be within the mean precision limits for the method.

*Use of blanks*

Method blanks and, where possible, field blanks should be analysed with each batch of samples. A method blank consists of reagent water, usually double-distilled water. A field blank is reagent water that has been bottled in the laboratory, shipped with sample bottles to the sampling site, processed and preserved as a routine sample and returned with the routine samples to the laboratory for analysis. The analysis of a blank should not yield a value higher than that allowed by the acceptance criteria. This procedure checks interference and the limit of detection of the assay.

*Recovery checking*

A specimen spiked with a known amount of the variable should be tested in each batch and the closeness of fit to the expected value calculated. In most cases this procedure provides a check on accuracy but, in assays where a variable is extracted from the original matrix (such as in many sample clean-up procedures used prior to chromatographic analysis), it can be used to monitor the extraction step. It is important that the matrix of the spiked specimen matches the real sample matrix as nearly as possible. Many laboratories use real samples with low natural values of the variable for this purpose, spiking them with known amounts of the variable and including both the spiked and natural samples in the same assay batch.

### 9.2.3 Precision and accuracy checks

Precision and accuracy checks are an extension of the validity checking described above. They have been dealt with separately because these checks allow the quality of the assay to be monitored over time using techniques such as control charting that will be described below. The validity checks described in the previous section only allow acceptance or rejection of the assay data to be decided. Precision and accuracy checking should allow slow deterioration of data quality to be identified and corrected before data have to be rejected. This results in increased efficiency and reduced costs for the laboratory.

*Control by duplicate analysis*

Use of duplicate analysis as a method of precision checking has two distinct advantages:

- quality control materials are matrix-matched, and
- the materials are readily available at no extra cost.

Since the samples are analysed using the same method, equipment and reagents, the same bias should affect all results. Consequently, duplicate analyses are only useful for checking precision; they provide no indication of the accuracy of the analyses. Results from duplicate analyses can be used to calculate a relative range value, $R$, by using the equation:

$$R = \frac{(X_1 - X_2)}{(X_1 + X_2)/2}$$

where $X_1$ and $X_2$ are the duplicate results from an individual sample and $X_1 - X_2$ is the absolute difference between $X_1$ and $X_2$. These values are then compared with the mean relative range values previously calculated for the assay during validation. The simplest method of assessment is to use the upper concentration limit (UCL), where UCL = 3.27 × mean $R$ value. When any value is greater than the UCL, the analytical procedure is out of control. This method, although statistically valid, provides no indication of deteriorating precision. A better and more sophisticated approach is to use acceptance criteria based on warning and action limits (as described below).

*Precision control using pooled reference material*

This method has the advantage of providing some monitoring of accuracy but is a viable control only if the material to be used will be stable in storage for a sufficiently long period of time. The reference material is normally prepared by taking previously analysed samples with known concentrations of the variable under investigation, mixing them and aliquoting the resultant pool. The aliquots are then stored in readiness for analysis. A small sample of the aliquots is analysed to determine the mean concentration of the variable, and the standard deviation and the coefficient of variance at that concentration level. Data may be used only if they come from analyses that are in control. This requires that the original pool materials must have been prepared during method validation, and that new materials must be prepared before the old ones are finished.

A typical precision control exercise would involve the analysis of four aliquots from each pool in each of five assays, thus obtaining 20 results. It is important that the material from the pool be analysed at several different times with different batches because between-batch variance is always slightly greater than within-batch variance.

Once 20 or more analyses have been made on this pool of material, the mean and standard deviations of the results are calculated. Any result that is more than three standard deviations from the mean is discarded and both of the statistics are recalculated. The mean is the "target" value and, ideally, will be a close approximation of the true concentration of the variable in the reference material. The mean and standard deviation become the basis of the acceptance criteria for the assay method and may be used to draw up control charts.

At least three separate reference materials with different mean values of variable concentration should be in use at any one time in order to provide control of the analytical method across a range of concentrations. If precision is checked at only one concentration of the variable, it is impossible to detect whether precision is deteriorating at other concentrations. Use of several reference materials also allows their preparation to be staggered so that they become exhausted at different times. This assures greater continuity of control because two or more old pools will still be in use during the first few assays of a new reference material.

Although the monitoring of accuracy by assessing deviation from the reference material mean (target value) is possible, care must be taken because the target value is only an approximation, albeit a good one, of the true value. As reference materials become exhausted and new ones are made, there will be a slow deterioration in accuracy. Accuracy can be safeguarded by regular participation in external quality control exercises and by the regular use of certified reference materials.

*Accuracy control using certified reference materials*

Certified reference materials (CRMs) are matrix-matched materials with assigned target values and assigned ranges for each variable, reliably determined from data obtained by repeated analysis. Target and range values may be generated from data produced by several laboratories using different analytical methods or calculated from data obtained by the use of one analytical method (usually a reference method). Consequently, there may be bias in the target value. The target values assigned to each variable in the matrix in certified reference materials are generally very close to the true value. For some variables, however, there is an appreciable difference in bias between different analytical methods and this may lead to wide assigned ranges. When a laboratory is not using one of the reference methods the "all method" range may be so wide that it is practically meaningless. Certified reference materials are also only practical for variables that are stable in long-term storage.

Since CRMs are prepared and checked under carefully controlled conditions, they are costly to produce and correspondingly expensive to purchase. Some authorities advocate the routine use of CRMs as precision control materials, rather like the pooled materials prepared in-house as described above, but it is more cost effective to use them for the periodic checking of accuracy, in combination with a rigorous IQC programme.

### 9.2.4 Use of control charts

Control charts have been used for recording internal quality control data for many years and are one of the most durable ways of using quality control data. The principle of control charts is that IQC data can be graphically plotted so that they can be readily compared and interpreted. Consequently, a control chart must be easy to use, easy to understand, and easy to act upon.

The Shewhart chart is the most widely used control chart. It is a graph with time (or assay batch) on the x-axis and the concentration of the variable in the reference material on the y-axis. Target, warning and action lines are marked parallel to the x-axis. Data obtained from precision control using reference materials (as described above) are usually plotted on a Shewhart chart. In this application the target line is at the mean concentration of the variable for that specific pool of material, warning lines are placed at two standard deviations to either side of the target line. Provided the distribution is normal, 95 per cent of results from assays in control will fall between the two warning lines. Action lines are normally placed at three standard deviations to either side of the target line and 99 per cent of normally distributed results should be between the action lines. Examples of typical Shewhart charts are shown in Figure 9.1.

In the regular day-to-day use of a Shewhart chart, an aliquot from an appropriate reference material is analysed with every batch of samples and the measured concentration of the variable in the aliquot is plotted on the chart. Normally, no more than 1 in 20 consecutive results should fall outside the warning lines. If this frequency is exceeded, or if a result falls outside the action lines, the method is out of control.

The scatter of the assay results for the reference material around the target line provides an indication of the precision of the method, while the mean of the assay results relative to the target value indicates whether there is any bias (consistent deviation) in the results.

The following general rules give guidance on the action to be taken should the analysis on one or more of the control specimens yield a result that is outside the warning or action lines on the chart.

- A single result outside the warning lines should lead to careful review of data from that analytical batch and two or three subsequent batches.
- Results outside the warning lines more frequently than once every 20 consecutive analyses of control specimens should prompt detailed checking of the analytical method and rejection of the assay data.
- A result outside the action limits should prompt detailed checking of the analytical method and rejection of the assay data.

### 9.2.5 Summary of an internal quality control programme

A summary of the internal quality control programme recommended by the GEMS/Water programme is given below. This programme offers a simple but effective introduction to IQC and is described in more detail in the *GEMS/WATER Operational Guide*.

For each variable:

1. Analyse five standard solutions at six different known concentrations covering the working range to develop a calibration curve or, when a calibration curve already exists, analyse two standard solutions at

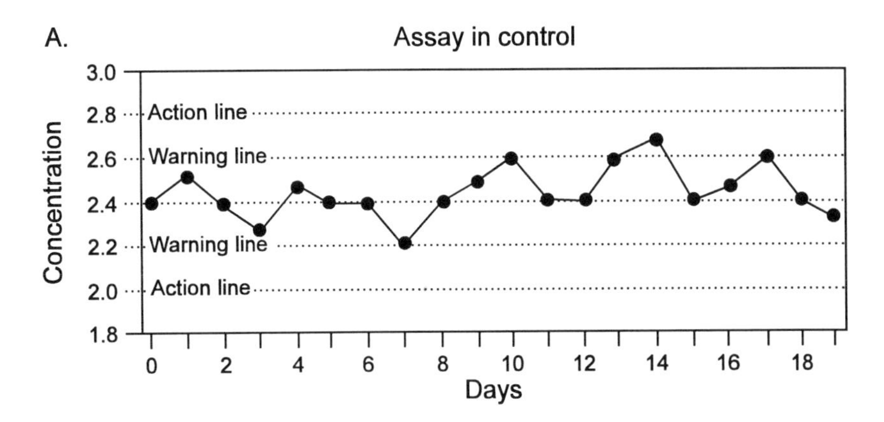


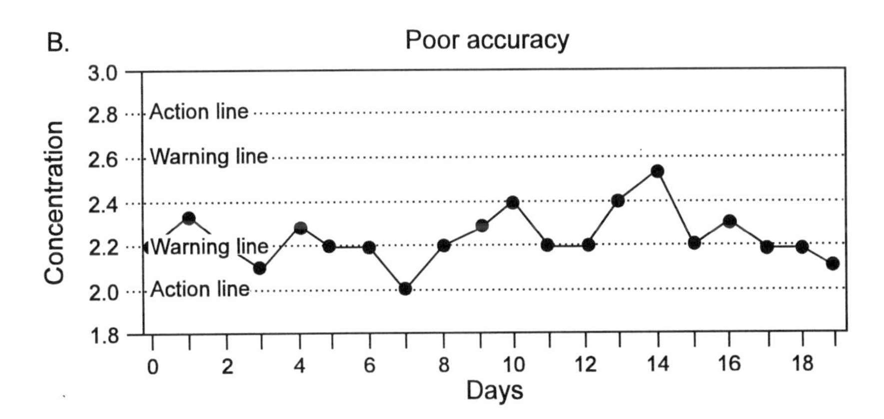


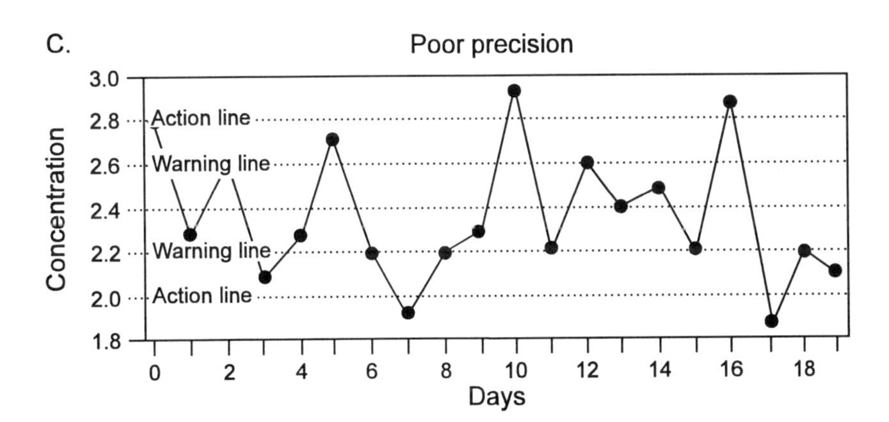


**Figure 9.1** Examples of typical Shewhart charts

different known concentrations covering the working range to validate the existing calibration curve.

2. Analyse one method blank per set of 20 samples.
3. Analyse one field blank per set of samples.
4. Analyse one duplicate of a sample chosen at random from each set of up to 20 samples. Interpret using the UCL method.
5. Analyse one specimen that has been spiked with a known amount of the variable as a recovery check. This specimen should have a matrix similar to those of the samples being processed.

### 9.2.6 Remedial action

If any of the quality control procedures indicates that a method is out of control or that a problem exists, corrective action must be taken. The main checks to make are calculations and records, standard solutions, reagents, equipment and quality control materials (Table 9.2).

## 9.3 External quality control

External quality control (EQC) is a way of establishing the accuracy of analytical methods and procedures by comparing the results of analyses made in one laboratory with the results obtained by others conducting the same analysis on the same material. This is usually accomplished by one laboratory, the reference laboratory, sending out sets of specimens with known and unknown concentrations of variables to all of the participating laboratories. Each participant analyses the specimens for the specified variables and reports the results to the reference laboratory.

The results from all participating laboratories are collated by the organisers of the EQC programme and then subjected to detailed statistical analysis. A report to each laboratory is generated, giving a target value for the reference sample or samples (usually consensus mean or median), a histogram illustrating distribution of results for each material, and an individual performance score relating the individual laboratory results to the target value. The calculations for performance indicators are often quite complex because multiple specimens have to be considered and the method variance varies with the concentration of the variable. However, the general principle of providing a method of performance comparison remains the same in all EQC exercises. An example of an external quality control report is shown in Figure 9.2.

External quality control reports should clearly indicate whether performance is satisfactory or not. If it is not satisfactory, two general actions must be taken. First, the analysis at fault must be examined to determine the cause of poor performance. Once found, the problem must be corrected. Secondly,

**Table 9.2** Necessary checks to be carried out when a problem is detected with an analytical method

| Item | Checks |
|---|---|
| Calculations and records | Check calculations for a transposition of digits or arithmetic errors. Confirm that results have been recorded in the proper units and that any transfer of data from one record to another has been made correctly. |
| Standard solutions | Check the standard solutions that are used for calibrating equipment. Old solutions may have deteriorated and errors may have occurred in the preparation of new ones. Check on storage conditions, the age of solutions and their expected shelf-life. |
| Reagents | Check whether old reagents have deteriorated.<br>Check fresh reagents to ensure that they have been properly prepared.<br>Check the storage conditions of reagents, especially those that must be stored away from the light or at a controlled temperature.<br>Check the shelf-life of reagents, discarding any that are outdated or have been improperly stored. |
| Equipment | Check calibration records and maintenance records for all reagent dispensers and measuring equipment used for the analysis of the variable where the method is out of control. Items such as automatic pipettes, balances and spectrophotometers should be checked and recalibrated if appropriate.<br>Ascertain that equipment is being properly used. |
| Quality control materials | Check on the storage conditions of quality control materials, ensuring that bottles are tightly sealed and that they are not being subjected to extremes of temperature.<br>Run analyses on several aliquots to determine whether the concentration of the variable remains within two standard deviations of the target value and close to the mean of the last 20 determinations. |

the internal quality control programme that allowed the deterioration to progress unchecked must be closely examined to establish where inadequacies exist. Any inadequacies must be corrected.

The general objective of EQC is to assess the accuracy of analytical results measured in participating laboratories and to improve inter-laboratory comparability. For an individual laboratory, participation in an EQC exercise is the only way to ensure that accuracy is independently monitored. Wherever possible, laboratories should participate in EQC programmes for each variable that is routinely analysed. This is only worthwhile where internal quality control is also part of a laboratory's normal procedures. Participation in relevant EQC programmes, and maintenance of adequate performance in those programmes, is often a requirement for laboratory accreditation (see section 9.1.2).

**Number of laboratories producing acceptable, flagged and double flagged results for Group 1 Hard Water.**

| Group 1 Hard Water | Total | Acceptable | Single Flagged | Double Flagged |
|---|---|---|---|---|
| Calcium | 88 | 85 | 1 | 2 |
| Magnesium | 86 | 78 | 5 | 3 |
| Total Hardness | 90 | 86 | 1 | 3 |
| Alkalinity | 91 | 89 | 0 | 2 |
| Potassium | 87 | 79 | 6 | 2 |
| Sodium | 86 | 79 | 4 | 3 |
| Chloride | 92 | 83 | 5 | 4 |
| Sulphate | 84 | 69 | 10 | 5 |
| Fluoride | 74 | 66 | 4 | 4 |
| Conductivity | 93 | 91 | 0 | 2 |
| Kjeldahl Nitrogen | 24 | 9 | 7 | 8 |
| Total Phosphorous | 50 | 38 | 6 | 6 |

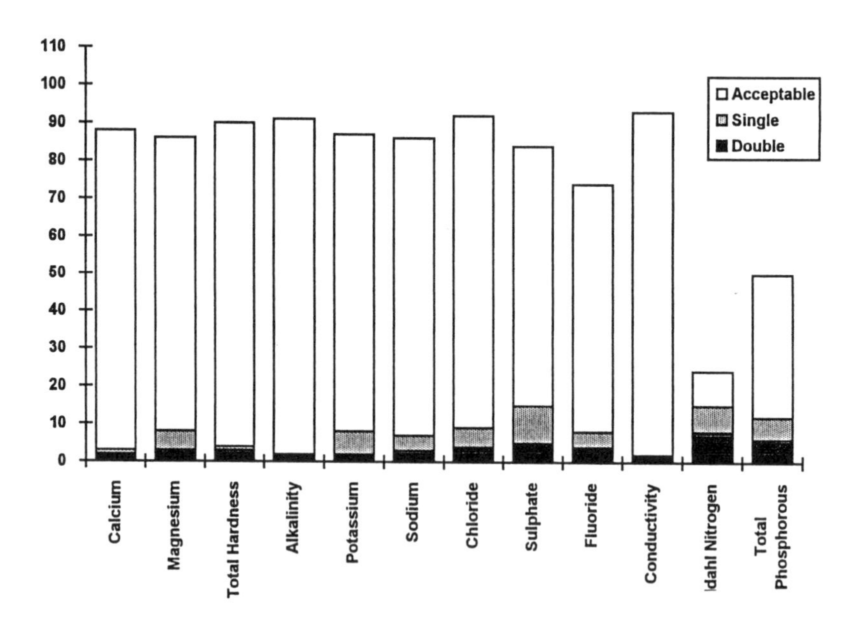


**Figure 9.2** Example of an external quality control report. Supplied by Dr Ian Taylor of Aquacheck, WRc, UK.

Distribution 89
Group - Hard Water - Sodium
Units - mgNa.l-1
Number of Laboratories Reporting: 86

Overall mean: 9.46
Reference Value: 9.46 - Mean of a
Relative Standard Deviation: 7.10 %
Range of reported concentrations: 1.46 to 16.05
Lower and Upper Flagging Limits: 7.85 to 11.07

| | Z-Score | Laboratory Numbers |
|---|---|---|
| Double Flagged | >2.00 | (153) (185) |
| | 1.80 to 2.00 | |
| | 1.60 to 1.80 | |
| Single | 1.40 to 1.60 | |
| Flagged | 1.20 to 1.40 | 10 54 |
| | 1.00 to 1.20 | |
| | **0.80 to 1.00** | **284** |
| | **0.60 to 0.80** | **359** |
| | **0.40 to 0.60** | **112 61 419 56** |
| | **0.20 to 0.40** | **65 25 34 55 50 464 45 8 140 286 <106414 421 416 182 46 478** |
| | **0.00 to 0.20** | **165 133 471 33 119 193 44 457 108 109 399 128 358 36 424 200 283 178** |
| | **0.00 to -0.20** | **101 134 35 60 143 455 332 96 214 164 197 21 395 272 23 63** |
| | **-0.20 to -0.40** | **466 355 9 47 472 113 418 213 275 216 177 154 422 202** |
| | **-0.40 to -0.60** | **100 120 42 62 222** |
| | **-0.60 to -0.80** | **38 339** |
| | **-0.80 to -1.00** | **141** |
| | -1.00 to -1.20 | 398 477 |
| Single | -1.20 to -1.40 | |
| Flagged | -1.40 to -1.60 | |
| | -1.60 to -1.80 | |
| | -1.80 to -2.00 | |
| Double Flagged | <-2.00 | (129) |

Laboratory number enclosed in brackets indicates that the lab's results were flagged as outliers.

The organisation of an EQC exercise requires substantial resources. Large quantities of stable reference materials must be prepared, these materials must be transported to the participating laboratories, data must be analysed and detailed reports on performance must be prepared. Consequently, these exercises are usually managed and run by large organisations which have adequate resources. Participating laboratories are usually charged for the service provided.

## 9.4 Source literature and further reading


Analytical Methods Committee, Royal Society of Chemistry 1992 Proficiency testing of analytical laboratories. *The Analyst,* **117**, 97–104.

Caulcutt, R. and Boddy, R. 1983 *Statistics for Analytical Chemists*. Chapman & Hall, London.

Duncan, A.J. 1986 *Quality Control and Industrial Statistics*. 5th edition, Irwin, Burr Ridge, IL.

Dux, J.P. 1990 *Handbook of Quality Assurance for the Analytical Chemistry Laboratory*. 2nd edition, Van Nostrand Reinhold, New York.

Garfield, F.M. 1991 *Quality Assurance Principles for Analytical Laboratories*. 2nd edition, AOAC International, Arlington, VA.

ISO 1984 *Development and Operations of Laboratory Proficiency Testing* Guide 43(E), International Organization for Standardization, Geneva.

ISO 1994 *Quality Management and Quality Assurance—a Vocabulary*. International Standard ISO 8402, International Organization for Standardization, Geneva.

Mesley, R.J., Pockington, W.D. and Walker, R.F. 1991 Analytical quality assurance—a review. *The Analyst*, **116**, 975–990.

Miller, J.C. and Miller, J.N. 1992 *Statistics for Analytical Chemistry*. 3rd edition, Prentice-Hall, Inglewood Cliffs, NJ.

Shewhart, W.A. 1986 *Statistical Method from the Viewpoint of Quality Control*. Dover Publications, New York.

WHO 1992 *GEMS/WATER Operational Guide*. Third edition, Unpublished WHO document GEMS/W.92.1, World Health Organization, Geneva.

# Chapter 10

# MICROBIOLOGICAL ANALYSES

The discharge of wastes from municipal sewers is one of the most important water quality issues world-wide. It is of particular significance to sources of drinking-water. Municipal sewage contains human faeces and water contaminated with these effluents may contain pathogenic (disease-causing) organisms and, consequently, may be hazardous to human health if used as drinking-water or in food preparation. Faecal contamination of water is routinely detected by microbiological analysis.

It is impractical to attempt the routine isolation of pathogens because they are present in relatively small numbers compared with other types of microorganism. Moreover, there are many types of pathogen and each requires a unique microbiological isolation technique. The approach that has been adopted is to analyse for indicator organisms that inhabit the gut in large numbers and are excreted in human faeces. The presence of these indicator organisms in water is evidence of faecal contamination and, therefore, of a risk that pathogens are present. If indicator organisms are present in large numbers, the contamination is considered to be recent and/or severe.

Bacteria in water are, in general, not present individually, but as clumps or in association with particulate matter. When enumerating bacteria in water it is not the number of individual bacteria present which are counted, but the number of clumps of bacteria or the particles and their associated bacteria. Each clump or particle may have many bacteria associated with it.

## 10.1 Characteristics of indicator organisms

*Total coliforms*

The term "total coliforms" refers to a large group of Gram-negative, rod-shaped bacteria that share several characteristics. The group includes thermotolerant coliforms and bacteria of faecal origin, as well as some bacteria that may be isolated from environmental sources. Thus the presence of total coliforms may or may not indicate faecal contamination. In extreme cases, a high count for the total coliform group may be associated with a low, or even zero, count for thermotolerant coliforms. Such a result would not necessarily indicate the presence of faecal contamination. It might be caused by

*This chapter was prepared by J. Bartram and S. Pedley*

entry of soil or organic matter into the water or by conditions suitable for the growth of other types of coliform. In the laboratory total coliforms are grown in or on a medium containing lactose, at a temperature of 35 or 37 °C. They are provisionally identified by the production of acid and gas from the fermentation of lactose.

*Thermotolerant (faecal) coliforms*

The term "faecal coliform" has been used in water microbiology to denote coliform organisms which grow at 44 or 44.5 °C and ferment lactose to produce acid and gas. In practice, some organisms with these characteristics may not be of faecal origin and the term "thermotolerant coliform" is, therefore, more correct and is becoming more commonly used. Nevertheless, the presence of thermotolerant coliforms nearly always indicates faecal contamination. Usually, more than 95 per cent of thermotolerant coliforms isolated from water are the gut organism *Escherichia coli*, the presence of which is definitive proof of faecal contamination. As a result, it is often unnecessary to undertake further testing to confirm the specific presence of *E. coli*.

In the laboratory thermotolerant coliforms are grown on media containing lactose, at a temperature of 44 or 44.5 °C. They are provisionally identified by the production of acid and gas from the fermentation of lactose.

Nutrient-rich environments may encourage the growth or persistence of some species of thermotolerant coliform other than *E. coli*. This possibility should be considered when, for example, an unusually high result is obtained from water that was thought to be relatively clean. In such a case, the advice of a microbiology laboratory should be sought for the determination of the more specific indicator, *E. coli*.

*Faecal streptococci*

The presence of faecal streptococci is evidence of faecal contamination. Faecal streptococci tend to persist longer in the environment than thermotolerant or total coliforms and are highly resistant to drying. It is, therefore, possible to isolate faecal streptococci from water that contains few or no thermotolerant coliforms as, for example, when the source of contamination is distant in either time or space from the sampling point. Faecal streptococci grow in or on a medium containing sodium azide, at a temperature of 37–44 °C. They are usually detected by the reduction of a dye (generally a tetrazolium-containing compound) or the hydrolysis of aesculin. Routine methods may give "false positives" and additional confirmatory tests may be required.

**Table 10.1** Comparison of methods for analysis of coliform bacteria

| Multiple fermentation tube technique | Membrane filter technique |
|---|---|
| Slower: requires 48 hours for a positive or presumptive positive | More rapid: quantitative results in about 18 hours |
| More labour-intensive | Less labour-intensive |
| Requires more culture medium | Requires less culture medium |
| Requires more glassware | Requires less glassware |
| More sensitive | Less sensitive |
| Result obtained indirectly by statistical approximation (low precision) | Results obtained directly by colony count (high precision) |
| Not readily adaptable for use in the field | Readily adapted for use in the field |
| Applicable to all types of water | Not applicable to turbid waters |
| Consumables readily available in most countries | Cost of consumables is high in many countries |
| May give better recovery of stressed or damaged organisms in some circumstances | |

*Heterotrophic plate count*

The heterotrophic plate count includes all of the micro-organisms that are capable of growing in or on a nutrient-rich solid agar medium. Two incubation temperatures and times are used: 37 °C for 24 hours to encourage the growth of bacteria of mammalian origin, and 22 °C for 72 hours to enumerate bacteria that are derived principally from environmental sources. The main value of colony counts lies in comparing the results of repeated samples from the same source. If levels increase substantially from normal values, there may be cause for concern.

## 10.2 Selecting a bacteriological analytical technique

Two techniques are commonly used to detect the presence of coliforms in water. The first of these is called the "multiple fermentation tube" or "most probable number" technique. In this method measured portions of a water sample are placed in test-tubes containing a culture medium. The tubes are then incubated for a standard time at a standard temperature. In the second technique, a measured volume of sample is passed through a fine filter that retains bacteria. The filter is then placed on culture medium and incubated. This is called the "membrane filter" technique. Features of the two techniques are compared in Table 10.1.

## 10.3 Multiple fermentation tube technique

The technique has been used for the analysis of drinking-water for many years with satisfactory results. It is the only procedure that can be used if water samples are very turbid or if semi-solids such as sediments or sludges are to be analysed. The procedure followed is fundamental to bacteriological analyses and the test is used in many countries.

It is customary to report the results of the multiple fermentation tube test for coliforms as a most probable number (MPN) index. This is an index of the number of coliform bacteria that, more probably than any other number, would give the results shown by the test. It is not a count of the actual number of indicator bacteria present in the sample.

*Principle*

Separate analyses are usually conducted on five portions of each of three serial dilutions of a water sample. The individual portions are used to inoculate tubes of culture medium that are then incubated at a standard temperature for a standard period of time. The presence of coliforms is indicated by turbidity in the culture medium, by a pH change and/or by the presence of gas. The MPN index is determined by comparing the pattern of positive results (the number of tubes showing growth at each dilution) with statistical tables. The tabulated value is reported as MPN per 100 ml of sample.

There are a number of variants to the multiple fermentation tube technique. The most common procedure is to process five aliquots of water from each of three consecutive 10-fold dilutions; for example, five aliquots of the sample itself, five of a 1/10 dilution of the sample and five of a 1/100 dilution. Aliquots may be 1-ml volumes, each added to 10 ml of single-strength culture medium, or 10-ml volumes, each added to 10 ml of double-strength medium. Results are compared with values such as those given in Table 10.5 (see later).

The use of one of the following variants of the technique may help to reduce the cost of analysis:

- A smaller number of tubes is incubated at each dilution, for example three instead of five. A different table must then be used for the MPN determination (see Table 10.6 later). Some precision is lost, but using 9 tubes instead of 15 saves materials, space in the incubator, and the analyst's time.
- For samples of drinking-water, one tube with 50 ml of sample and five tubes with 10 ml of sample are inoculated and incubated. The results are compared with the values such as those given in Table 10.7 (see later) to obtain the MPN.

### 10.3.1 Culture media and buffered dilution water

Each part of the test requires a different type of medium. For example, when enumerating coliforms, lauryl tryptose (lactose) broth is used in the first (isolation or presumptive) part of the test. In the second (confirmation) part, brilliant green lactose bile (BGLB) broth is used to confirm total coliforms and *E. coli* medium to confirm faecal coliforms. Some of the characteristics of these and other media suitable for use in most probable number analyses are described in Table 10.2. Media can be made from primary ingredients but are also available in the following forms:

- Dehydrated powder, packaged in bulk (200 g or more), to be weighed out when the medium is prepared and dissolved in an appropriate volume of distilled water, dispensed to culture tubes and sterilised before use.
- Dehydrated powder, packaged in pre-weighed amounts suitable for making one batch of medium, to be dissolved in an appropriate volume of distilled water, dispensed to culture tubes and sterilised before use.
- Ampoules of solution, ready to use.

The ampoules of ready-to-use media are the most convenient form but are the most expensive and have the shortest shelf-life. Pre-weighed packages are easy to use and reduce the risk of error in making up a batch of medium. Media are inevitably expensive when packaged in small quantities. However, they are not a major component of the cost of bacteriological analysis, and the extra cost may be negligible when compared with the greater convenience and reduced wastage. Large bottles containing dehydrated media must be tightly resealed after use to prevent spoilage; this is especially important in humid environments.

Media should be stored in a cool, dark, dry place. After a medium has been prepared by dissolving the powder in distilled water, it should be distributed into culture tubes or bottles and sterilised. Batches of media should be tested before use, using a known positive and negative control organism. If the appropriate reactions are not observed, the media and the control organisms should be investigated and the tests repeated. Media should be used immediately but may be stored for several days provided that there is no risk of their becoming contaminated.

A stock solution of buffered dilution water is prepared by dissolving 34.0 g of potassium dihydrogen phosphate, $KH_2PO_4$, in 500 ml of distilled water. The pH is checked and, if necessary, adjusted to 7.2 by the addition of small quantities of 1 mol $l^{-1}$ NaOH solution. Distilled water is added to bring the final volume to 1 litre. The buffered water is stored in a tightly stoppered bottle in the refrigerator.

**Table 10.2** Culture media for most probable number (MPN) analyses

| Medium | Uses | Incubation temperature | Remarks |
|---|---|---|---|
| *Isolation media* | | | |
| Lactose broth | Total or thermotolerant coliforms | 48 hours at 35 ± 0.5 °C or 37 ± 0.5 °C for total coliforms and 24 hours at 44 ± 0.25 °C or 44.5 ± 0.25 °C for thermotolerant coliforms | Prepare single strength medium by diluting double strength medium with distilled water. Each tube or bottle should contain an inverted fermentation (Durham) tube |
| MacConkey broth | Total or thermotolerant coliforms | 48 hours at 35 ± 0.5 °C or 37 ± 0.5 °C for total coliforms and 24 hours at 44 ± 0.25 °C or 44.5 ± 0.25 °C for thermotolerant coliforms | |
| Improved formate lactose glutamate medium | Total or thermotolerant coliforms | 48 hours at 35 ± 0.5 °C or 37 ± 0.5 °C for total coliforms and 24 hours at 44 ± 0.25 °C or 44.5 ± 0.25 °C for thermotolerant coliforms | Available commercially in dehydrated form as Minerals Modified Glutamate Medium |
| Lauryl tryptose (lactose) broth | Total or thermotolerant coliforms | 48 hours at 35 ± 0.5 °C or 37 ± 0.5 °C for total coliforms and 24 hours at 44 ± 0.25 °C or 44.5 ± 0.25 °C for thermotolerant coliforms | |
| *Confirmatory media* | | | |
| Brilliant green lactose bile broth | Total or thermotolerant coliforms (gas production) | 44.5 ± 0.25 °C for thermotolerant coliforms | |
| EC medium | Thermotolerant coliforms (indole production) | 44.5 ± 0.25 °C for thermotolerant coliforms | Addition of 1 % (m/m) L- or DL-tryptophan may improve performance of the medium |
| Tryptone water | Thermotolerant coliforms (gas + indole production) | 44.5 ± 0.25 °C for thermotolerant coliforms | |
| Lauryl tryptose mannitol broth with tryptophan | Thermotolerant coliforms (gas + indole production) | 44.5 ± 0.25 °C for thermotolerant coliforms | |

Source: Adapted from ISO, 1990b

To prepare bottles of dilution water, 1.25 ml of stock solution is added to 1 litre of distilled water, mixed well and dispensed into dilution bottles in quantities that will provide, after sterilisation, 9 or 90 ml. The bottles are loosely capped, placed in the autoclave, and sterilised for 20 minutes at 121 °C. After the bottles have been removed from the autoclave, the caps should be tightened and the bottles stored in a clean place until needed.

### 10.3.2 Procedure

The procedure described below is for five tubes at each of three sample dilutions and provides for confirmation of both total and thermotolerant (faecal) coliforms. If fewer tubes (e.g. three of each of three sample dilutions or one 50-ml and five 10-ml portions) are inoculated, the MPN index must be determined from tables specific to the combination of tubes and dilutions used.

*Apparatus*

- ✓ Incubator(s) or water-baths capable of maintaining a temperature to within ± 0.5 °C of 35 and 37 °C and to within ± 0.25 °C of 44 and 44.5 °C. The choice of temperature depends on the indicator bacteria and the medium.
- ✓ Autoclave for sterilising glassware and culture media. The size required depends on the volume of work to be undertaken. A capacity of 100–150 litres would be required for a medium-size laboratory undertaking work on a routine basis.
- ✓ Distillation apparatus, with storage capacity for at least 20 litres of distilled water.
- ✓ Laboratory balance, accuracy ± 0.05 g, with weighing scoop. This may be omitted if culture media and potassium dihydrogen phosphate are available in pre-weighed packages of the proper size.
- ✓ Racks for tubes and bottles of prepared culture media and dilution water. These must fit into the autoclave.
- ✓ Pipettes, reusable, glass, 10-ml capacity graduated in 0.1-ml divisions, and 1-ml capacity graduated in 0.01-ml divisions.
- ✓ Test-tubes, 20 x 150 mm for 10 ml of sample + 10 ml of culture medium, with metal slip-on caps.
- ✓ Bottles, with loose-fitting caps, calibrated at 50 and 100 ml, for 50 ml of sample + 50 ml of culture medium.
- ✓ Measuring cylinders, unbreakable plastic or glass, capacity 100, 250, 500 and 1,000 ml.
- ✓ Test-tube racks to hold tubes in incubator and during storage.
- ✓ Thermometer for checking calibration of incubator or water-bath.
- ✓ Refrigerator for storage of prepared culture media.
- ✓ Hot-air steriliser for sterilising pipettes.
- ✓ Bunsen burner or alcohol lamp.
- ✓ Durham tubes, 6 x 30 mm.
- ✓ Pipette cans for sterilising pipettes.
- ✓ Flasks for preparation of culture media.
- ✓ Wash-bottle.
- ✓ Pipette bulbs.
- ✓ Wire loops for inoculating media, and spare wire.
- ✓ Spatula.
- ✓ Container for used pipettes.
- ✓ Brushes for cleaning glassware (several sizes).

**Table 10.3** Typical sample volumes and number of tubes for multiple fermentation tube analysis

| | Sample volume (ml) | | | | |
|---|---|---|---|---|---|
| Sample type | 50 | 10 | 1 | 0.1 | 0.01 |
| Treated drinking-water | 1 | 5 | | | |
| Partially treated drinking-water | | 5 | 5 | 5 | |
| Recreational water | | 5 | 5 | 5 | |
| Protected-source water | | 5 | 5 | 5 | |
| Surface water | | | 5 | 5 | 5 |

Volumes of 0.1 and 0.01 ml of sample are obtained by addition of 1 ml of a 1/10 and 1/100 dilution, respectively, of the sample to 10 ml of single-strength culture medium.

✓ Fire extinguisher and first-aid kit.
✓ Miscellaneous tools.
✓ Waste bin.

*Consumables*
✓ Culture media: for example lauryl tryptose broth, brilliant green lactose bile (BGLB) broth, and *E. coli* medium.
✓ Disinfectant for cleaning laboratory surfaces and the pipette discard container.
✓ Detergent for cleaning glassware and equipment.
✓ Phosphate-buffered dilution water.
✓ Autoclave tape.

*Procedure*
*Note:* Aseptic technique must be used.

1. Prepare the required number of tubes of culture medium. The volume and strength (single or double) of medium in the tubes will vary depending on the expected bacteriological density in the water and the dilution series planned. For most surface waters, 10 ml volumes of single-strength medium are appropriate.

2. Select and prepare a range of sample dilutions; these will normally be suggested by experience. Recommended dilutions for use when there is no experience with samples from that station are given in Table 10.3. To prepare a 1/10 dilution series, mix the sample bottle well. Pipette 10 ml of sample into a dilution bottle containing 90 ml of phosphate-buffered dilution water. To prepare a 1/100 dilution, mix the 1/10 dilution bottle well and pipette 10 ml of its contents into a bottle containing 90 ml of dilution water. Subsequent dilutions are made in a similar way. Alternatively, 1 ml of sample may be added to a bottle containing 9 ml of dilution water.

3. Pipette the appropriate volumes of sample and diluted sample into the tubes of medium, as shown in Figure 10.1a.

4. Label the tubes with the sample reference number, the dilution and the volume of sample (or dilution) added to the tube. Shake gently to mix the sample with the medium. Place the rack in an incubator or water-bath for 48 hours at 35 ± 0.5 °C or 37 ± 0.5 °C.

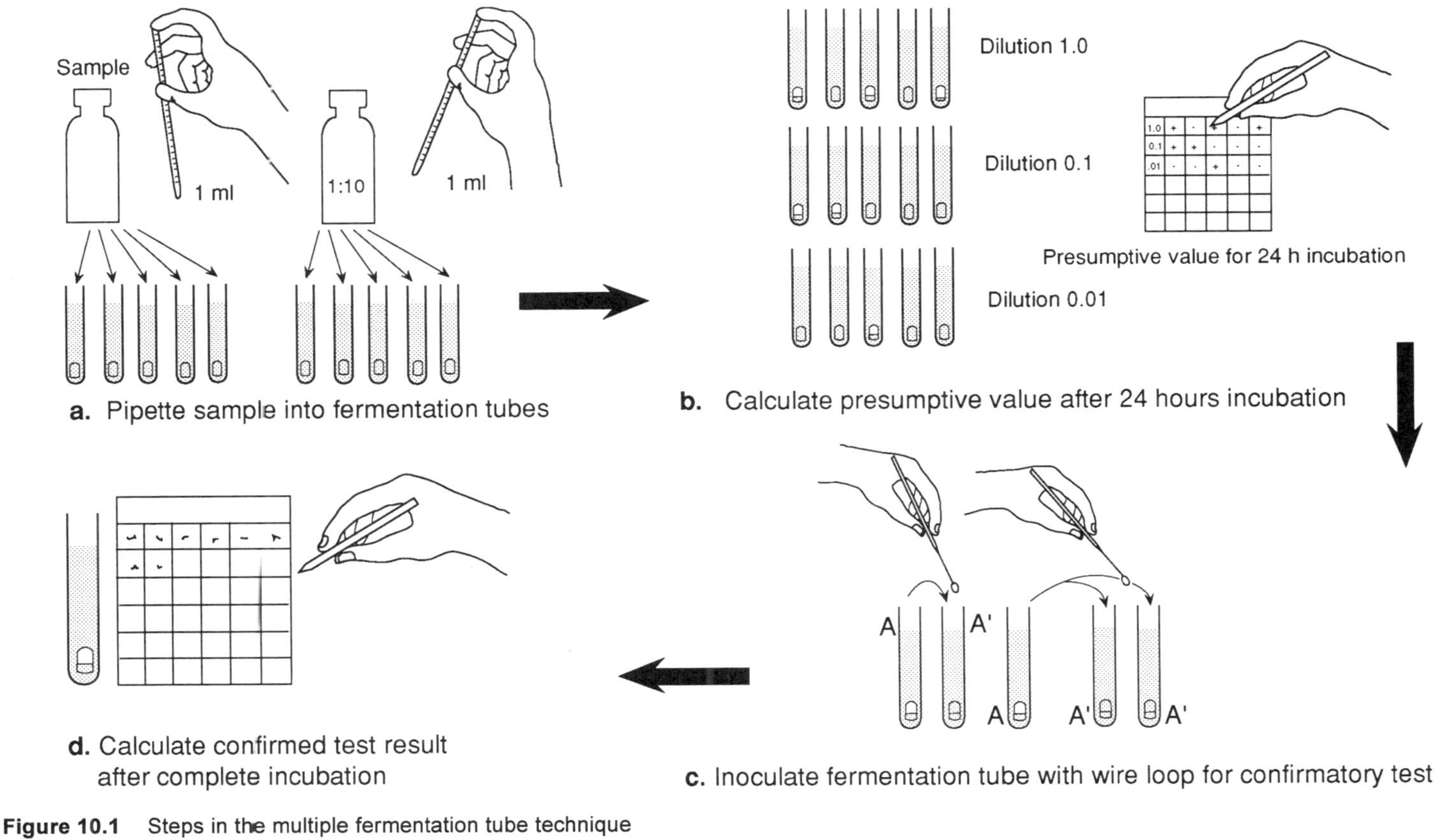


**Figure 10.1** Steps in the multiple fermentation tube technique

**Table 10.4** Reactions following analysis by the MPN method

| | Reactions | |
|---|---|---|
| Medium | Total coliforms at 35 or 37 °C | Thermotolerant coliforms at 44 or 44.5 °C |
| *Isolation media* | | |
| Lactose broth | Gas visible in the inverted fermentation (Durham) tube plus turbidity of the medium | Same as total coliforms at 35 or 37 °C |
| MacConkey broth | Gas visible in the inverted fermentation (Durham) tube plus turbidity of the medium | Same as total coliforms at 35 or 37 °C |
| Improved formate lactose glutamate medium | Gas visible in the inverted fermentation (Durham) tube plus turbidity of the medium | Same as total coliforms at 35 or 37 °C |
| Lauryl tryptose (lactose) broth | Gas visible in the inverted fermentation (Durham) tube plus turbidity of the medium | Same as total coliforms at 35 or 37 °C |
| *Confirmatory media* | | |
| Brilliant green lactose bile broth | Gas visible in the inverted fermentation (Durham) tube plus turbidity of the medium | Same as total coliforms at 35 or 37 °C |
| EC medium | Gas visible in the inverted fermentation (Durham) tube plus turbidity of the medium | Same as total coliforms at 35 or 37 °C |
| Tryptone water | | Add KOVACS' reagent to tube; a red colour denotes presence of indole |
| Lauryl tryptose mannitol broth with tryptophan | | Allows detection of gas + indole production in same tube |

Source: Adapted from ISO, 1990b

5. After 18 or 24 hours, note which tubes show growth. The reactions are listed in Table 10.4. Tubes that show turbidity and gas production, or a colour change indicating the production of acid (if the medium contains a pH indicator), are regarded as positive. Record the number of positive tubes at each dilution, as shown in Figure 10.1b. Return the tubes to the incubator and re-examine after a total of 48 hours of incubation. Continue with the next step of the procedure.

6. Prepare the required number of tubes of confirmation culture medium (BGLB broth for total coliforms and *E. coli* medium for faecal coliforms). Using a sterile wire loop, transfer inocula from positive tubes into the confirmation medium, as shown in Figure 10.1c. Sterilise the loop between successive transfers by heating in a flame until it is red hot. Allow it to cool before use. If confirmation of both total and faecal coliforms is required, a BGLB and an *E. coli* medium tube should be inoculated from each presumptive positive. Label these tubes carefully with the same code used in the presumptive test and incubate them for 48 hours at 35 ± 0.5 °C or 37 ± 0.5 °C for total coliforms (BGLB broth) or for 24 hours at 44 ± 0.5 °C for faecal coliforms (*E. coli* medium).

**Table 10.5** MPN index and 95 per cent confidence limits for various combinations of positive results when five tubes are used per dilution (10 ml, 1.0 ml, 0.1 ml portions of sample)

| Combination of positives | MPN index per 100 ml | 95 % confidence limits | | Combination of positives | MPN index per 100 ml | 95 % confidence limits | |
|---|---|---|---|---|---|---|---|
| | | Upper | Lower | | | Upper | Lower |
| 0–0–0 | < 2 | – | – | 4–2–0 | 22 | 9.0 | 56 |
| 0–0–1 | 2 | 1.0 | 10 | 4–2–1 | 26 | 12 | 65 |
| 0–1–0 | 2 | 1.0 | 10 | 4–3–0 | 27 | 12 | 67 |
| 0–2–0 | 4 | 1.0 | 13 | 4–3–1 | 33 | 15 | 77 |
| | | | | 4–4–0 | 34 | 16 | 80 |
| 1–0–0 | 2 | 1.0 | 11 | 5–0–0 | 23 | 9.0 | 86 |
| 1–0–1 | 4 | 1.0 | 15 | 5–0–1 | 30 | 10 | 110 |
| 1–1–0 | 4 | 1.0 | 15 | 5–0–2 | 40 | 20 | 140 |
| 1–1–1 | 6 | 2.0 | 18 | 5–1–0 | 30 | 10 | 120 |
| 1–2–0 | 6 | 2.0 | 18 | 5–1–1 | 50 | 20 | 150 |
| | | | | 5–1–2 | 60 | 30 | 180 |
| 2–0–0 | 4 | 1.0 | 17 | 5–2–0 | 50 | 20 | 170 |
| 2–0–1 | 7 | 2.0 | 20 | 5–2–1 | 70 | 30 | 210 |
| 2–1–0 | 7 | 2.0 | 21 | 5–2–2 | 90 | 40 | 250 |
| 2–1–1 | 9 | 3.0 | 24 | 5–3–0 | 80 | 30 | 250 |
| 2–2–0 | 9 | 3.0 | 25 | 5–3–1 | 110 | 40 | 300 |
| 2–3–0 | 12 | 5.0 | 29 | 5–3–2 | 140 | 60 | 360 |
| 3–0–0 | 8 | 3.0 | 24 | 5–3–3 | 170 | 80 | 410 |
| 3–0–1 | 11 | 4.0 | 29 | 5–4–0 | 130 | 50 | 390 |
| 3–1–0 | 11 | 4.0 | 29 | 5–4–1 | 170 | 70 | 480 |
| 3–1–1 | 14 | 6.0 | 35 | 5–4–2 | 220 | 100 | 580 |
| 3–2–0 | 14 | 6.0 | 35 | 5–4–3 | 280 | 120 | 690 |
| 3–2–1 | 17 | 7.0 | 40 | 5–4–4 | 350 | 160 | 820 |
| 4–0–0 | 13 | 5.0 | 38 | 5–5–0 | 240 | 100 | 940 |
| 4–0–1 | 17 | 7.0 | 45 | 5–5–1 | 300 | 100 | 1,300 |
| 4–1–0 | 17 | 7.0 | 46 | 5–5–2 | 500 | 200 | 2,000 |
| 4–1–1 | 21 | 9.0 | 55 | 5–5–3 | 900 | 300 | 2,900 |
| 4–1–2 | 26 | 12.0 | 63 | 5–5–4 | 1,600 | 600 | 5,300 |
| | | | | 5–5–5 | > 1,600 | – | – |

Source: After APHA, 1992

7. After the prescribed incubation time, note which tubes show growth with the production of gas, and record the number of positives for each sample dilution as shown in Figure 10.1d.
8. Compare the pattern of positive results with a most probable number table such as one of those given in Tables 10.5, 10.6 and 10.7.

**Table 10.6** MPN index for various combinations of positive results when three tubes are used per dilution (10 ml, 1.0 ml and 0.1 ml portions of sample)

| Number of tubes giving a positive reaction from | | | |
|---|---|---|---|
| 3 of 10 ml each | 3 of 1 ml each | 3 of 0.1 ml each | MPN |
| 0 | 0 | 0 | < 3 |
| 0 | 0 | 1 | 3 |
| 0 | 1 | 0 | 3 |
| 1 | 0 | 0 | 4 |
| 1 | 0 | 1 | 7 |
| 1 | 1 | 0 | 7 |
| 1 | 1 | 1 | 11 |
| 1 | 2 | 0 | 11 |
| 2 | 0 | 0 | 9 |
| 2 | 0 | 1 | 14 |

Source: After WHO, 1985

**Table 10.7** MPN index and 95 per cent confidence limits for various combinations of positive results for a set of one 50 ml and five 10 ml portions of sample

| Number of tubes giving positive reaction | | | 95 % confidence limits | |
|---|---|---|---|---|
| 1 x 50 ml | 5 x 10 ml | MPN Index per 100 ml | Lower | Upper |
| 0 | 0 | < 1 | | |
| 0 | 1 | 1 | 0.5 | 4 |
| 0 | 2 | 2 | 0.5 | 6 |
| 0 | 3 | 4 | 0.5 | 11 |
| 0 | 4 | 5 | 1 | 13 |
| 0 | 5 | 7 | 2 | 17 |
| 1 | 0 | 2 | 0.5 | 6 |
| 1 | 1 | 3 | 0.5 | 9 |
| 1 | 2 | 6 | 1 | 15 |
| 1 | 3 | 9 | 2 | 21 |
| 1 | 4 | 16 | 4 | 40 |
| 1 | 5 | > 18 | | |

Source: After Department of Health and Social Security, 1982

## 10.4 Membrane filter technique

The membrane filter technique can be used to test relatively large numbers of samples and yields results more rapidly than the multiple fermentation tube technique. It was originally designed for use in the laboratory but portable equipment is now available that permits use of the technique in the field. The names and addresses of several manufacturers of portable equipment are given in Appendix 1.

*Principle*

The membrane filter method gives a direct count of total coliforms and faecal coliforms present in a given sample of water. A measured volume of water is filtered, under vacuum, through a cellulose acetate membrane of uniform pore diameter, usually 0.45 μm. Bacteria are retained on the surface of the membrane which is placed on a suitable selective medium in a sterile container and incubated at an appropriate temperature. If coliforms and/or faecal coliforms are present in the water sample, characteristic colonies form that can be counted directly.

The technique is unsuitable for natural waters containing very high levels of suspended material, sludges and sediments, all of which could block the filter before an adequate volume of water has passed through. When small quantities of sample (for example, of sewage effluent or of grossly polluted surface water) are to be tested, it is necessary to dilute a portion of the sample in sterile diluent to ensure that there is sufficient volume to filter across the entire surface of the membrane.

Suggested volumes to be filtered for water from different sources are listed in Table 10.8. If the quality of water is totally unknown, or there is doubt concerning the probable bacterial density, it is advisable to test two or more volumes in order to ensure that the number of colonies on the membrane will be in the optimum range for counting (i.e. 20–80 colonies per membrane). If a suitable volume of sample cannot be filtered through a single membrane, the sample may be filtered through two or more and the numbers of colonies on the membranes added to give the total count for the sample.

Membrane filtration and colony count techniques assume that each bacterium, clump of bacteria, or particle with bacteria attached, will give rise to a single visible colony. Each of these clumps or particles is, therefore, a colony forming unit (cfu) and the results are expressed as colony forming units per unit volume. In the case of thermotolerant coliform bacteria the result should be reported as thermotolerant coliforms [No.] cfu per 100 ml.

**Table 10.8** Typical sample volumes for membrane filtration analysis

| Sample type | Sample volume (ml) | | | | | |
|---|---|---|---|---|---|---|
| | 100 | 10 | 1[1] | 0.1[1,2] | 0.01[1,2] | 0.001[1,2] |
| Treated drinking water | x | | | | | |
| Partially treated drinking water | x | x | | | | |
| Recreational water | | | x | x | | |
| Protected source water | | x | x | | | |
| Surface water | | | x | x | | |
| Wastewater | | | x | x | x | |
| Discharge from sewage treatment plant | | | x | x | x | |
| Ponds, rivers, stormwater runoff | | | | x | x | x |
| Raw sewage | | | | x | x | x |

1 Small volumes should be added to the filtration apparatus together with a minimum of 9 ml of sterile diluent to ensure adequate dispersal across the surface of the filter membrane

2 1.0, 0.1, 0.01 and 0.001-ml volumes are filtered after first preparing serial dilutions of the sample. To filter:
1.0 ml of sample, use 10 ml of 1/10 dilution
0.1 ml of sample, use 10 ml of 1/100 dilution
0.01 ml of sample, use 10 ml of 1/1,000 dilution
0.001 ml of sample, use 10 ml of 1/10,000 dilution

### 10.4.1 Culture media and buffered dilution water

The general comments on media in the section devoted to the multiple fermentation tube technique (section 10.3.1) are also relevant here. The media that may be used are described in Table 10.9. A stock solution of buffered dilution water is prepared as described in section 10.3.1 and stored in a tightly stoppered bottle in the refrigerator.

To prepare bottles of dilution water, 1.25 ml of stock solution is added to 1 litre of distilled water, mixed well and dispensed into dilution bottles in quantities that will provide, after sterilisation, 9 ± 0.5 ml. The bottles are loosely capped, placed in the autoclave and sterilised for 20 minutes at 121 °C. After the bottles have been removed from the autoclave, the caps should be tightened and the bottles stored in a clean place until they are needed.

A supply of good-quality distilled water may not be available if membrane filter equipment is being used in the field. Some alternatives are:

- Rainwater, which is normally low in dissolved solids and, if filtered through the membrane filter apparatus, should be free of suspended solids and bacteria. In some areas rainwater may be acidic and, therefore, pH should be checked before the water is used for the preparation of culture media.

**Table 10.9** Culture media for membrane filtration

| Medium | Uses | Incubation temperature | Remarks |
|---|---|---|---|
| Lactose TTC agar with Tergitol 7 | Total or thermotolerant coliforms | 18–24 hours at 35 ± 0.5 °C or 37 ± 0.5 °C for total coliforms and 18–24 hours at 44 ± 0.25 °C or 44.5 ± 0.25 °C for thermotolerant coliforms | Adjust pH before sterilisation. Filter TTC supplement to sterilise. Tergitol supplement sterilised by autoclaving. Supplements of Tergitol and TTC to be added aseptically. Prepared plates have max. shelf-life of 10 days. Store in dark. |
| Lactose agar with Tergitol 7 | Total or thermotolerant coliforms | 18–24 hours at 35 ± 0.5 °C or 37 ± 0.5 °C for total coliforms and 18–24 hours at 44 ± 0.25 °C or 44.5 ± 0.25 °C for thermotolerant coliforms | Prepared plates have max. shelf-life of 10 days. Store prepared plates at 4 °C. |
| Membrane enrichment with Teepol broth | Total or thermotolerant coliforms | 18–24 hours at 35 ± 0.5 °C or 37 ± 0.5 °C for total coliforms and 18–24 hours at 44 ± 0.25 °C or 44.5 ± 0.25 °C for thermotolerant coliforms | Check pH before sterilisation |
| Membrane lauryl sulphate broth | Total or thermotolerant coliforms | 18–24 hours at 35 ± 0.5 °C or 37 ± 0.5 °C for total coliforms and 18–24 hours at 44 ± 0.25 °C or 44.5 ± 0.25 °C for thermotolerant coliforms | Check pH before sterilisation |
| Endo medium | Total coliforms only | 35–37 °C | Basic fuchsin may be a carcinogen. Also requires ethanol. Do not autoclave. Prepared medium has a shelf-life of 4 days. Store prepared medium at 4 °C in the dark. |
| LES Endo medium | Total coliforms only | 35–37 °C | Basic fuchsin may be a carcinogen. Also requires ethanol. Do not autoclave. Prepared medium has a shelf-life of 2 weeks. Store prepared medium at 4 °C in the dark. |
| MFC | Thermotolerant coliforms | 44 °C | Do not autoclave. Discard unused medium after 96 hours. Rosalic acid stock solution has a maximum shelf-life of 2 weeks. Check pH before sterilisation. Store prepared medium at 2–10 °C. |

Source: Adapted from ISO, 1990a

- De-ionising packs, which are available from commercial suppliers and, if used in accordance with manufacturers' instructions, will supply good-quality water. They may be expensive and are not readily available in some countries.
- Water for car batteries, which is often available at stations where vehicle fuel is sold. However, it may be of lower purity than desired and should probably be filtered through the membrane filter apparatus.

The quality of distilled water should be checked before use. Conductivity should be less than 10 µmhos $cm^{-1}$ and pH should be close to 10.

Culture media and buffered dilution water may be prepared in the field, but this requires the transport of all necessary equipment, which may include measuring cylinders, beakers, distilled water, autoclavable bottles, a large pressure-cooker and a gas burner or other source of heat. Pre-weighed portions of the dehydrated medium should be used. The medium should be mixed with water and dispensed into bottles, each of which should contain enough medium for one day's work. These bottles should be sterilised in the pressure-cooker for 20 minutes. Buffered dilution water should be prepared, dispensed into bottles and sterilised.

*Note:* Whenever a pressure-cooker is used, care must be taken to ensure that the pressure has dropped to atmospheric pressure before the lid is removed.

### 10.4.2 Procedure

*Apparatus*

- ✓ Incubator(s) or water-bath(s) capable of maintaining a temperature to within ± 0.5 °C of 35 and 37 °C and to within ± 0.25 °C of 44 and 44.5 °C. Choice of temperature depends on the indicator bacteria and the medium.
- ✓ Membrane filtration apparatus, complete with vacuum source (electrically operated pump, hand-pump or aspirator) and suction flask.
- ✓ Autoclave for sterilising prepared culture media. A pressure-cooker, heated on a hot-plate or over a Bunsen burner, may be substituted in some circumstances.
- ✓ Boiling-pan or bath (if filtration apparatus is to be disinfected in boiling water between uses).
- ✓ Laboratory balance, accurate to ± 0.05 g, and with weighing scoop. This may be omitted if media and potassium dihydrogen phosphate are available in pre-weighed packages of the correct size.
- ✓ Racks for bottles of prepared culture media and dilution water. These must fit into the autoclave or pressure-cooker.
- ✓ Distilling apparatus with storage capacity for at least 5 litres of distilled water.
- ✓ Refrigerator for storage of prepared culture media.
- ✓ Hot-air steriliser for sterilising pipettes and glass or metal Petri dishes.
- ✓ Thermometer for checking calibration of incubator or water-bath.
- ✓ Pipette cans for sterilising pipettes.
- ✓ Boxes for Petri dishes for use in hot-air steriliser.
- ✓ Reusable bottles for culture media.
- ✓ Reusable glass or metal Petri dishes.

- ✓ Measuring cylinders, capacity 100 ml and 250 ml.
- ✓ Reusable pipettes, glass, capacity 1 ml and 10 ml.
- ✓ Bottles to contain 9-ml volumes of buffered dilution water.
- ✓ Flasks for preparation of culture media.
- ✓ Wash-bottle.
- ✓ Blunt-edged forceps.
- ✓ Pipette bulbs.
- ✓ Spatula.
- ✓ Container for used pipettes.
- ✓ Brushes for cleaning glassware (several sizes).
- ✓ Fire extinguisher and first-aid kit.
- ✓ Miscellaneous tools.
- ✓ Waste bin.

Filtration apparatus should be disinfected between analyses of consecutive samples and sterilised at intervals. The choice of method will depend on the type of apparatus, where it is being used, the equipment available and cost considerations. Sterilising is generally achieved by autoclaving but any of the following methods is suitable for disinfection between analyses of consecutive samples:

- —immersion of components in boiling water for at least 1 minute,
- —rinsing in methanol followed by rinsing in distilled water,
- —flaming with methanol, or
- —exposure to formaldehyde gas generated by burning methanol in the absence of oxygen (generally preferred as a field technique).

*Consumables*

- ✓ Methanol for disinfecting filtration apparatus using formaldehyde gas (unnecessary in the laboratory, but essential if analyses are done in the field). It is essential to use methanol. Ethanol or methylated spirits cannot be substituted.
- ✓ Membrane filters, 0.45 µm pore size and of diameter appropriate for the filtration apparatus being used and complete with absorbent pads.
- ✓ Disinfectant for cleaning laboratory surfaces and a container for discarded pipettes.
- ✓ Culture media (options are listed in the section on media).
- ✓ Phosphate-buffered dilution water.
- ✓ Petri dishes, glass or aluminium (reusable) or plastic (disposable).
- ✓ Polyethylene bags for wrapping Petri dishes if dry incubator is used.
- ✓ Magnifying lens (as an aid to counting colonies after filters are incubated).
- ✓ Wax pencils for labelling Petri dishes.
- ✓ Autoclave tape.
- ✓ Detergent for cleaning glassware and equipment.

*Procedure*

1. Add absorbent pads to sterile Petri dishes for the number of samples to be processed. Sterile pads may be placed in the Petri dishes with sterile forceps or with an automatic dispenser as shown in Figure 10.2a.

2. Soak the pads with nutrient medium. Nutrient medium may be dispensed with a sterile pipette or by carefully pouring from an ampoule or bottle, as shown in Figure 10.2b. In all cases, a slight excess of medium should be added (e.g. about 2.5 ml). Immediately before processing a sample, drain off most of the excess medium, but always ensure that a slight excess remains to prevent the pad drying during incubation. *Note:* Absorbent pads soaked in liquid medium may be replaced by medium solidified by agar. In this case, Petri dishes should be prepared in advance and stored in a refrigerator.

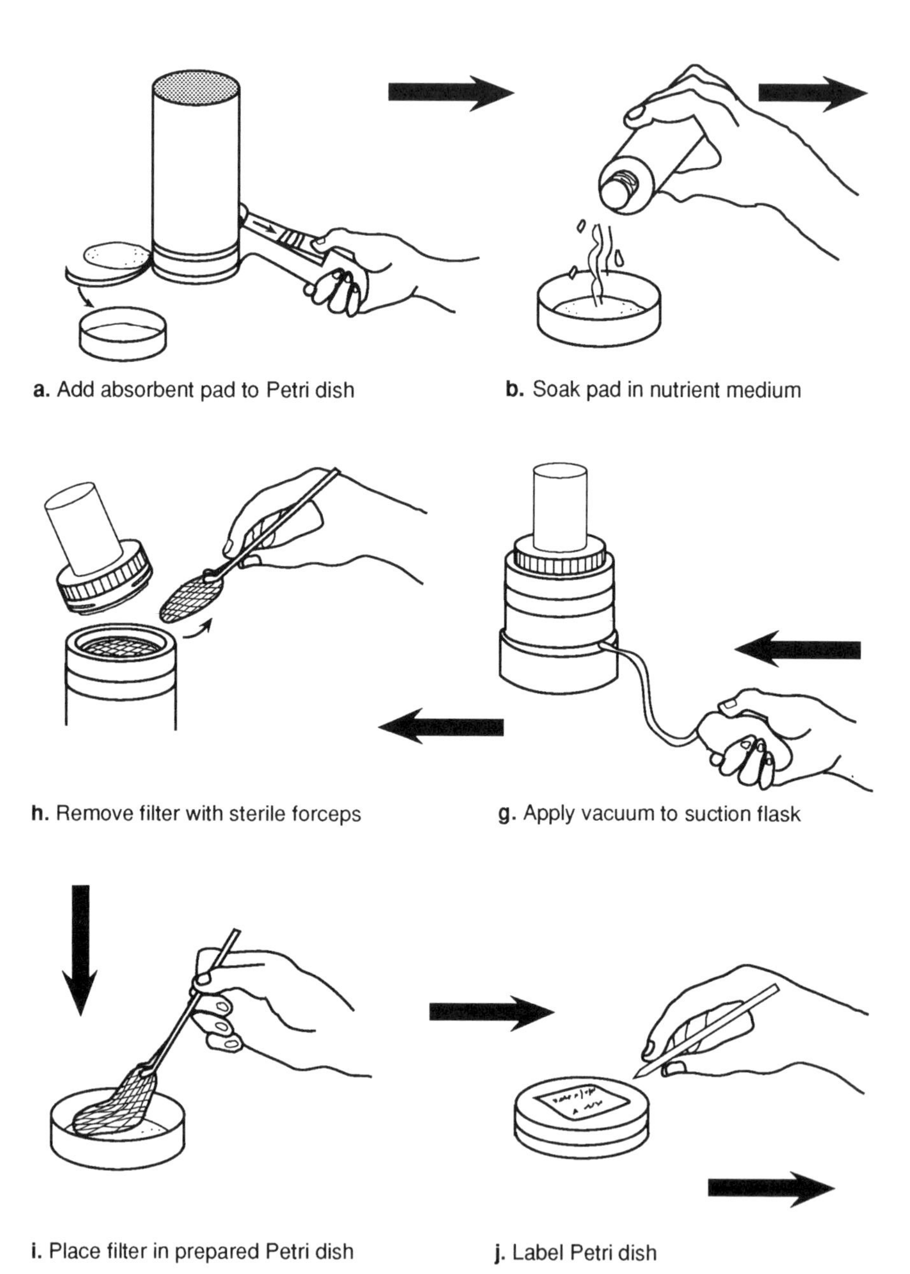


**Figure 10.2** Steps in the membrane filter technique

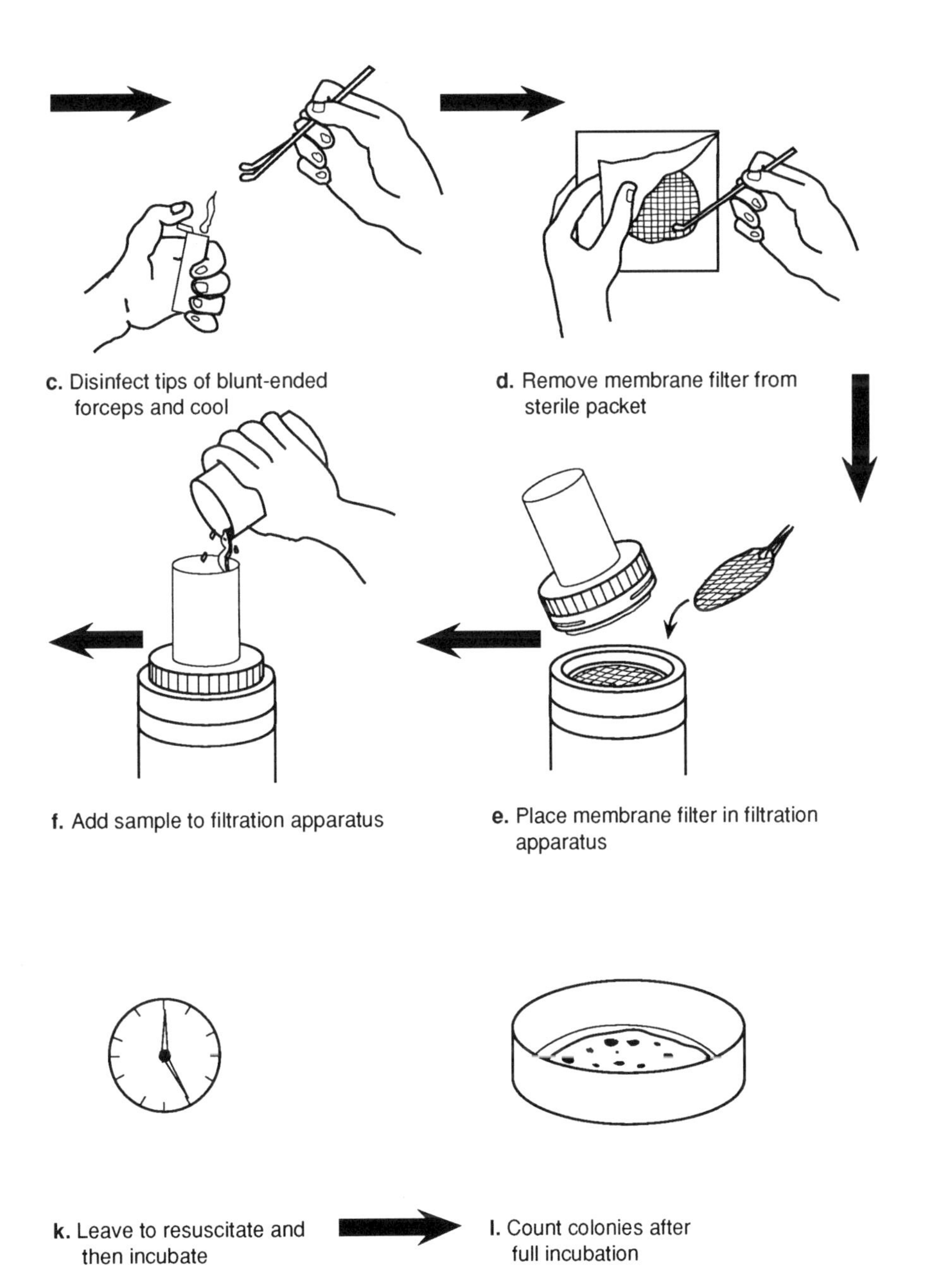
c. Disinfect tips of blunt-ended forceps and cool
d. Remove membrane filter from sterile packet
e. Place membrane filter in filtration apparatus
f. Add sample to filtration apparatus
k. Leave to resuscitate and then incubate
l. Count colonies after full incubation

3. Sterilise the tips of the blunt-ended forceps in a flame and allow them to cool (Figure 10.2c).
4. Carefully remove a sterile membrane filter from its package, holding it only by its edge as shown in Figure 10.2d.
5. Place the membrane filter in the filter apparatus as shown in Figure 10.2e, and clamp it in place. If the apparatus has been disinfected by boiling, ensure that it has cooled down before inserting the membrane filter.
6. Mix the sample by inverting its container several times. Pour or pipette the desired volume of sample into the filter funnel (see Figure 10.2f). This volume should normally be chosen in the light of previous experience, but suggested volumes are given in Table 10.8. If the volume to be filtered is less than 10 ml, it should be made up to at least 10 ml with sterile diluent so that the sample will be distributed evenly across the filter during filtration. Alternatively, the sample may be diluted as suggested in footnote 2 to Table 10.8.
7. Apply a vacuum to the suction flask and draw the sample through the filter; disconnect vacuum (Figure 10.2g).
8. Dismantle the filtration apparatus and remove the membrane filter using the sterile forceps, taking care to touch only the edge of the filter (Figure 10.2h).
9. Remove the lid of a previously prepared Petri dish and place the membrane, grid side uppermost, onto the pad (or agar). Lower the membrane, starting at one edge in order to avoid trapping air bubbles. The procedure is shown in Figure 10.2i.
10. Replace the lid of the Petri dish and mark it with the sample number or other identification (see Figure 10.2j). The sample volume should also be recorded. Use a wax pencil or waterproof pen when writing on Petri dishes.
11. If membranes are going to be incubated at 44 or 44.5 °C, the bacteria on them may first require time to acclimatise to the nutrient medium. After processing samples from areas of temperate climate, leave each Petri dish at environmental temperature for 2 hours before placing it in the incubator. Samples from areas of tropical climate may be incubated immediately.
12. Maintain the Petri dish in a humid atmosphere (e.g. in a plastic bag or in a small container with a moist pad in the base) and incubate it either in an incubator or in a weighed canister in a water bath. This ensures that the pad does not dry out during the incubation period.
13. The incubation periods and temperatures required for each culture medium are listed in Table 10.9. Characteristics of total coliform and thermotolerant coliform colonies grown on the various culture media are described in Table 10.10.
14. After incubation, count the colonies. Express the results as number of colonies per 100 ml of sample. Where smaller volumes have been used, results are calculated from the following formula:

    No. of colonies per 100 ml = [(No. of colonies)/(volume filtered)] x 100

    The colonies counted at this stage are presumed to be coliform bacteria (presumptive results).

**Table 10.10** Colony characteristics following analysis by the membrane filtration method

| Medium | Colony characteristics: Total coliforms at 35 or 37 °C | Thermotolerant coliforms at 44 or 44.5 °C |
|---|---|---|
| Lactose TTC agar with Tergitol 7 | Yellow, orange or brick red coloration with yellow central halo in the medium under the membrane | Same as total coliforms at 35 or 37 °C |
| Lactose agar with Tergitol 7 | Yellow central halo in the medium under the membrane | Same as total coliforms at 35 or 37 °C |
| Membrane enriched Teepol broth | Yellow colour extending on to the membrane | Same as total coliforms at 35 or 37 °C |
| Membrane lauryl sulphate broth | Yellow colour extending on to the membrane | Same as total coliforms at 35 or 37 °C |
| Endo agar or broth | Dark red colour with golden-green metallic sheen | (not applicable) |
| LES Endo agar | Dark red colour with golden-green metallic sheen | (not applicable) |
| MFC medium | (not applicable) | Blue colonies |

Source: Adapted from ISO, 1990a

### 10.4.3 Confirmatory tests

For the examination of raw or partly treated waters, presumptive results may be adequate but, in certain other circumstances, it is important to carry out confirmatory tests on pure subcultures.

To confirm the membrane results for total coliforms, each colony (or a representative number of colonies) is subcultured to tubes of lactose peptone water and incubated at 35 or 37 °C for 48 hours. Gas production within this period confirms the presence of total coliforms.

To confirm thermotolerant coliforms and *E. coli* on membranes, whether incubated at 35, 37 or 44 °C, each colony (or a representative number of colonies) is subcultured to a tube of lactose peptone water and a tube of tryptone water. Tubes are incubated at 44 °C for 24 hours. Growth with the production of gas in the lactose peptone water confirms the presence of thermotolerant coliforms. Confirmation of *E. coli* requires the addition of 0.2–0.3 ml of Kovac's reagent to each tryptone water culture. Production of a red colour indicates the synthesis of indole from tryptophan and confirms the presence of *E. coli*.

*Note:* Use of lauryl tryptose mannitol broth with tryptophan allows both gas production and indole synthesis to be demonstrated in a single tube.

## 10.5 Quality assurance

Quality assurance is discussed in detail in Chapter 9, which should be read in conjunction with this section. In this section, guidance is given on those aspects of analytical quality control that apply only to microbiological laboratories, particularly the preparation and control of laboratory consumables (media and dilution solutions, membrane filters and pads, plastic- and glassware). The monitoring of laboratory equipment is discussed briefly. Readers who wish to develop more rigorous quality control procedures should refer to the appropriate literature cited in section 10.6, such as the *Standard Methods for the Examination of Water and Wastewater*, which provides detailed guidelines for microbiological quality control.

### 10.5.1 Laboratory equipment

Chapter 9 deals with monitoring and control of laboratory equipment, maintenance of operational records and use of calibrated instruments for taking measurements. The principles discussed there also apply to equipment used in microbiological laboratories. Particular attention should be paid to incubators, water-baths, refrigerators and freezers, since correct operation is vital to the reliability of test results. It is recommended that temperature measurements are made twice daily: first thing in the morning before equipment is used, and again at the end of the working day.

Incubators, refrigerators and freezers should be cleaned at least once a month. The manufacturers' instructions should include advice on cleaning and may recommend suitable detergents and disinfectants. Water-baths may need more frequent cleaning to control bacterial growth.

### 10.5.2 Glassware and plasticware

Some detergents and other agents used for the cleaning of laboratory glassware may contain substances that will either inhibit or stimulate the growth of bacteria. Samples of glassware should be checked regularly, for example once a month, if washing procedures and products are always the same. If procedures change, however, or new products are introduced, additional checks should be made. Laboratory plastics may also contain inhibitory residues and each new batch of plasticware should be checked.

*Procedure*

1. Use a type strain of *E. coli* or a well characterised laboratory isolate. Culture the organism overnight in a nutrient broth at 37 °C, and dilute in quarter-strength Ringer's solution to a concentration of 50–200 colony-forming units per ml.
2. Wash and rinse six glass Petri dishes (glass Petri dishes may also be used to simulate washed glassware) according to usual laboratory practice. Designate these as group A.

3. Wash a further six Petri dishes and rinse 12 times with successive portions of reagent-grade water. Designate these as group B.
4. Wash another six Petri dishes with detergent wash water and dry without further rinsing. Designate these as group C.
5. Sterilise the Petri dishes of groups A, B and C by the usual procedure.
6. To test pre-sterilised plasticware, designate six plastic Petri dishes as group D.
7. Add 1 ml of the *E. coli* culture from step 1 to each dish and mix thoroughly with 15 ml of molten yeast-extract agar, cooled to 50 °C (nutrient agar is acceptable if yeast extract agar is not available).
8. Incubate the dishes at 37 °C for 24 hours. Count the number of colonies on each dish and calculate the mean count for each group.
9. The difference in the mean count between the groups should be less than 15 per cent if there are no toxic or inhibitory effects. Differences in mean counts of less than 15 per cent between groups A and B, or of more than 15 per cent between groups A and C show that the detergent has inhibitory properties that are eliminated during routine washing.

### 10.5.3 Media

Every new batch of media should be tested for sterility and for its ability to support the growth of the test organism with the production of its characteristic biochemical reactions.

Sterility should be tested by incubating a sample volume of the medium, at an appropriate temperature, for 48 hours and observing for growth.

The ability of a medium to support the growth of a test organism and to differentiate the test organisms from other organisms is tested by inoculating a sample of the medium with positive and negative control organisms. The positive control must include a type strain of *E. coli* or a well characterised laboratory isolate. The negative control may include one or more of the following: *Staphylococcus aureus*, *Pseudomonas* sp., *Streptococcus faecalis*.

Results of the control tests should be entered in a media log-book. If the positive control does not grow or does not give the correct growth characteristics, the purity and the identification of the test culture should be checked and the test repeated with a fresh culture prepared from a frozen stock. If there are no problems with the test culture, a new batch of medium should be prepared and glassware records checked to ensure that no inhibitory substances have been found during quality control checks. The pH of the medium should also be checked, since pH changes can affect the growth characteristics of an organism. Each new batch of media should be tested alongside the media currently in use.

### 10.5.4 Evaluation of reagents, media and membranes

Before new batches of reagents, media, membranes and membrane pads are released for routine use, they should be compared with those currently in use. Only one variable should be changed with each comparison.

*Procedure*

1. Assemble at least five positive water samples (samples that have been shown to be contaminated). The use of more samples will increase the sensitivity of the test.
2. Process the samples using the test batch of materials and the batch currently in use.
3. Incubate the plates (or tubes).
4. Compare the growth characteristics of the contaminating organism on the two batches of materials. Note any atypical results.
5. Count or calculate the number of colonies per 100 ml, or determine the MPN.
6. Transform the counts to logarithms and enter the results for the two batches of materials in parallel columns.
7. Calculate the difference $d$ between the two transformed results for each sample (include the + or – sign).
8. Calculate the mean of the differences $\bar{d}$ and the standard deviation.
9. Perform a Student's $t$-test, using the number of samples as $n$.
10. Use a Student's table to determine the critical value of $t$ at the 0.05 significance level (two-tailed test). Some critical values are given below.

| No. of samples ($n$) | Degrees of freedom ($n$ – 1) | Critical value of $t$ at 0.05 significance level |
|---|---|---|
| 5 | 4 | 2.78 |
| 6 | 5 | 2.57 |
| 7 | 6 | 2.45 |
| 8 | 7 | 2.37 |
| 9 | 8 | 2.31 |
| 10 | 9 | 2.26 |

11. If the calculated value of $t$ exceeds the critical value, the two batches of materials give significantly different results.

If this test indicates a problem with the new batch of materials, the test conditions and procedure should be carefully reviewed and the batch retested. The batch should be rejected as unsatisfactory only if the problems are confirmed by this second test.

### 10.5.5 Precision testing

Precision testing is important in the microbiological laboratory because test results can reveal procedural problems and problems with the materials. The principles of precision testing are discussed in detail in Chapter 9.

Satisfactory results must be obtained from precision tests before the results of monitoring tests are reported.

*Procedure*

1. At the beginning of each month, or at the earliest convenient time, collect 15 samples that are likely to be positive by the first procedure, with a range of positive results.
2. Make duplicate analyses of each sample. The same analyst should do the tests, but all analysts should be included, on a rota basis.
3. Record the results of the duplicate tests as $D_1$ and $D_2$. Calculate the logarithm of each result. If either of a set of duplicate results is zero, add 1 to both values before calculating the logarithms.
4. Calculate the difference $R$ between each pair of transformed duplicates, and the mean of these differences $\overline{R}$.
5. Calculate the precision criterion as $3.27\overline{R}$.
6. Thereafter, analyse 10 per cent of routine samples, or a minimum of two samples per day, in duplicate. Calculate the logarithm of each result and the difference between the logarithms. If the difference is greater than the calculated precision criterion, analyst variability is excessive and the analytical procedure should be reviewed. The laboratory manager should decide whether or not to release monitoring test results in the light of past performance and other mitigating factors.

## 10.6 Source literature and further reading

APHA 1992 *Standard Methods for the Examination of Water and Wastewater*. 18th edition, American Public Health Association (APHA), American Water Works Association (AWWA) and Water Pollution Control Federation (WPCF), Washington, D.C.

Department of Health and Social Security 1982 *The Bacteriological Examination of Drinking Water Supplies*. Her Majesty's Stationery Office (HMSO), London.

Environment Canada 1981 *Analytical Methods Manual*. Water Quality Branch, Environment Canada, Ottawa.

ISO 1984 *Water Quality — Detection and Enumeration of Faecal Streptococci. Part 1: Method by Enrichment in a Liquid Medium*. International Standard ISO 7899-1, International Organization for Standardization, Geneva.

ISO 1984 *Water Quality — Detection and Enumeration of Faecal Streptococci. Part 2: Method by Membrane Filtration*. International Standard ISO 7899-2, International Organization for Standardization, Geneva.

ISO 1988 *Water Quality — General Guide to the Enumeration of Microorganisms by Culture*. International Standard ISO 8199, International Organization for Standardization, Geneva.

ISO 1988 *Water Quality — Enumeration of Viable Microorganisms. Colony Count by Inoculation in or on a Solid Medium*. International Standard ISO 6222, International Organization for Standardization, Geneva.

ISO 1990a *Water Quality — Detection and Enumeration of Coliform Organisms, Thermotolerant Coliform Organisms and Presumptive* Escherichia coli. *Part 1: Membrane Filtration Method.* International Standard ISO 9308-1, International Organization for Standardization, Geneva.

ISO 1990b *Water Quality — Detection and Enumeration of Coliform Organisms, Thermotolerant Coliform Organisms and Presumptive* Escherichia coli. *Part 2: Multiple Tube (Most Probable Number) Method.* International Standard ISO 9308-2, International Organization for Standardization, Geneva.

Mara, D. and Cairncross, A. 1989 *Guidelines for the Safe Use of Wastewater and Excreta in Agriculture and Aquaculture.* World Health Organization, Geneva.

WHO 1985 *Guidelines for Drinking-water Quality. Volume 3: Drinking-water Control in Small Community Supplies.* World Health Organization, Geneva.

WHO 1989 *Health Guidelines for the Use of Wastewater in Agriculture and Aquaculture. Report of a WHO Scientific Group.* WHO Technical Report Series, No. 778, World Health Organization, Geneva.

# Chapter 11

# BIOLOGICAL MONITORING

Water quality can be described in terms of physical, chemical and biological characteristics (see Chapter 2). Although biologists have been studying the effects of human activities on aquatic systems and organisms for decades, their findings have only relatively recently been translated into methods suitable for monitoring the quality of water bodies. Artificial (and in some cases natural) changes in the physical and chemical nature of freshwaters can produce diverse biological effects ranging from the severe (such as a total fish kill) to the subtle (for example changes in enzyme levels or sub-cellular components of organisms). Changes like these indicate that the ecosystem and its associated organisms are under stress or that the ecosystem has become unbalanced. As a result there could be possible implications for the intended uses of the water and even possible risks to human health.

The responses of biological communities, or of the individual organisms, can be monitored in a variety of ways to indicate effects on the ecosystem. The co-existence and abundance of certain species at particular locations can indicate, for example, whether that habitat has been adversely altered. The reactions of individual organisms, such as behavioural, physiological or morphological changes, can also be studied as responses to stress or adverse stimuli (for example caused by the presence of contaminants). Some approaches are suitable for field use and some have been developed specifically for use in the laboratory (particularly toxicity tests and bioassays).

Certain contaminants, particularly metals and organic compounds, may be accumulated in the tissues of organisms. Therefore, chemical analysis of the appropriate biological tissues can be used to show that the organism has been exposed to contaminants and, in some cases, to monitor the spatial distribution, or accumulation, of that contaminant in the aquatic ecosystem.

When designing a monitoring programme biological methods should be considered along with other approaches (see Chapter 3). They should be used only if appropriate to the objectives of the programme and should always be accompanied by the appropriate physical and chemical measurements that are necessary for proper interpretation of results. Biological monitoring

*This chapter was prepared by D. Chapman and J. Jackson with contributions from F. Krebs*

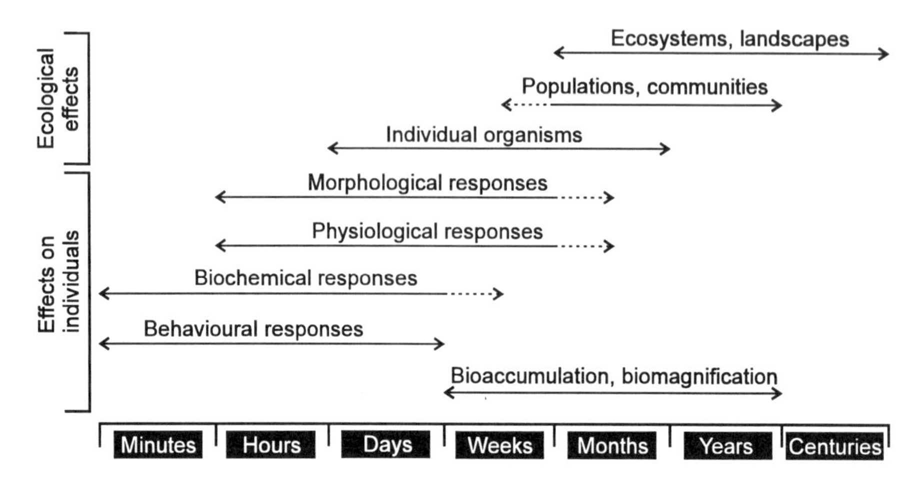


**Figure 11.1** The duration of environmental effects which can be monitored by different biological approaches (Modified from de Zwart, 1995)

should not be seen as an alternative to physical and chemical monitoring but as a useful complementary approach. Although physical and chemical analyses can identify that many contaminants may be present, biological methods can integrate responses to combinations of all contaminants and to other sources of environmental stress, thereby indicating overall effects in a water body. Physical and chemical analyses give a measurement which is valid only for the instance in time when the sample was collected, whereas some biological methods reflect the effects of the physical and chemical conditions to which the organisms were exposed over a period of time (see Figure 11.1). Many biological approaches can be cheaper than chemical methods in terms of equipment, but would normally place heavy demands on field and laboratory personnel. Financial savings can sometimes be made in a monitoring programme by using biological methods to "trigger" the need for intensive and sensitive chemical analyses (see section 11.5).

A disadvantage of biological methods is that it can be difficult to relate observed effects to specific aspects of environmental disturbance, such as contamination or natural changes. For example, methods do not always provide precise information on the identity of a contaminant unless supplementary information from chemical analyses is available. In addition, the response of organisms may be affected by their natural cycles, such as life stage and reproductive condition. Consequently, like other techniques, biological monitoring methods should be developed and interpreted by experienced biologists.

Within the context of this chapter it is not possible to describe all biological methods which may be useful in detail. Emphasis is placed here on describing standardised methods or methods which are less costly or can be relatively easily developed for use in local circumstances. Although the methods are described principally with respect to their use for monitoring water quality for intended uses, or in relation to actual or anticipated effects, some methods (particularly bioassay techniques) can also be applied to monitoring liquid effluents prior to their discharge.

Examples of the application of some popular methods in monitoring and assessment programmes, together with techniques for the interpretation of results are given in the companion guidebook, *Water Quality Assessments*, and in other literature describing general, specific and standardised methods (see section 11.9 for some key references and widely available texts).

## 11.1 Selection of appropriate methods and organisms

The decision to use a biological method will be based on a number of factors:

- The objectives of the monitoring programme.
- The availability of a suitable method in relation to the objectives.
- The financial and personnel resources available.

The selection of the most suitable technique or approach is very important if objectives are to be met and the value of the biological method is to be realised. There are several important factors which must be considered during the selection process. The ability of any biological method to indicate the state of the environment is dependent on the degree and duration of exposure of the disturbance being matched with the sensitivity and response rate of the biological process (Figure 11.1). A trained biologist is normally necessary to select and develop biological methods and to interpret their results.

The most widely applied biological method is the monitoring of bacteria associated with faecal contamination. This approach is used to monitor a very specific water quality issue and gives a direct indication of risk to human health. Microbiological methods are treated in detail in Chapter 10.

Assessments of long-term water quality variations on a regional basis often include biological methods as a means of indicating the overall condition of ecosystems. Baseline surveys of aquatic habitats and their communities provide a reference point against which future monitoring programme results can be evaluated. Specific objectives, such as early-warning of environmental degradation close to an important drinking water intake, require sensitive biological methods which give clear, unambiguous results, immediately the adverse environmental conditions occur. Where existing standardised methods are not available it may be necessary to adapt a technique to local conditions, using organisms found naturally in the aquatic habitat to be monitored.

If the financial resources and the technical capacity for extensive chemical monitoring are limited, it may be desirable to use a biological method in order to determine the presence and severity of an environmental effect or to trigger the need for more complex techniques which are not used routinely. Nevertheless, it should always be borne in mind that meaningful interpretation of the results of biological methods usually requires some basic physical and chemical measurements in order to help the biologist to identify natural influences on aquatic organisms and communities.

### 11.1.1 Selection of techniques

Once it has been established that biological methods will provide useful information in a monitoring programme, an appropriate technique must be selected from the many methods available. The principal biological methods can be divided into five categories:

- Ecological methods: based on community structure and presence or absence of species.
- Physiological and biochemical methods: based on community metabolism (such as oxygen production or consumption, growth rates) or biochemical effects in individuals or communities (such as enzyme inhibition).
- Controlled biotests: based on measuring toxic (or beneficial) effects (death, growth rates, reproductive capacity) on organisms under defined laboratory conditions or the effects on behaviour *in situ* or in controlled environments.
- Contaminants in biological tissues: based on measurements of the accumulation of specific contaminants in the tissues of organisms living in the environment (passive monitoring) or deliberately exposed in the environment (active monitoring).
- Histological and morphological methods: based on the observation of cellular changes or morphological changes such as gill damage or skin lesions in fish.

Most of the approaches listed above have one or more specific methods which are widely accepted as useful and practicable for environmental monitoring (Table 11.1).

The global diversity of aquatic species has led to the adaptation and modification of several standard biological methods (see section 11.5), for use with an enormous range of organisms. Many of these methods are useful for indicating adverse effects in aquatic environments, but often only in specific situations. It has been estimated that there are about 120 different laboratory-based toxicity tests, and about 100 different field methods for

**Table 11.1** Examples of biological monitoring methods which can be applied in freshwaters

| Biological approach | Organisms or methods used | Organism response or observation criterion |
|---|---|---|
| Single species, acute toxicity tests in the laboratory | Fish | Death |
| | *Daphnia* | Death, immobilisation |
| | Bacterial luminescence | Light emission |
| | Daphnia IQ test | Enzyme inhibition |
| | Rotoxkit F | Death |
| | Thamnotoxkit F | Death |
| | Toxichromotest | Enzyme inhibition |
| | Ames test, SOS chromotest, Mutatox test | Bacterial mutagenicity |
| Single species, sub-lethal toxicity tests in the laboratory | Protozoa/bacteria | Population growth |
| | Algae | Population growth |
| | *Daphnia* | Reproduction |
| | Fish | Early life stage, growth, chromosome aberration |
| | *Lemna* test | Colony growth |
| Sub-organismal toxicity test in the laboratory | *In vitro* tissue test | Growth, death histopathology |
| Early warning or semi-continuous field toxicity tests | Fish | Ventilation |
| | | Rheotaxis |
| | | Swimming behaviour |
| | Algae | Productivity |
| | Bacteria | Luminescence |
| | | Respiration |
| | *Daphnia* | Swimming activity |
| | Mussels | Valve movement |
| Field toxicity test (active monitoring) | Caged organisms | Death, growth, reproduction, bioaccumulation, scope for growth |
| | | Biomarkers: metallothioneine formation, lysosome stability, mixed function oxidase (MFO) induction |
| Observation of effects in the field (passive monitoring) | Eco-epidemiology in selected species: fish, *Chironomus* | Incidence of diseases and morphological variations |
| | Indicator species | Presence or absence |
| | Colonisation of artificial substrates | Species composition, diversity abundance |
| | Community structure: benthic macrofauna, diatoms | Species composition, diversity abundance |
| | Ecological function | Primary productivity, respiration, biomass, turnover, degradation, material cycling |

Source: After de Zwart, 1995

assessing aquatic community effects, described in the international literature (de Zwart, 1995). The suitability of any particular method must be evaluated with respect to:

- the information it will generate (ecological as well as in relation to environmental problems),
- the specificity of the method and the selected organisms,
- the ability of the habitat or the organisms to return to their natural condition (i.e. recolonise, regenerate) once the environmental problem is removed,
- the sensitivity and rate and range of responses of the organisms or methods in relation to the environmental stress,
- the cost and practicality of the method, and
- the ability of the method to provide information that can be translated into useful management information and to help achieve the monitoring programme objectives.

As with all other components of a monitoring programme, it is important to assess the suitability of biological methods against the specified objectives of the whole programme and not only against the biological component of the programme.

The effects demonstrated by organisms experiencing environmental stress range from death or migration from the affected area to subtle biochemical changes at the sub-cellular level. Many of these effects have been incorporated and developed into possible monitoring methods, but most are only applicable to certain types of environmental stress, to particular time scales of stress effects or to specific habitats or localities.

The most widely applicable methods (i.e. geographically) are those which are laboratory based, using samples of water collected in the field, such as toxicity tests and bioassays. However, the direct relevance of such methods to the field situation is sometimes questioned and not always clear. Conversely, field-based methods which integrate and accumulate direct effects in the water body or aquatic habitat are usually very specific to geographic regions and sometimes to specific aquatic habitats within those regions. Care must be taken to apply these methods only in appropriate locations (unless they have been specifically adapted) and to choose monitoring sites which are directly comparable with each other.

### 11.1.2 Choice of organisms

In order to detect environmental disturbance using biota it is important that the organisms reflect the situation at the site from which they are collected, i.e. they must not migrate. As a result, the most widely used group of aquatic organisms are the benthic macroinvertebrates. These organisms are usually relatively immobile, thereby indicating local conditions and, since many have life spans covering a year or more, they are also good integrators of environmental conditions. They are ubiquitous and abundant in aquatic ecosystems and relatively easy to collect and identify (at least to family level). No expensive equipment is necessary for their collection. It is sometimes possible to get a qualitative impression of water quality in the field because the presence or absence of certain groups is related to their tolerance to specific environmental conditions (such as organic pollution) and may be well documented (Figure 11.2).

For evaluation of long-term (i.e. several years) environmental conditions over a broader area, such as a river stretch or small lake, fish have several advantages. They are long lived and mobile and, because different species cover the full range of trophic levels (omnivores, herbivores, insectivores, planktivores and piscivores), fish community structure is a good indicator of overall environmental condition. Fish also have the advantage that they are an important source of food for humans which gives them a direct relevance for assessing potential contamination. They are also important for recreational and commercial activities, and monitoring them gives an evaluation of water quality which is readily understood by non-specialists. In addition, fish often form the basis of ecological water quality standards. Basic sampling methods can be cheap and easy (such as netting techniques) and a trained fish biologist can make a field evaluation of the condition of the habitat without harming the fish.

The other group of organisms which has been fairly widely used for water quality monitoring is the algae. They are widely used in the evaluation of eutrophication and turbidity. Algae are primary producers and are, therefore, mostly affected by physical and chemical variations in their environment. Their pollution tolerances are also fairly well documented. Simple, non-taxonomic methods are well developed, such as the determination of chlorophyll (see sections 11.3 and 7.9), and sampling methods are simple and cheap. Although other groups of organisms have been found to be suitable for particular biological methods they are less suitable for indicating the integrated effects of environmental disturbance.

The following sections present details of the basic procedures for a few selected methods which are widely applicable, relatively less expensive to perform and most easily adapted to local conditions. Examples and results of using some biological methods, together with the manipulation and interpretation of the resultant data, are given in the companion guidebook *Water Quality Assessments*.

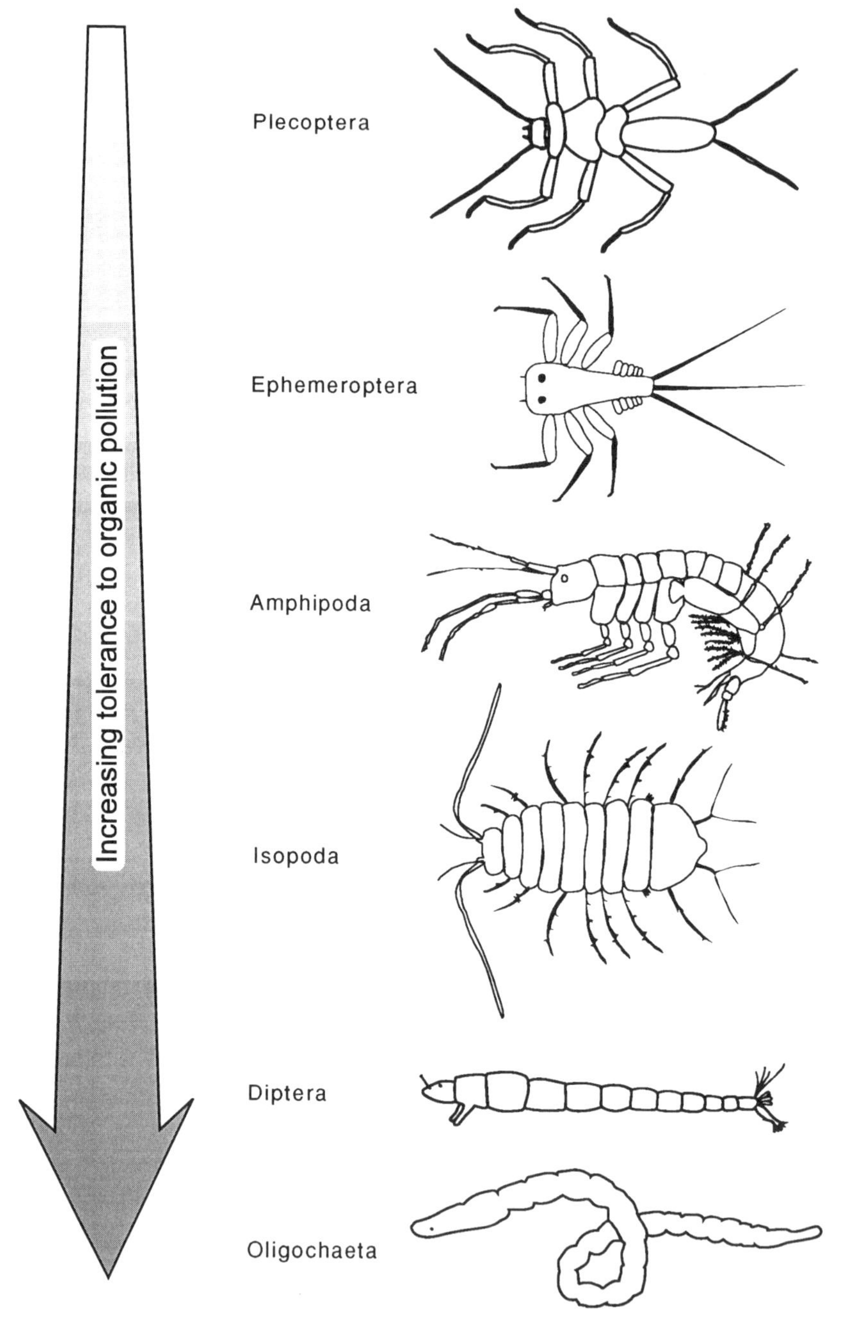


**Figure 11.2** The relative tolerance to organic pollution of some key groups of aquatic macroinvertebrates (Adapted from Mason, 1981)

## 11.2 Ecological methods

Each aquatic organism has particular requirements with respect to the physical, chemical and biological condition of its habitat. Changes in these conditions can result in reduction in species numbers, a change in species dominance or total loss of sensitive species by death or migration. The presence or absence of certain species in relation to particular water quality characteristics has been exploited in the development of ecological methods based on "indicator species". These methods are frequently referred to as biotic indices and require a good knowledge of the organisms in the specific environments to which the methods are applied. Information on the physical and chemical status of the aquatic habitats in which these methods are used is also essential in order to determine whether certain species could survive there, even under undisturbed conditions. The fluctuations in diversity and numerical abundance of species have also been developed into a variety of community structure indices. These methods often require a less detailed knowledge of the species in a particular habitat and have, as a result, been rather widely applied without adequate investigation into their biological relevance. They are, nevertheless, very useful while a (possibly) more sensitive method is being developed or tested. Macroinvertebrates are particularly suitable for both of these approaches.

At the most basic level, ecological monitoring involves determining the species and abundance of the fauna and flora present and observing any changes in the community structure over time. However, other methods have been developed which present the information such that non-biologists can interpret the observed differences between sampling sites. These approaches produce numerical indices in which the magnitude of the index value is related to a qualitative assessment of ecosystem or water quality (e.g. from clean to polluted). Unfortunately, such simplified interpretations are open to mis-use and the importance of interpretation of the results by a trained biologist cannot be over-emphasised.

A preliminary survey or sampling programme, using the proposed methods at the intended sites, can help to eliminate possible problems and to save time and effort during the main programme. The survey can evaluate the most suitable organisms, test the efficiency of the sampling device and determine the size and number of samples required in relation to the size, density and spatial distribution of the organisms being sampled.

### 11.2.1 Indicator organisms

The most studied and monitored degradation of water quality is due to organic pollution arising from sewage discharges. In northern temperate regions the typical assemblages of organisms associated with a certain degree of organic pollution (or the gradual recovery from the effects of an organic discharge) are well documented and numerous biotic indices have been developed. In Central Europe, the association of particular (easily identified) organisms from all trophic levels with defined physical and chemical characteristics of the water, have been combined with the abundance of the organisms to calculate their saprobic value (a number between 1 and 20 representing the association of the organisms with clean to severely polluted conditions). The saprobic value is then combined with index values related to the abundance and sensitivity of the organisms to calculate the Saprobic Index (ranging from 1.0 to 4.0, indicating very clean to severely polluted waters). This index was first devised at the beginning of the twentieth century and, although it has been improved and standardised, it is time consuming and costly to conduct frequently or intensively.

Many simplified biotic indices are based on the Trent Biotic Index which was originally devised in 1964 for use in the River Trent in England. Standardised collection methods (see section 11.2.3) are used for benthic macroinvertebrates and the indices are derived from scores allocated according to the presence or absence of indicator groups and/or indicator species. The Trent Biotic Index is based on the number of groups (specified taxa) in relation to the presence of six key organisms found at the sample site. This type of simplified index is often most suited to fast flowing, upland stretches of rivers. In order to use or develop this basic approach for other geographic areas it is necessary to have detailed knowledge of the distribution of local species in relation to different environmental conditions (e.g. clean and organically contaminated). It is important that biotic indices are used together with all other available information to ensure correct interpretation of the biological information.

Both of the methods mentioned above are described in more detail in the companion guidebook *Water Quality Assessments*.

*The Biological Monitoring Working Party (BMWP) score*

In order to reduce the effort and taxonomic expertise necessary for routine biological assessments based on indicator organisms, there is much interest in developing score systems which rely only on identification to the family level and which are not specific to any single river catchment or geographical area. One such method, the Biological Monitoring Working Party score (BMWP) has been standardised by the International Organization for Standardization (ISO). It can be used to reflect the impact of organic pollution, such as results from sewage disposal or farm waste.

**Table 11.2** The biological scores allocated to groups of organisms by the Biological Monitoring Working Party (BMWP) score

| Score | Groups of organisms |
|---|---|
| 10 | Siphlonuridae, Heptageniidae, Leptophlebiidae, Ephemerellidae, Potamanthidae, Ephemeridae<br>Taeniopterygidae, Leuctridae, Capniidae, Perlodidae, Perlidae, Chloroperlidae<br>Aphelocheiridae<br>Phryganeidae, Molannidae, Beraeidae, Odontoceridae, Leptoceridae, Goeridae, Lepidostomatidae, Brachycentridae, Sericostomatidae |
| 8 | Astacidae<br>Lestidae, Agriidae, Gomphidae, Cordulegasteridae, Aeshnidae, Corduliidae, Libellulidae<br>Psychomyiidae (Ecnomidae), Phylopotamidae |
| 7 | Caenidae<br>Nemouridae<br>Rhyacophilidae (Glossosomatidae), Polycentropodidae, Limnephilidae |
| 6 | Neritidae, Viviparidae, Ancylidae (Acroloxidae)<br>Hydroptilidae<br>Unionidae<br>Corophiidae, Gammaridae (Crangonyctidae)<br>Platycnemididae, Coenagriidae |
| 5 | Mesovelidae, Hydrometridae, Gerridae, Nepidae, Naucoridae, Notonectidae, Pleidae, Corixidae<br>Haliplidae, Hygrobiidae, Dytiscidae (Noteridae), Gyrinidae, Hydrophilidae (Hydraenidae), Clambidae, Scirtidae, Dryopidae, Elmidae<br>Hydropsychidae<br>Tipulidae, Simuliidae<br>Planariidae (Dogesiidae), Dendrocoelidae |
| 4 | Baetidae<br>Sialidae<br>Pisicolidae |
| 3 | Valvatidae, Hydrobiidae (Bithyniidae), Lymnaeidae, Physidae, Planorbidae, Sphaeriidae<br>Glossiphoniidae, Hirudinidae, Erpobdellidae<br>Asellidae |
| 2 | Chironomidae |
| 1 | Oligochaeta |

Groups in brackets are new groups of organisms that were previously contained in the group immediately before it in the list. These new groups are the result of developments in the taxonomic system since the BMWP score was originally prepared.

*Principle:* Invertebrates are collected from different habitats (e.g. gravel, silt, weed beds) at representative sites on river stretches and identified to the required taxonomic level (normally family level). Each group or family is allocated a score between 1 and 10, according to their sensitivity to environmental disturbance (Table 11.2). The most sensitive organisms, such as stoneflies, score 10 and the least sensitive, such as oligochaete worms, score 1. The scores for each family represented in the sample are then summed to give the BMWP score. In order to reduce the effects of sample size, sampling effort and sampling efficiency on the results obtained by this method, the Average Score Per Taxon (ASPT) should also be taken into consideration. This is obtained by dividing the BMWP score by the total number of taxa (families) in the sample. The number of taxa present is indicative of the diversity of the community (see below). A BMWP score greater than 100, together with an ASPT value greater that 4, generally indicates good water quality.

*Procedure*

1. Use a standardised collection technique (normally a pond net, although dredges or grabs can be used in deeper water, see section 11.2.3) to collect macroinvertebrates from each of the major habitat types at the sample site (e.g. gravel, silt, weed beds). If using a pond net collect organisms for a fixed period of time, or if using a dredge or grab collect a standard volume of substrate.
   *Note:* Each habitat type represents a part of the complete sample for the site.
2. Remove large pieces of organic debris or stones from the sample and empty it into a suitable, labelled, container.
3. If sorting and identification cannot be carried out in the field, the sample may be preserved with formaldehyde or alcohol (see Table 11.3) for transportation and storage at the laboratory.
4. Empty the sample into a white tray and sort the macroinvertebrates present into the groups identified in Table 11.2.
5. Tick off the groups present on a sample record sheet. Note that even if more than one species occurs for a particular group that group is only recorded once.
6. Add the scores for all groups ticked on the record sheet to give the BMWP score (e.g.if Oligochaeta, Assellidae, Sphaeriidae and Baetidae were present the score would be 11).
7. Add up the total number of groups occurring in the sample (for the example given in step 6 above the total number is 4).
8. Divide the BMWP score by the total number of groups present to give the ASPT (for the example above 11/4 = 2.75)
9. Record the result as BMWP..... ASPT..... (for the example above the result of BMWP 11 ASPT 2.75 would suggest very poor water quality).

### 11.2.2 Community structure

Community structure methods are based on the numerical abundance of each species rather than relying on particular indicator species. Some of the resultant indices are derived from mathematical principles, such as information theory, and their direct relevance to, and suitability for, the environmental situations in which they may be used should be thoroughly tested. Although a knowledge of taxonomy is required to sort and count samples of organisms, the indices can be useful to non-specialists as an indicator of whether environmental conditions are changing. Community structure indices should only

**Table 11.3** Selected techniques for the preservation of biological samples

| Organisms | Container type | Preservation method | Storage of preserved samples |
|---|---|---|---|
| *Identification and counting* | | | |
| Benthic macro-invertebrates | P or G | 10 % Ethanol. 5 % formaldehyde | 1 year |
| Fish | P or G | Addition of 10 % (m/m) formaldehyde. 3 g of sodium borate decahydrate and 50 ml of glycerol per litre | Short as possible but < 1 year |
| Macrophytes | P or OG | Addition of 5 % (m/m) formaldehyde | 1 year |
| Periphyton | P or OG | Addition of 5 % (m/m) neutral formaldehyde | 6 months in the dark |
| Phytoplankton | P or OG | Addition of 5 % (m/m) neutral formaldehyde or mentholate Lugol's iodine solution | 6 months in the dark |
| Zooplankton | P or G | Addition of 5 % (m/m) formaldehyde Lugol's solution | 1 year |
| *Fresh weight* | | | |
| Fish | P or G | Not applicable | Not applicable Weigh on site |
| All other groups | P or G | Cool to 2–5 °C | 24 hours |
| *Dry or ash weight* | | | |
| All groups | P or G | Filter (where appropriate) Freeze to –20 °C | 6 months |
| *Chlorophyll analyses* | | | |
| Phytoplankton | P | Filter immediately and add 0.2 ml of $MgCO_3$ suspension (1.0 g $MgCO_3$ in 100 ml water) as last of sample is being filtered Freeze filters to –20 °C | In the dark. Minimum possible but up to a few weeks if frozen and desiccated |
| *Chemical analysis* | | | |
| Animal or plant tissue | [1] | Freeze to –20 °C | 1 year |

A preservation method should be chosen which is suitable for all the determinations which may be necessary on a single sample. Ideally, all analyses should be carried as soon as possible after sample collection, and preservation should be avoided where feasible.

P Polyethylene

G Glass

OG Opaque glass

[1] Storage and transport containers depend on the analyses to be performed on the tissues. For most metal analyses new polythene bags are suitable

Source: After WMO, 1988.

be used to study changes at the same sites over time or to compare sites with similar natural physical and chemical features. They are normally applied to samples of organisms of the same type, e.g. benthic macroinvertebrates, diatoms or fish.

*Diversity indices*

Diversity indices are best applied to situations of toxic or physical pollution which impose general stress on the organisms. Stable ecosystems are generally characterised by a high species diversity, with each species represented by relatively few individuals. Although diversity can be reduced by anthropogenic disturbance or stress, some natural conditions can also lead to reduced diversity (such as nutrient poor headwaters) and it is very important that diversity indices are only used to compare sites of similar physical and chemical characteristics. A widely used diversity index is the Shannon Index which combines data on species or taxa richness with data on individual abundance. The species number indicates the diversity of the system and the distribution of numbers of individuals between species indicates the evenness.

*Procedure for Shannon Index*

1. Sort organisms into particular taxa (species, genus or family). All organisms should be identified to the same taxonomic level, e.g. genus level.
2. Count organisms in each taxonomic group.
3. Total the number of organisms in the whole sample.
4. Calculate the Shannon Index $H'$ from the following formula:

$$H' = \sum_{i=1}^{s} \frac{n_i}{n} \ln \frac{n_i}{n}$$

where $s$ = the number of taxa in the sample
$n_i$ = the number of individuals in the $i$th taxa
$n$ = the total number of individuals in the sample

The numerical values generated cannot be taken to be indicative of any particular water quality unless extensively tested and related to physical, chemical and biological conditions in the specific water bodies from which they have been obtained (see example in Figure 14.2). They can be used, however, to show relative differences from one site to another within the same aquatic system or at the same site over time.

### 11.2.3 Sampling methods and sample handling

Some common sampling methods for different aquatic organisms are compared in Table 11.4. Ecological methods can use a wide range of sampling techniques ranging from qualitative collection (such as selection of macrophytes by hand), to semi-quantitative methods (such as collection of

**Table 11.4** Comparison of sampling methods for aquatic organisms

| Sampler/ sampling mechanism | Most suitable organisms | Most suitable habitats | Advantages | Disadvantages |
|---|---|---|---|---|
| Collection by hand | Macrophytes, attached or clinging organisms | River and lake margins, shallow waters, stony substrates | Cheap — no equipment necessary | Qualitative only. Some organisms lost during disturbance. Specific organisms only collected |
| Hand net on pole (c. 500 µm mesh) | Benthic invertebrates | Shallow river beds, lake shores | Cheap, simple | Semi-quantitative. Mobile organisms may avoid net |
| Plankton net | Phytoplankton (c. 60 µm mesh), zooplankton (c. 150–300 µm mesh) | Open waters, mainly lakes | Cheap and simple. High density of organisms per sample. Large volume or integrated samples possible | Qualitative only (unless calibrated with a flow meter). Selective according to mesh size. Some damage to organisms possible |
| Bottle samplers (e.g. Friedinger, Van Dorn, Ruttner) | Phytoplankton, zooplankton (inc. protozoa), micro-organisms | Open waters, groundwaters | Quantitative. Enables samples to be collected from discrete depths. No damage to organisms | Expensive unless manufactured "in house". Low density of organisms per sample. Small total volume sampled |
| Water pump | Phytoplankton, zooplankton (inc. protozoa), micro-organisms | Open waters, groundwaters | Quantitative if calibrated. Rapid collection of large volume samples. Integrated depth sampling possible | Expensive and may need power supply. Sample may need filtration or centrifugation to concentrate organisms. Some damage to organisms possible |

Continued

**Table 11.4** Continued

| Sampler/ sampling mechanism | Most suitable organisms | Most suitable habitats | Advantages | Disadvantages |
|---|---|---|---|---|
| Grab (e.g. Ekman, Peterson, Van Veen) | Benthic invertebrates living in, or on, the sediment. Macrophytes and associated attached organisms | Sandy or silty sediments, weed zones | Quantitative sample. Minimum disturbance to sample | Expensive. Requires winch for lowering and raising |
| Dredge-type | Mainly surface living benthic invertebrates | Bottom sediments of lakes and rivers | Semi-quantitative or qualitative analysis depending on sampler | Expensive. Mobile organisms avoid sampler. Natural spatial orientation of organisms disturbed |
| Corer (e.g. Jenkins or made in-house) | Micro-organisms and benthic inverts living in sediment | Fine sediments, usually in lakes | Discrete, quantitative samples possible with commercial corers | Expensive unless made in-house. Small quantity of sample |
| Artificial substrates (e.g. glass slides plastic baskets) | Epiphytic algae, attached invertebrate species, benthic invertebrates | Open waters of rivers and lakes, weed zones, bottom substrates | Semi-quantitative compared to other methods for similar groups of organisms. Minimum disturbance to community on removal of sampler. Cheap | "Unnatural habitat", therefore, not truly representative of natural communities. Positioning in water body important for successful use |
| Poisons (e.g. rotenone) | Fish | Small ponds or river stretches | Total collection of fish species in area sampled | Destructive technique |
| Fish net/trap | Fish | Open waters, river stretches, lakes | Cheap. Non-destructive | Selective. Qualitative unless mark recapture techniques used |
| Electro-fishing | Fish | Rivers and lake shores | Semi-quantitative. Non-destructive | Selective technique according to current used and fish size. Expensive. Safety risk to operators if not carried out carefully |

Source: After Friedrich *et al.*, 1996

benthic organisms using a standardised handnet technique), to fully quantitative techniques (such as bottle samples for plankton or grab samples for benthic organisms). Ecological methods based on biotic or community structure indices require the use of quantitative or semi-quantitative methods.

The simplest and cheapest method of collecting benthic invertebrates in shallow, flowing waters is by means of a standard handnet (Figure 11.3) as described below.

*Procedure*

1. The net is held vertically against the river or stream bed, with the mouth facing upstream.
2. In shallow water the operator turns over the stones by hand for a defined area immediately upstream of the net. Dislodged organisms are carried into the net by the water current. In deeper water the toe or heel of the operator's boot can be used to kick the substrate in a defined area for a fixed period of time in order to dislodge organisms.
3. Stones from the defined area are examined and any attached organisms dislodged into the net.
4. The fine substrate from the sample area is also disturbed.
5. The contents of the net are gently washed into a corner of the net using flowing water and then the net turned inside out and gently shaken (and/or washed) into a sample container. Organisms clinging to the net should be removed to the sample container by hand.
6. Ideally, organisms should be identified and enumerated live but samples can be preserved for later sorting, identification and counting (see Table 11.3 for various methods of preservation).
7. To reduce the volume of sample requiring preservation and storage, net samples can be gently washed through a 500 μm sieve and placed in a white dish containing water. The organisms can then be sorted from the stones and debris with a wide mouth pipette and placed in a labelled sample jar.
8. The net should be thoroughly washed before taking the next sample.

Quantitative samples may be obtained in water with a flow rate greater than 10 cm $s^{-1}$ using a Surber sampler (Figure 11.3) following the same basic procedure given above for the handnet. The hinged metal frame quadrat is placed flat on the substrate with the mouth of the sample net facing upstream. The substrate within the square of the frame is stirred and lifted and organisms are dislodged into the net. The vertical side-flaps help reduce the loss of organisms around the sides of the net. When the stream bed is uneven, it may be necessary to use a foam rubber strip on the lower edge of the metal frame to close any gaps through which organisms may escape.

Slow flowing or static water, such as lakes and ponds, should be sampled using grabs, corers or dredges. Where the water is shallow and the substrate is soft, a semi-quantitative benthic sample can be obtained by pushing one end of a 25 cm diameter plastic pipe vertically into the substrate. The

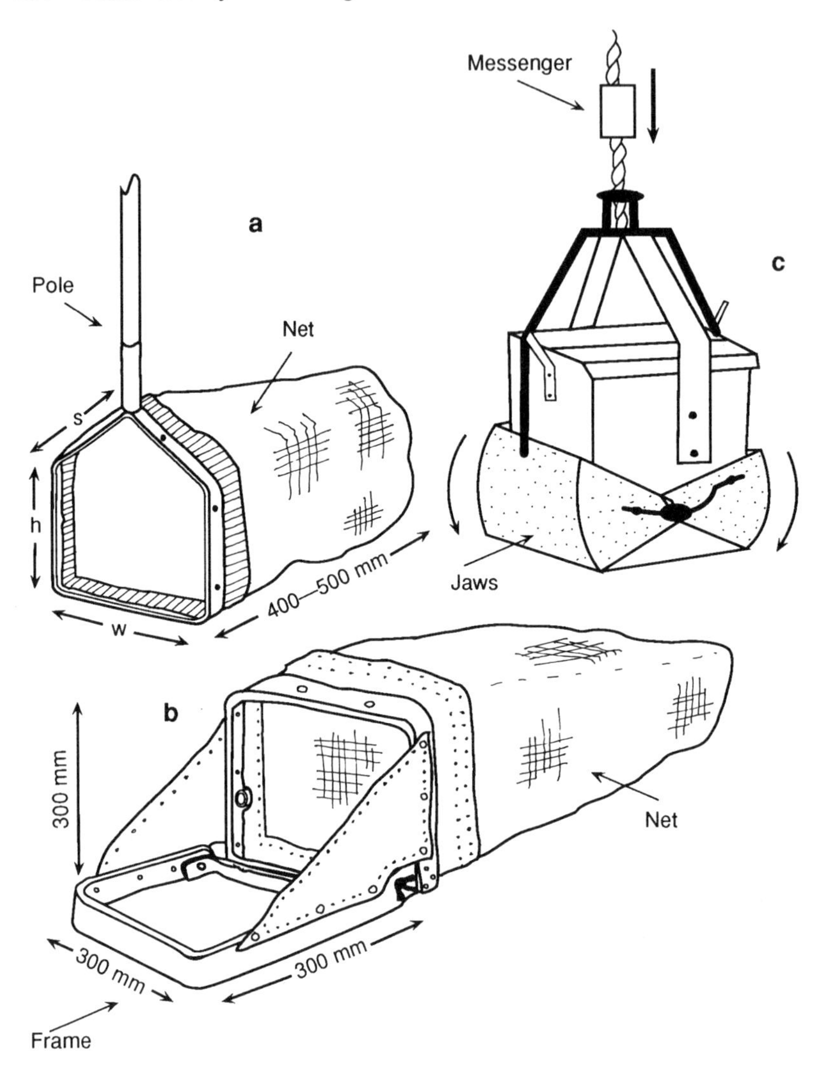


**Figure 11.3** Various methods of sampling benthic invertebrates: a. handnet, b. Surber sampler, c. grab, e.g. Ekman grab

sediment inside the tube is then removed with a plastic beaker or fine-mesh net. In deeper water, unless access is available from a bridge or pier, a grab or corer must be used from a boat. A grab (Figure 11.3) has jaws which close beneath a known area of sediment when triggered by a messenger released down the suspension wire from the boat. The entire content of the grab represents a quantitative sample from a known area. It is usually necessary to use a corer in very fine sediments. A corer consists of a perspex cylinder (usually

15 cm diameter) with end flaps which close when triggered automatically, or by a messenger, so that the mud and the water above it are held firmly in the tube. A dredge scrapes the surface layer collecting dislodged organisms as it is towed along.

*Size of sample*

The size of the sample required depends on the requirements of the ecological method and the statistical techniques which will eventually be applied to the data. Some methods require that several samples (sample units) are taken in different habitats from the sample area and pooled to create the representative sample. In other cases the single sample unit (e.g. one handnet kick sample or one grab sample) may yield insufficient numbers for statistical analysis and several units may be required for a single sample.

If the method requires an estimate of the populations of organisms from a sample site, the dimension of the sampling unit, the number of units within the sample and the location of the sampling units are all important. It is usually acceptable, for ecological surveys, to estimate the population with a precision of 20 per cent (i.e. $D = 0.2$ in the formula below). Using the results obtained from a preliminary survey, an estimate of the required number of sample units for a given precision can be calculated from:

$$n = \left( \frac{s}{D \times \bar{x}} \right)^2$$

where $D$ = index of precision
$\bar{x}$ = mean
$s$ = standard deviation
$n$ = number of sample units

The sample units should normally be selected at random within the sample site. A grid of numbered sampling units can be allocated to the sampling site and the units selected by reference to a random number table. Further detail on the selection of sample numbers and location for statistical treatment of data is available in the companion guidebook *Water Quality Assessments*.

*Artificial substrates*

Where the collection of samples using traditional methods is difficult, such as in deep, fast-flowing rivers, or where it is difficult to find sites with similar physical characteristics, artificial substrates may be suitable. There are many types of artificial substrates designed for different groups of organisms. Two types of artificial substrates for invertebrates are illustrated in Figure 11.4, but other substrates include synthetic foam blocks and glass slides for

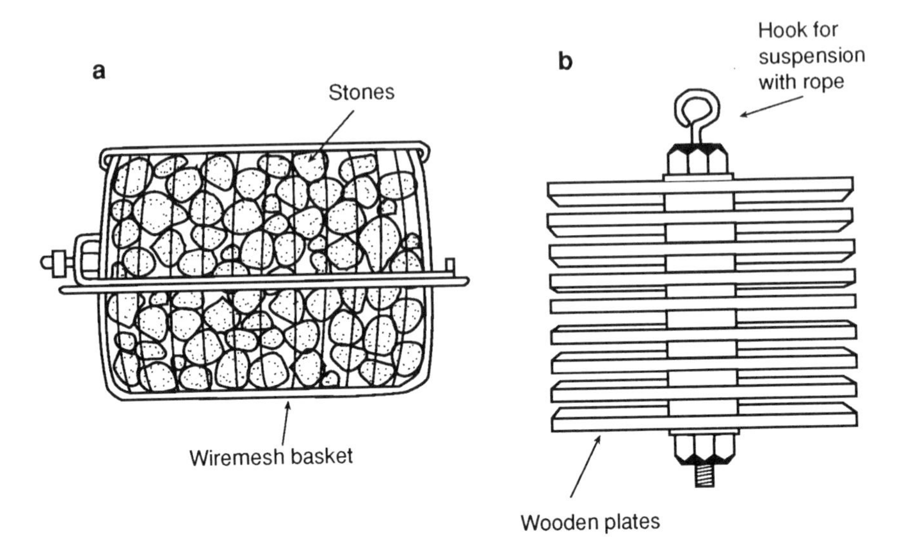


**Figure 11.4** Two types of artificial substrates for sampling aquatic organisms (Modified from Mason, 1981)

micro-organisms or periphytic algae, and rope or plastic strips to imitate plants. The size of the artificial unit can be selected according to the habitat it is intended to represent and the kind of organisms that will colonise it. Some preliminary experimentation may be required to ensure that sufficient numbers of organisms can be collected from the size of artificial unit in use.

An artificial substrate represents a standardised sampling unit (provided it is deployed according to standardised procedures), allowing greater precision when comparing sites and reducing the natural patchiness that tends to occur within habitats. Less skill and training is required for routine collection of artificial substrates, although an experienced specialist is necessary for selecting the sample sites at which they are to be deployed.

Although artificial substrates offer several advantages, their suitability should be thoroughly tested before routine use. It takes several weeks for effective colonisation and the species composition may change if the substrate is left in place for too long. A reasonably representative community of invertebrates generally develops in about six weeks. It must be recognised that the artificial substrate will be selective for certain types of organisms and the results are not, therefore, directly comparable with other sampling techniques. However, they can be used to assess the potential of a site to support certain aquatic life purely on the basis of its water quality (and not its habitat availability). Unfortunately, artificial substrates are subject to theft or vandalism and they may be swept away in extremely high flows.

### 11.2.4 Field data and habitat assessment

Habitat assessment forms an important aspect of the field work component of ecological methods because any change (natural or unnatural) in the habitat will lead to a change in the ecological balance (species, abundance and diversity of organisms present). Consequently, a field record sheet, which describes the habitat condition and indicates any signs of contamination, should be completed for each sampling site. An example of a field record sheet is given in Figure 11.5. Details relating to the specific samples taken should also be recorded in the field notebook as described in Chapter 5 (see Figure 5.7).

## 11.3 Measurement of chlorophyll *a*

A relatively simple estimate of the total mass (i.e. biomass) of algal material present can be obtained from the concentration of the photosynthetic pigment, chlorophyll. There are three main chlorophyll pigments commonly found in phytoplankton: *a, b* and *c*, but chlorophyll *a* is the one measured most often. This pigment aids the assimilation of nutrients into cell biomass by harnessing the energy of sunlight. Its concentration is related to the quantity of cell carbon. Phytoplankton chlorophyll is one of the most commonly used biological measurements in water quality monitoring and assessments, particularly in relation to the effects of increasing nutrients or in relation to the operation of water treatment processes.

The analytical procedure for chlorophyll *a* is given in section 7.9. Water samples are usually collected from the same sites and at the same time as other measurements, using a standard water sampling technique. Chlorophyll *a* concentrations may vary with water depth depending on the penetration of light (necessary for algal photosynthesis) and whether there is sufficient turbulence to mix the algae within the water column. An integrated measure for the productive water layer (euphotic zone) can be obtained by taking depth samples and combining them or by using an integrated sampling technique, such as a hose-pipe sampler or pump sampler (see section 5.4.1). Ideally, water samples should be filtered immediately (in the field) through a glass fibre (GF/C Whatman) filter but, if necessary, they can be stored for a few hours in a cool place (out of direct sunlight) in polyethylene bottles. Zooplankton should be removed from bottled samples by passing the sample through a suitable net (e.g. 300 μm mesh).

## FIELD RECORD

Water body________________________ Date__________________ Time__________

Name and number of sample site__________________________ No.______________

| Weather | | | | |
|---|---|---|---|---|
| Rain | None ☐ | Last 24 hours ☐ | Duration ________ | |
| Cloud cover | None or little ☐ | Variable ☐ | Extensive ☐ | |

| Water conditions | | | | | | |
|---|---|---|---|---|---|---|
| Width (m) | <1 ☐ | 1–2 ☐ | 3–5 ☐ | 6–10 ☐ | 11–20 ☐ | >20 ☐ |
| Depth (m) | <0.1 ☐ | 0.1–0.3 ☐ | 0.4–0.6 ☐ | 0.7–1.0 ☐ | >1.0 ☐ | |
| Currents | No turbulence ☐ | Still ☐ | Turbulent ☐ | Strong turbulence ☐ | | |
| Estimated velocity (m $s^{-1}$) | <0.2 ☐ | 0.2–0.4 ☐ | 0.5–0.8 ☐ | >0.8 ☐ | | |
| Turbidity | None ☐ | Little ☐ | Moderate ☐ | Strong ☐ | | |
| Colour | Coloured ☐ | Uncoloured ☐ | | | | |
| Odour | Absent ☐ | Present ☐ | Chemical ☐ | $H_2S$ ☐ | Other ☐ | |

| Substrate | | | | Score: | |
|---|---|---|---|---|---|
| Boulders | 1 2 3 | Silt | 1 2 3 | | |
| Large stones | 1 2 3 | Detritus | 1 2 3 | Rare <25% | 1 |
| Small stones | 1 2 3 | Mosses | 1 2 3 | Intermediate 25–50% | 2 |
| Filamentous algae | 1 2 3 | Submerged plants | 1 2 3 | Dominant >50% | 3 |
| Emergent plants | 1 2 3 | Marginal plants | 1 2 3 | | |

| Habitat | | | | |
|---|---|---|---|---|
| % still water | Very little ☐ | <25% ☐ | 26–50% ☐ | >50% ☐ |
| % flowing water | Very little ☐ | <25% ☐ | 26–50% ☐ | >50% ☐ |
| Shade | None ☐ | Little ☐ | Moderate ☐ | Heavy ☐ |
| River protection measures | None ☐ | Concrete banks ☐ | | Others ☐ |

| Reducing conditions | | |
|---|---|---|
| Reducing conditions absent ☐ | | Reducing conditions present ☐ |
| Bubble-forming sediments | Yes ☐ | No ☐ |
| Black sediment | Superficial ☐ | Below sediment surface ☐ |
| Undersides of stones | None ☐ | Partial ☐ | Total ☐ |

| Signs of pollution | | | | |
|---|---|---|---|---|
| Open water | None ☐ | Sewage ☐ | Oil ☐ | Others ☐ |
| Margins | None ☐ | Sewage ☐ | Oil ☐ | Others ☐ |

Other comments________________________

________________________________________

________________________________________

Observer's signature________________

Physical-chemical measurements:

pH__________ Air Temp °C__________

Diss. $O_2$ (mg $l^{-1}$)________ Water Temp °C________

Conductivity (μS $cm^{-1}$)________

**Figure 11.5** Typical field record sheet for biological sampling and ecological surveys (Based on an example used by Fundaçao de Tecnologia de Saneamento Ambiental (FEEMA), Rio de Janiero)

*Procedure*

1. Gently mix the sample by inverting the bottle (avoiding bright light) and measure a suitable volume into a measuring cylinder (between 100 ml and 1 litre depending on the turbidity of the sample).
2. Using a gentle vacuum (i.e. a hand-operated vacuum pump), pass as much of the measured sample through a fresh GF/C filter paper as possible.
3. Add 0.2 ml of $MgCO_3$ suspension (1.0 g $MgCO_3$ in 100 ml $H_2O$) to the last of the sample, as it is being filtered, to preserve it on the filter.
4. Do not allow the filter to dry whilst adding the sample to the filter cup.
5. Once the sample has passed through the filter rinse the sides of the filter cup with about 50 ml of distilled water.
6. Allow the filter to dry for a few seconds and then fold it in half, with the sample folded inside.
7. Place the filter in a Petri dish or small polythene bag labelled with the sample identifier and volume filtered and, if storage is necessary, place the filter in its container in the dark, in a freezer at –20 °C. If the sample cannot be frozen until several hours later keep it cool and in the dark (note the length and conditions of storage).
8. Do not store frozen samples for more than a few months.
9. Analyse the sample as described in section 7.9.

In addition to the extraction method described above, chlorophyll can be measured by fluorescence techniques. Fluorescence measurements can be made in the field using samples immediately after removal from the water body or, with a suitable instrument, they can be made *in situ*, underwater. Extensive surveys or detailed depth profiles can be made rapidly in this way. For proper interpretation, fluorescence measured in water samples should be correlated to other variables such as extracted chlorophyll (see Figure 14.2B) or cell numbers, and instruments should be calibrated frequently against extracted chlorophyll measurements.

## 11.4 Physiological techniques

For the purposes of water quality monitoring, the most widely exploited physiological responses of aquatic organisms to environmental stress are production, respiration and growth rates. Most of these responses have been developed for biological monitoring under controlled conditions, such as during bioassays (see section 11.5). The growth criteria (light, nutrients, temperature) for some common freshwater algal species have been well studied and documented and several methods based on algal growth rates have now been standardised. In addition to being themselves affected by variations in water quality, phytoplankton can also directly affect water quality, particularly oxygen concentrations, turbidity and even toxicity to consumers (some algal species release chemicals which are toxic to livestock and humans when in sufficiently high concentrations). Measures of algal growth have been incorporated into some water quality management models.

For most physiological methods the results can only be considered as relative. Nevertheless, such methods are useful for monitoring large areas, along long river stretches, or for short, intensive programmes. In addition, some methods are particularly useful for monitoring the effects of effluents, where measurements can be made upstream and downstream of the discharge. An example of a relatively simple method using biological samples from the sites of interest is presented below.

Care must be taken to follow the specified procedures accurately if physiological methods are to be used successfully. Unless the method specifies a particular technique for collecting water samples, standard water samplers can be used (see section 5.5). If incubations are to be carried out *in situ*, care must be taken to avoid subjecting the samples and controls to adverse or unusual environmental conditions which might affect respiration or photosynthesis rates (such as sudden or extreme changes in temperature or light conditions).

### 11.4.1 Oxygen production and consumption

The net production of oxygen by phytoplankton can be used as an indicator of the activity of algae (see Figure 14.4) and of possible toxic inhibition, especially when correlated with the concentration of chlorophyll pigments. The Oxygen Production Potential has been developed as a standardised method and consists of incubating water samples containing native phytoplankton in light and dark bottles for a given time (usually 24 hours) and measuring the oxygen production. The incubations can be carried out in the bottles *in situ* (suspended in the water body) or in an incubator with heat and light. Toxic inhibition should be suspected where the Oxygen Production Potential decreases per unit of algal biomass (measured as chlorophyll). The Oxygen Production Potential can also be expressed as the equivalent concentration of chlorophyll required to generate 10 mg $l^{-1}$ of oxygen.

An alternative method is to measure oxygen consumption resulting from respiration by bacteria in the test water samples. Substrates which stimulate bacterial growth (e.g. peptone or glucose) are added to the water samples and the concentration of oxygen after incubation is compared with controls without growth stimulators. Greater oxygen consumption in the test samples suggests bacterial activity is normal, but if oxygen consumption ceases or is very low, the samples may contain a substance which is toxic to the bacteria.

## 11.5 Controlled biotests

Bioassay methods can be used to reveal or confirm the presence of toxic conditions in water bodies as well as to provide information on the toxicity of effluents. Bioassay methods can be used to demonstrate the presence of "unknown" contaminants, to locate the position of diffuse or point discharges of contaminants or to monitor the dispersion of known toxic discharges. In addition, such methods are useful for evaluating persistence and the combined effects of several contaminants or effluents.

Toxicity tests and bioassays can be used to evaluate the necessity or urgency for chemical analysis when many samples have been taken, for example, following an accidental pollution incident. A rapid evaluation of toxicity can highlight the samples requiring immediate chemical analysis, thereby focusing the attention of the chemical analysis laboratory on samples from areas where the greatest environmental effects, and possible risks to water users, may be found.

Acute toxicity is usually indicated by death and standard laboratory toxicity tests use this response to assess the lethal concentration of a sample or compound. Bioassays usually measure more subtle, sub-lethal effects in organisms under defined conditions (for a particular temperature or time interval). Bioassays can be carried out *in situ* in the water body, or in the laboratory using samples of water collected from the field. When compared with acute tests, sub-lethal methods are generally found to be more sensitive to the dilutions of contaminants expected in water bodies after discharge and dispersion. Many sub-lethal responses can be detected by physiological measurements (see also previous section), such as growth rates or oxygen production, while others can be manifest by behavioural, biochemical or mutagenic changes.

Test organisms can be exposed *in situ* and the response measured at regular time intervals or they can be placed at selected sites over a study area and the results evaluated from all sites for the same time interval. This is known as active biomonitoring. Responses such as mortality, growth and reproduction are commonly used in these situations. Some physiological or behavioural responses, such as swimming activity, algal fluorescence or oxygen consumption, are suitable for continuous monitoring in early warning systems (see below). Physical and chemical variables should be monitored simultaneously during any *in situ* exposure in order to help eliminate effects on the test organisms other than those due to the contaminant under investigation.

It is very important, when conducting or developing bioassays, to select organisms which are compatible with the water samples being tested, i.e.

**Table 11.5** Examples of standardised bioassay techniques

| Organism | Response | Exposure time | Standard method |
|---|---|---|---|
| Bacteria | | | |
| *Pseudomonas putida* | Inhibition of oxygen consumption | 30 minutes | DIN 38 412 Part 8 (1991)<br>DIN 38 412 Part 27 (1993) |
| *Vibrio fischeri*[1] | Inhibition of luminescence | 30 minutes | DIN 38 312 Part 34 (1991)<br>DIN 38 312 Part 341 (1993) |
| Invertebrates | | | |
| *Daphnia magna* | Immobilisation | 24 hours | DIN 38 412 Part 11 (1982)<br>DIN 38 412 Part 30 (1989)<br>OECD 202 Part I (1984) |
| | Reproduction | 14 days | OECD 202 Part II (1993) |
| Fish | | | |
| *Brachydanio rerio* | Death | 1–2 days | ISO 7346/1/2/3 (1984)<br>OECD 203 (1992) |
| | Sub-lethal effects | 14 days | OECD 204 (1984) |
| | Death | 28 days | DIN 38 412 Part 15 (1982)<br>DIN 38 412 Part 20 (1981)<br>DIN 38 412 Part 31 (1989) |
| *Oncorhynchus mykiss* | Sub-lethal effects | 14 days | OECD 204 (1984) |
| Algae | | | |
| *Scenedesmus subspicatus* | Growth rate, reproduction | 3 days | ISO 8692 (1989)<br>DIN 38 412 Part 33 (1991) |
| *Selenastrum capricornutum* | Growth rate, reproduction | 3 days | ISO 8692 (1989)<br>OECD 201 (1984) |

[1] *Phytobacterium phosphoreum*

organisms which tolerate the normal physical and chemical conditions of the test water (such as salinity, temperature, hardness). There are now many standardised bioassays using widely available organisms which are easily maintained in the laboratory (Table 11.5). However, the relevance of the chosen tests to the individual water body and its associated water quality problems needs to be carefully evaluated before use. For tests carried out *in situ* it is important to use indigenous organisms.

The following section describes a rapid test for determining the effects on the microcrustacean organism *Daphnia magna* resulting from chemicals or substances in a water sample or from municipal or industrial wastewater samples. The precise details of other standardised tests are available in the relevant documentation (see section 11.9). Some general guidance on site selection and water sampling is given in section 11.7.

### 11.5.1 *Daphnia magna* immobilisation test

*Daphnia magna* Straus are members of the zooplankton of still waters and obtain their nutrition by filter feeding on particulate organic matter. The effects of a water sample or an effluent on the *Daphnia* can be determined and expressed as the concentration at the start of the test which results in 50 per cent of the test organisms becoming immobile and incapable of swimming during the 24 hours of the test period. This concentration is known as the effective inhibitory concentration and is designated as the 24h—$EC50_i$. If it is not possible to record the 24h—$EC50_i$, or even a 48h—$EC50_i$, useful information can be gained from the lowest concentration tested which immobilises all the *Daphnia* and the highest concentration tested which does not immobilise any of the organisms.

A preliminary test using a single series of concentrations can be carried out which determines the range of concentrations for use in the final toxicity test and gives an approximate value of the 24h—$EC50_i$ (or, where appropriate, the 48h—$EC50_i$). When the approximate value obtained in the preliminary test is not adequate a definitive test is carried out. This allows calculation of the 24h—$EC50_i$, the 48h—$EC50_i$ and determination of the concentrations corresponding to 0 per cent and 100 per cent immobilisation. The range of concentrations of the test solution must be chosen to give at least three percentages of immobilisation between 10 per cent and 90 per cent.

The test should be carried out on samples within 6 hours of collection whenever possible. Samples can be stored for up to 48 hours if cooled to 4 °C immediately on collection, or they can be frozen and used for testing within 2 months.

*Apparatus*

- ✓ pH meter.
- ✓ Oxygen meter.
- ✓ Glass beakers, 50-ml and 2,000-ml, washed and rinsed first with distilled water and then with dilution water.
- ✓ Culture tubes, 18 mm diameter, 180 mm high, washed and rinsed first with distilled water and then with dilution water.
- ✓ Petri dishes, washed and rinsed first with distilled water and then with dilution water.
- ✓ Matt black surface.
- ✓ Sieves, nylon meshes of 0.65 mm and 0.20 mm apertures.
- ✓ Graduated flasks, 100-ml and 1,000-ml.
- ✓ Graduated pipettes, 1, 10 and 25 ml.
- ✓ Pasteur pipettes.

*Consumables*

Analytical reagent grade chemicals and pure (e.g. deionised or distilled) water must be used.

- ✓ Calcium chloride solution: dissolve 11.76 g of calcium chloride dihydrate, $CaCl_2.2H_2O$, in water and make up to 1,000 ml.
- ✓ Magnesium sulphate solution: dissolve 4.93 g magnesium sulphate heptahydrate, $MgSO_4.7H_2O$, in water and make up to 1,000 ml.
- ✓ Sodium bicarbonate solution: dissolve 2.59 g of sodium bicarbonate, $NaHCO_3$, in water and make up to 1,000 ml.
- ✓ Potassium chloride solution: dissolve 0.23 g of potassium chloride, KCl, in water and make up to 1,000 ml.

✓ Dilution water prepared by transferring 25 ml, by pipette, from each of the above four solutions into a graduated flask and making the volume up to 1,000 ml. Dilution water should be made with water of maximum conductivity 10 μS $cm^{-1}$. The final pH must be 7.8 ± 0.2 and hardness 250 mg $l^{-1}$ (as $CaCO_3$) with a molar Ca:Mg ratio close to 4:1. The dissolved oxygen concentration must be greater than 7 mg $l^{-1}$. Natural water of similar pH and hardness may be used for culture. Dilution water must be aerated until the pH has stabilised and the dissolved oxygen concentration has reached saturation. The pH can be adjusted with sodium hydroxide or hydrochloric acid.
✓ Potassium dichromate, $K_2Cr_2O_7$, of analytical grade.
✓ *Daphnia magna* Straus of at least third generation. A clone culture can be purchased or raised in the laboratory.
✓ Food for *Daphnia*, e.g. single-celled green algae.

*Procedure*

1. Start the culture by adding no more than 100 adult *Daphnia* per litre of culture water in large glass beakers. Add a quantity of food which can be consumed within a few days (the water in the beaker will go from green to clear as the algae are consumed).
2. Keep the culture either in the dark or under a 16 h/8 h light/dark photoperiod at 20±2°C and change the culture water once or twice a week. When changing the water separate adults from young by passing the culture through a fine and a coarse mesh sieve. The adults are retained on the coarse sieve and the younger ones on the fine sieve.
3. To obtain *Daphnia* less than 24 hours old for the test, select gravid females from the adults retained on the coarse sieve and transfer into a fresh culture vessel. Collect newly released neonates with a fine sieve within 24 hours.
4. Using dilution water, prepare the desired dilutions or concentrations of the test sample or effluent immediately before conducting the test.
5. Pour 10 ml of each test solution into a series of culture tubes or 20 ml into a series of beakers. For samples with a high oxygen consumption use 20 ml volumes in Petri dishes.
6. Add no more than 20 *Daphnia* to each container of test concentration or dilution (a maximum of 5 *Daphnia* per 10 ml of test solution is recommended). For each test series prepare one control of dilution water only (using the same volume of solution and number of *Daphnia* as used for the test solutions).
7. Place control and test samples in the dark or 18 h/8 h light/dark at 20 ± 2 °C for 24 hours without food.
8. After 24 hours gently agitate the liquid in each beaker for 15 seconds and place it on a matt black surface. Count and record the number of *Daphnia* in each beaker which do not immediately exhibit swimming behaviour when the water is agitated.
9. Determine the concentration range giving 0 per cent to 100 per cent immobilisation and note abnormalities in the behaviour of the *Daphnia*.
10. Record the dissolved oxygen concentration in the test container with the solution of lowest concentration at which all the *Daphnia* were immobilised.
11. Periodically determine the 24h—$EC50_i$ (following the procedure above) of the potassium dichromate reference solution (by making up a range of concentrations with dilution water). This verifies the sensitivity of the *Daphnia*. If the 24h—$EC50_i$ falls outside the concentration range 0.6–1.7 mg $l^{-1}$ the application of the test procedure

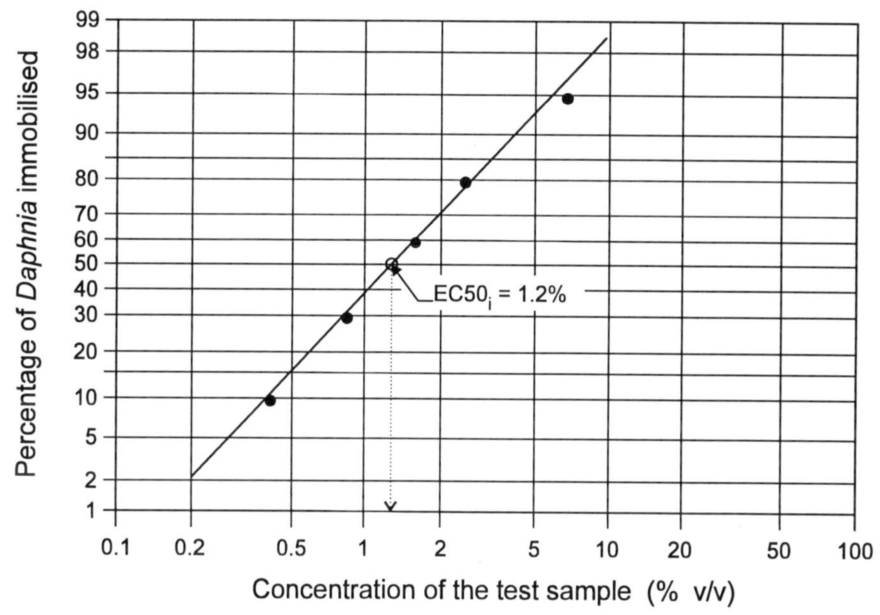


**Figure 11.6** Graphical representation of the results from an acute *Daphnia* test from which the $EC50_i$ value can be interpolated

should be checked, as well as the culture technique for the *Daphnia*. A new strain of *Daphnia magna* may be necessary.

*Evaluation and expression of results*

Calculate the percentage immobilisation for each test concentration. The $EC50_i$ value can then be obtained by calculation or by interpolation from a graph such as that in Figure 11.6, where the concentration or dilution of the test sample is plotted on the x-axis (logarithmic values on probability paper can be used) and the corresponding percentage of the *Daphnia* incapable of swimming are plotted on the y-axis. The highest concentration of the sample in which all *Daphnia* remained capable of swimming (0 per cent immobilisation) and the lowest concentration at which all *Daphnia* were incapable of swimming (100 per cent immobilisation) should also be recorded, particularly if data are insufficient to calculate the 24h—$EC50_i$.

The test is only valid a) if less than 10 per cent of the *Daphnia* in the control are incapable of swimming, b) if the $EC50_i$ value of the potassium dichromate lies in the range 0.6–1.7 mg $l^{-1}$ and c) if the dissolved oxygen concentration at the end of the test was $\geq$ 2 mg $l^{-1}$.

Results are reported as a percentage dilution in the case of test samples or as mg $l^{-1}$ for specified chemical substances. The 24h—$EC50_i$ and the concentrations resulting in 0 per cent and 100 per cent immobilisation should be quoted, together with any other relevant information.

## 11.5.2 Early warning methods

Biological early warning systems (also referred to as biomonitors and dynamic tests) rely on a biological response in the test organisms placed *in situ* triggering further action when defined thresholds (expressed in terms of the biological response) are exceeded. The further action usually takes the form of a more detailed investigation of the causes, including chemical monitoring. If these systems are used close to important water intakes, such as for major drinking water supplies, the response can be used to trigger the temporary shut down of the intake while further investigations are carried out. When used close to major effluent discharges, the biological responses can be used to signal a sudden change in the nature of the discharge, such as may occur due to a treatment plant malfunction. Rapid measures to reduce environmental effects can then be taken without waiting for extensive chemical monitoring results.

There are many commercially available biological early warning systems and some can be fairly easily constructed to suit specific requirements. They are, however, expensive to set-up and to run efficiently. The most common organisms employed are fish, the crustacean *Daphnia* sp., algae and bacteria. The basic principle usually involves diverting water from the water body of interest, as a continuous flow, through special tanks containing the test organisms. The response of the organisms is continuously measured by an appropriate sensor device and recorded, for example by computer or by paper and pen recorder. Some semi-continuous flow tests take water into the tank containing the test organisms and then close the intake for a defined period during which the responses are recorded. The test water is then renewed for a new test cycle.

Certain organisms have been shown to be particularly suitable for indicating the presence (although not the nature or concentration) of certain chemicals. Fish and *Daphnia* are able to indicate the presence of many insecticides, and algae are able to indicate the presence of herbicides (through photosynthetic inhibition). In order to act as a general early warning mechanism for unspecified contaminants, several tests using organisms of different trophic levels should be used simultaneously.

Biomonitors using algae rely on detecting relative changes in *in-vivo* chlorophyll fluorescence at 685 nm. High energy lights (435 nm) are used to activate the chlorophyll *a* molecules of the algae and the energy which is not used for photosynthesis is emitted as fluorescence. Damaged algae (for example by the presence of toxic compounds) show increased fluorescence.

Methods using bacteria are based on the measurement of oxygen depletion in the test tank arising from normal bacterial respiration. If the oxygen concentrations are not reduced in the test water after contact with the bacteria the test sample is assumed to have had a toxic effect on the bacteria. The test water is first enriched with oxygen and oxygen electrodes are used to

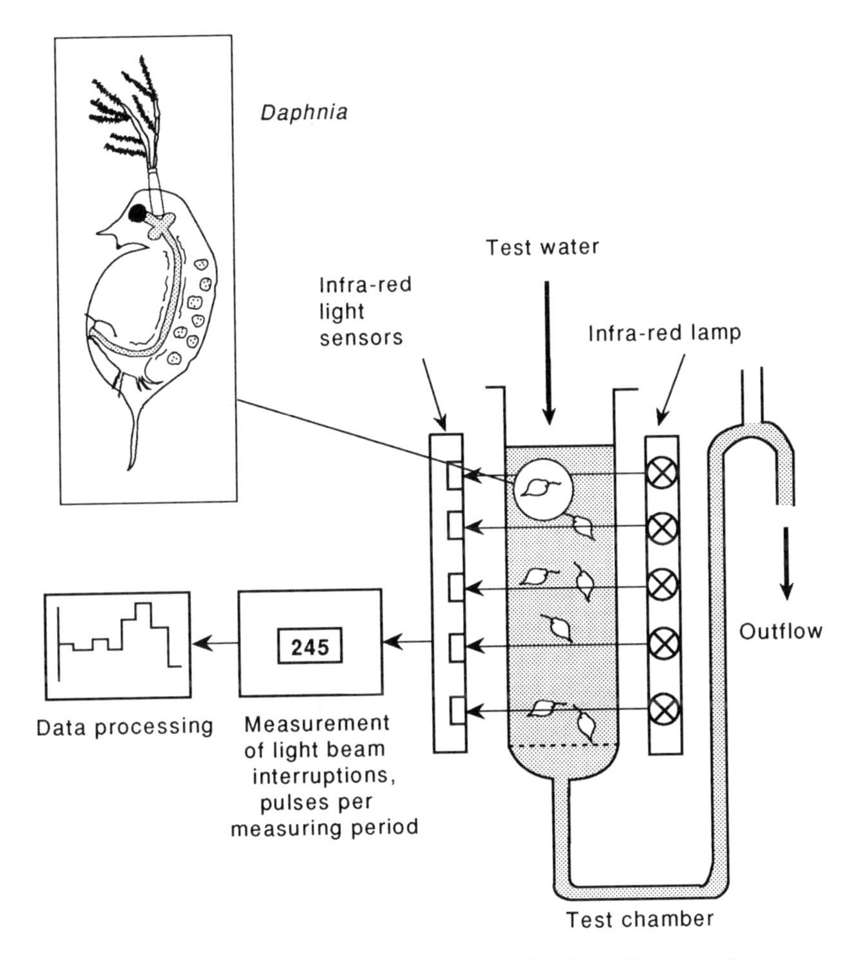


**Figure 11.7** A biological early warning system using *Daphnia* in a continuous flow-through chamber with light sensors to detect variations in motility (Details provided by the State Agency for Water and Waste, North-Rhine/Westphalia, Germany)

measure the initial oxygen concentration in the test sample and the residual oxygen in the contact tank containing the test bacteria.

The planktonic crustacean *Daphnia* sp. is an important member of many aquatic communities, where it forms a link in the food chain by grazing on algae and forming a food source for small and juvenile fish. The difference between the swimming activity of *Daphnia* in a control tank and a tank receiving test water is used as a measurable response to the presence of contaminants causing inhibition of movement. The swimming activity is measured by an array of light beams and light sensors (Figure 11.7). This dynamic *Daphnia* test has been shown to be highly sensitive and reliable over many years of operation.

Although the use of fish has the advantage that the test organisms are vertebrates (and hence closer to humans in their physiological and biochemical responses than other test organisms), their biological responses which are sufficiently sensitive to be used in early warning tests for unknown contaminants require complex measurement techniques. One approach uses a computer-controlled image processing system to detect and analyse the behaviour in a pair of continuous flow-through tanks each containing a shoal of six small fish. The activity of the fish in each tank is recorded by two video cameras and six behavioural parameters are analysed by the computer for each tank: i.e. motility, irregularity of swimming velocity, number of turnings, swimming height, mean horizontal position and shoaling behaviour. The results are compared statistically with "normal" behaviour patterns.

## 11.6 Contaminants in biological tissues

Many organisms have been found to accumulate certain contaminants in their tissues during their life time without detectable effects on normal physiological functions. This phenomenon is often known as bioaccumulation. The contaminants are effectively detoxified and/or tolerated within the organism, often as a result of binding into particular tissues in the body. Some organisms accumulate particular contaminants in this way until a critical body burden is reached which then triggers a physiological effect, other organisms regulate the body burden by excreting the contaminant. In some organisms, a sudden change in metabolic pattern (such as during breeding) causes the remobilisation of the accumulated contaminants, subsequently producing adverse physiological effects. The transfer of contaminants from one trophic level to another, through ingestion of organisms which have already bioaccumulated a contaminant, is known as food-chain transfer. This may lead to organisms higher in the food chain accumulating contaminants to concentrations much higher than occur in the ambient surroundings.

If the correlation between the concentration of a contaminant in a water body and the concentration in the tissues of an aquatic organism is good, the organism may be used for chemical monitoring of the contaminant in the water body as an alternative monitoring medium to water or particulate matter. Biological tissues sometimes have the advantage that the contaminant concentrations being analysed are much higher than in water or particulate matter, thus requiring lower detection limits and, possibly, less sophisticated analytical techniques. In addition, biological tissues can be bulked together to give a greater total concentration.

At present there are few species which can be used world-wide for chemical monitoring, although many indigenous organisms have been tested and found suitable for local use. Analysis of organisms such as fish and shellfish, which are consumed by human populations, provides information of direct relevance to human health. Active monitoring of the distributions of contaminant concentrations can be carried out by deliberately placing organisms at specified locations in water bodies and collecting them for analysis after suitable time intervals have elapsed during which bioaccumulation may have occurred. Examples of the use of organisms in chemical monitoring are given in the companion guidebook *Water Quality Assessments*.

### 11.6.1 Organism selection

Organisms suitable for chemical monitoring should be immobile (such as bivalve molluscs) or should remain or should be restrained (e.g. in a cage) within the area which is being studied so that that they are not reflecting the accumulation of contaminants from another area from which they may have migrated. Plants and animals have been found to be useful for monitoring contamination by metals, organic chemicals and radioisotopes, and many examples of organisms and methods can be found in the published literature.

When investigating suitable local organisms certain criteria should be tested and met:

- The organism should accumulate the contaminant of interest when exposed to the environmental concentrations present without experiencing lethal toxic effects.
- There should be a simple correlation between the concentration of the contaminant in the organism and the average concentration in the water body.
- Bioaccumulation should reach concentrations which allow direct analysis of organism tissues.
- The organisms should be abundant in, and representative of, the water body being monitored.
- The organism should be easily sampled.
- The organism should survive in the laboratory long enough for contaminant uptake studies to be performed.

Assuming all these criteria can be met for the chosen organism, it is also important that they are fulfilled at each sampling site under all the environmental conditions that may be encountered during the monitoring programme. In order to make valid comparisons between sites in a water body using this technique, the correlation between contaminant concentrations in the biological tissues and the water body must be the same for all sampling sites.

### 11.6.2 Procedures

Most chemical analyses of biological tissues are performed using standard techniques, such as atomic absorption spectrophotometry (AAS) with flame or graphite furnace for metals (see section 8.1), gas chromatography–mass spectrometry for organic micropollutants (see section 8.2) or high pressure liquid chromatography with fluorescence for polyaromatic hydrocarbons. Acid digests of the biological tissues are used for metal analyses and organic solvent extracts are used for organic contaminants. Usually between 10 g and 15 g of dry tissue is required and results are expressed as $\mu g\ g^{-1}$ dry weight or $mg\ kg^{-1}$ wet weight or dry weight. For contaminants with a strong affinity for lipids the results may be expressed as $mg\ kg^{-1}$ fat content.

The handling of tissues for chemical analysis requires the same degree of cleanliness as for water sample analysis in order to avoid contamination. Great care must be taken during all stages of sample handling, including collection in the field, transportation to the laboratory and during laboratory preparation. For example, where tissues are to be analysed for metals, only acid washed plastic or glass samplers, containers and instruments should be used and all procedures, such as dissecting, drying and weighing, should be carried out in a clean, preferably dust-free, environment. At least duplicate analyses should be made for each sample and sample blanks should be prepared with each batch of tissues in order to check for contamination (see Chapter 9).

There are many detailed descriptions, available in the published literature and specialised texts, of satisfactory techniques which have been developed for common contaminants in many different biological tissues. The basic procedure for metal analysis is outlined here as a basis for developing, or selecting methods, appropriate to the specific objectives of a monitoring programme.

*Procedure*

1. Select organisms of a similar species, sex, size (weight or length) and stage of life cycle. If whole organisms are to be analysed a period of depuration in uncontaminated water may be necessary (during which the organisms flush out the contents of their guts).

2. Sampled organisms should be transported and/or stored in specially cleaned containers appropriate to the future analyses, e.g. acid-cleaned plastic pots or new polythene bags. If analyses will not be performed for several days or weeks, samples (whole organisms or dissected tissues) should be frozen immediately and stored at –20 °C.

3. Weigh whole organisms or selected tissues if wet weight is to be used or oven- or freeze-dry to constant weight and then weigh if dry weights are to be used.

4. Add a small volume of high grade concentrated nitric acid to the dry tissue in an appropriately cleaned container (e.g. glass test tube with reflux bulb or PTFE beakers with covers). If the whole sample is too large for digestion about 10–15 g of dry tissue should be weighed out.

5. Heat the samples gently (e.g. on a hot plate at 70 °C) until digestion of the tissue is complete and no residue remains. The digest should be clear and not cloudy.
6. Make up the digest to a known volume (e.g. 50 ml) with analytical-grade water.
7. Analyse the digest, or suitable dilutions of it, using an appropriate technique (such as AAS, see section 8.1) and calculate the concentration of the metal in terms of the wet or dry weight of tissue (mg $kg^{-1}$ or µg $kg^{-1}$). Allowance must be made for any measurements obtained from the blank digests.

## 11.7 Site selection and sampling frequency

Selection of sampling site location and sampling frequency depend on the objectives of the monitoring programme, together with any special requirements of the biological techniques being used. Consideration should be given to sites where other physical and chemical measurements are made. The criteria for site selection are, therefore, fundamentally the same as has been discussed in Chapter 3. Reference can be made to any earlier biological monitoring results as an aid to site selection.

Most of the general principles for site selection in relation to the programme objectives described in Chapter 3 are applicable for ecological methods. However, the suitability of the site for the anticipated sampling method also has to be tested. It is preferable that ecological samples are taken at the same sites as hydrological measurements and samples for physical and chemical analysis, in order to have the maximum information possible available for interpretation of any ecological effects observed.

Typically, ecological methods are used for long-term evaluation of biological water quality by sampling once a year in the same place and at the same time of year. Intensive surveys to assess baseline ecological quality may be conducted several times a year in the first instance in order to determine seasonal variations in the organisms present. The life cycles of organisms and the ability for their populations to recover from sampling activities which totally remove all species, must be taken into consideration.

Since many bioassay and physiological techniques rely on detecting a biological response which differs from normal, it is usually essential to find additional sites which can act as controls, i.e. the water is uncontaminated and the organisms found there are not subject to the same environmental stress (such as upstream of discharges to a river). Recovery sites (beyond the mixing zone of contaminants) should also be selected, as well as sites which cover the full range of the suspected pollution gradient, e.g. at intervals downstream of a discharge to a river, or in a grid radiating outwards from a point source in a lake. Water samples for bioassay techniques should be taken, whenever possible, when the environmental dilutions of contaminant inputs are at their lowest, such as during low flow periods in rivers.

Discrete or composite water samples can be taken with most water samplers as described in section 5.5. The frequency of sampling is determined by the objectives, as for all kinds of monitoring. Where accidental contamination is being monitored, the frequency of sampling will be high (e.g. daily) and determined by the severity and duration of the effects, although the activity might only continue for a total period of weeks. Early warning systems are usually in operation, *in situ,* continuously.

The same principles apply to sampling biological materials for chemical analysis as for taking water samples for similar analyses. For passive monitoring, collection of organisms of a particular growth stage or period of the life cycle may require sampling at a particular time of year. Active monitoring requires placement of the organisms in carefully selected control and exposure sites which are readily accessible and environmentally suitable for the test organisms.

## 11.8 Quality assurance

The general principles of quality assurance described in Chapter 9 should also be applied to all biological methods. The precise details of the methods adopted for a monitoring programme should be recorded in the standard operating procedures and made available to all personnel involved. This includes the details of field sampling techniques (e.g. length of time required for an invertebrate kick sample, precise orientation of an artificial substrate in relation to flow direction, etc.) as well as the laboratory-based procedures. This ensures that all personnel use exactly the same procedures for a particular monitoring programme, regardless of techniques they may have used elsewhere. Relevant observations should be recorded at all steps in the procedure, from the field record to the laboratory worksheets and the final calculations, so that discrepancies or unusual results can be checked back against all the raw data. Any deviations from the standard operating procedure should be recorded.

In order to compare the results of biological monitoring obtained by different laboratories or organisations, it is necessary to use generally accepted and standardised methods. With the increasing interest in, and use of, biological approaches an increasing number of methods are being tested and standardised at national and international level. The precision of the methods should be known and, where possible, acceptable limits of accuracy should be set according to the requirements of the monitoring programme rather than defined by the easiest method available.

## 11.9 Source literature and further reading

Abel, P.D. 1989 *Water Pollution Biology*. Ellis Horwood Limited, Chichester.

APHA 1992 *Standard Methods for the Examination of Water and Wastewater*. 18th edition. American Public Health Association, American Water Works Association and Water Pollution Control Federation, Washington, DC.

Boudou, A. and Ribeyre, F.V. [Eds] 1989 *Aquatic Ecotoxicology: Fundamental Concepts and Methodologies*. Volumes 1 and 2, CRC Press, Boca Raton.

Calow, P. [Ed.] 1993 *Handbook of Ecotoxicology*. Blackwell Scientific Publications, Oxford.

Chapman, D. [Ed.] 1996 *Water Quality Assessments. A Guide to the Use of Biota, Sediments and Water in Environmental Monitoring*. Second edition. Chapman & Hall, London.

de Zwart, D. 1995 *Monitoring Water Quality in the Future. Volume 3, Biomonitoring*. Ministry of Housing, Spatial Planning and the Environment (VROM), The Hague.

DIN 1981 *Determination of the Effect of Waste Water and Industrial Effluents on Fish (L20) (Fish-Test)*. DIN 38 412 Part 20, Deutsches Institut für Normung e.V. (German Institute for Standardization), Berlin.

DIN 1982 *Determination of the Effect of Substances in Water on Microcrustaceans (*Daphnia*-Short-Term Test)*. DIN 38 412 Part 11, Deutsches Institut für Normung e.V. (German Institute for Standardization), Berlin.

DIN 1982 *Determination of the Effect of Substances in Water on Fish (L15) (Fish-Test)*. DIN 38 412 Part 15, Deutsches Institut für Normung e.V. (German Institute for Standardization), Berlin.

DIN 1989 *Determination of the Non-acute Poisonous Effect of Waste Water to* Daphnia *by dilution limits (L30)*. DIN 38 412 Part 30, Deutsches Institut für Normung e.V. (German Institute for Standardization), Berlin.

DIN 1989 *Determination of the Non-acute Poisonous Effect of Waste Water to Fish by Dilution Limits (L31)*. DIN 38 412 Part 31, Deutsches Institut für Normung e.V. (German Institute for Standardization), Berlin.

DIN 1991 *Determination of the Inhibitory Effect of Water Constituents on Bacteria (L8) (Pseudomonas Cell Multiplication Inhibition Test)*. DIN 38 412 Part 08, Deutsches Institut für Normung e.V. (German Institute for Standardization), Berlin.

DIN 1991 *Determination of the Non-acute Poisonous Effect of Waste Water to Green Algae by Dilution Limits (L33) (Scenedesmus Chlorophyll Fluorescence Test)*. DIN 38 412 Part 33, Deutsches Institut für Normung e.V. (German Institute for Standardization), Berlin.

DIN 1991 *Determination of the Inhibitory Effect of Waste Water on the Light Emission of* Photobacterium phosphoreum — *Luminescent Bacteria Waste Water Test Using Conserved Bacteria (L34).* DIN 38 412 Part 34, Deutsches Institut für Normung e.V. (German Institute for Standardization), Berlin.

DIN 1993 *Determination of the Inhibitory Effect of Waste Water on the Oxygen Consumption of* Pseudomonas putida *(L27).* DIN 38 412 Part 27, Deutsches Institut für Normung e.V. (German Institute for Standardization), Berlin.

DIN 1993 *Determination of the Inhibitory Effect of Waste Water on the Light Emission of* Photobacterium phosphoreum — *Luminescent Bacteria Waste Water Test, Extension of the Method DIN 38 412, Part 34 (L341).* DIN 38 412 Part 341, Deutsches Institut für Normung e.V. (German Institute for Standardization), Berlin.

Elliott, J.M. 1977 *Some Methods for the Statistical Analysis of Samples of Benthic Invertebrates.* 2nd edition, Scientific Publication No. 25, Freshwater Biological Association, Ambleside.

Elliott, J.M. and Tullett, P.A. 1978 *A Bibliography of Samplers for Benthic Invertebrates.* Occasional Publication No. 4, Freshwater Biological Association, Ambleside.

EN ISO 1995 *Water Quality — Sampling in Deep Waters for Macro-invertebrates — Guidance on the Use of Colonization, Qualitative and Quantitative Samplers.* European Standard EN ISO 9391:1995 E. European Committee for Standardization, Paris.

Friedrich, G., Chapman, D. and Beim, A. 1996 The use of biological material. In: D. Chapman [Ed.] *Water Quality Assessments. A Guide to the Use of Biota, Sediments and Water in Environmental Monitoring.* Second edition. Chapman & Hall, London.

Gilbert, R.O. 1987 *Statistical Methods for Environmental Pollution Monitoring.* Van Nostrand Reinhold Company, New York.

Gruber, D.S. and Diamond, J.M. 1988 *Automated Biomonitoring: Living Sensors and Environmental Monitors.* Ellis Horwood Publishers, Chichester.

Hellawell, J.M. 1986 *Biological Indicators of Freshwater Pollution and Environmental Management.* Pollution Monitoring Series, K. Mellanby [Ed.], Elsevier Applied Science, Barking.

ISO 1984 *Determination of the Acute Lethal Toxicity of Substances to a Freshwater Fish* Brachydanio rerio *Hamilton-Buchanan (Teleostei, Cyprinidae). Part 1: Static Methods. Part 2: Semi-static Method. Part 3: Flow-Through Method.* International Standards 7346-1, 7346-2, 7346-3. International Organization for Standardization, Geneva.

ISO 1985 *Sampling Part 3: Guidance on the Preservation and Handling of Samples*. International Standard 5667-3. International Organization for Standardization, Geneva.

ISO 1985 *Water Quality — Methods of Biological Sampling — Guidance on Handnet Sampling of Aquatic Benthic Macro-invertebrates*. International Standard 7828. International Organization for Standardization, Geneva.

ISO 1988 *Water Quality — Design and Use of Quantitative Samplers for Benthic Macro-invertebrates on Stony Substrata in Shallow Freshwaters*. International Standard 8265. International Organization for Standardization, Geneva.

ISO 1989 *Water Quality — Fresh Water Algae Growth Inhibition Test with* Scenedesmus subspicatus *and* Selenastrum capricornutum. International Standard 8692. International Organization for Standardization, Geneva.

ISO 1992 *Spectrometric Determination of Chlorophyll* a *Concentrations*. International Standard 10260. International Organization for Standardization, Geneva.

ISO 1995 *Determination of the Inhibition of the Mobility of* Daphnia magna *Straus (Cladocera, Crustacea) — Acute Toxicity Test*. Draft International Standard ISO/DIS 6341.2 International Organization for Standardization, Geneva.

ISO-BMWP 1979 *Assessment of the Biological Quality of Rivers by a Macro-invertebrate Score*. ISO/TC147/SC5/WG6/N5, International Organization for Standardization, Geneva.

Mason, C.F. 1981 *Biology of Freshwater Pollution*. Longman, Harlow (Second edition, 1991).

Newman, P.J., Piavaux, M.A. and Sweeting, R.A. [Eds] 1992 *River Water Quality — Ecological Assessment and Control*. Commission of the European Communities, Brussels.

OECD 1984 *Algal Growth Inhibition Test*. OECD Guidelines 201, Organisation for Economic Cooperation and Development, Paris.

OECD 1984 *Daphnia, Acute Immobilisation Test*. OECD Guidelines 202 Part I, Organisation for Economic Cooperation and Development, Paris.

OECD 1984 *Fish, Prolonged Toxicity Test, 14-day Study*. OECD Guidelines 204, Organisation for Economic Cooperation and Development, Paris.

OECD 1987 *The Use of Biological Tests for Water Pollution Assessment and Control*. Environment Monographs, No. 11. Organisation for Economic Cooperation and Development, Paris.

OECD 1992 *Fish, Acute Toxicity Test*. OECD Guidelines 203, Organisation for Economic Cooperation and Development, Paris.

OECD 1993 *Daphnia Reproduction Test — Draft 6/93*. OECD Guidelines 202 Part II, Organisation for Economic Cooperation and Development, Paris.

Philips, D.J.H. 1980 *Quantitative Aquatic Biological Indicators*. Applied Science Publishers Ltd, London.

Richardson, M. [Ed.] 1993 *Ecotoxicology Monitoring*. VCH Verlagsgesellschaft, Weinheim.

Turner, A.P.F., Krube, J. and Wilson, G.S. 1987 *Biosensors — Fundamentals and Applications*. Oxford University Press, Oxford.

Washington, H.G. 1984 Diversity, biotic and similarity indices. A review with special relevance to aquatic ecosystems. *Water Research,* **18**(6), 653–694.

WMO 1988 *Manual on Water Quality Monitoring*. WMO Operational Hydrology Report No. 27, WMO Publication No. 680. World Meteorological Organization, Geneva.

# Chapter 12

# HYDROLOGICAL MEASUREMENTS

Hydrological measurements are essential for the interpretation of water quality data and for water resource management. Variations in hydrological conditions have important effects on water quality. In rivers, such factors as the discharge (volume of water passing through a cross-section of the river in a unit of time), the velocity of flow, turbulence and depth will influence water quality. For example, the water in a stream that is in flood and experiencing extreme turbulence is likely to be of poorer quality than when the stream is flowing under quiescent conditions. This is clearly illustrated by the example of the hysteresis effect in river suspended sediments during storm events (see Figure 13.2). Discharge estimates are also essential when calculating pollutant fluxes, such as where rivers cross international boundaries or enter the sea. In lakes, the residence time (see section 2.1.1), depth and stratification are the main factors influencing water quality. A deep lake with a long residence time and a stratified water column is more likely to have anoxic conditions at the bottom than will a small lake with a short residence time and an unstratified water column.

It is important that personnel engaged in hydrological or water quality measurements are familiar, in general terms, with the principles and techniques employed by each other. This chapter provides an introduction to hydrological measurements for personnel principally concerned with water quality monitoring. More detailed information on hydrological methods is available in the specialised literature listed in section 12.5 and from specialised agencies such as the World Meteorological Organization. Further detail and examples of the use of hydrological measurements in water quality assessments in rivers, lakes, reservoirs and groundwaters are available in the companion guidebook *Water Quality Assessments*.

## 12.1 Rivers

Proper interpretation of the significance of water quality variables in a sample taken from a river requires knowledge of the discharge of the river at the time and place of sampling (for further discussion see also section 2.1). In order to calculate the mass flux of chemicals in the water (the mass of a

*This chapter was prepared by E. Kuusisto*

chemical variable passing a cross-section of the river in a unit of time), a time series of discharge measurement is essential.

The flow rate or discharge of a river is the volume of water flowing through a cross-section in a unit of time and is usually expressed as $m^3 s^{-1}$. It is calculated as the product of average velocity and cross-sectional area but is affected by water depth, alignment of the channel, gradient and roughness of the river bed. Discharge may be estimated by the slope–area method, using these factors in one of the variations of the Chezy equation. The simplest of the several variations is the Manning equation which, although developed for conditions of uniform flow in open channels, may give an adequate estimate of the non-uniform flow which is usual in natural channels. The Manning equation states that:

$$Q = \frac{1}{n} A R^{2/3} S^{1/2}$$

where $Q$ = discharge ($m^3 s^{-1}$)
$A$ = cross-sectional area ($m^2$)
$R$ = hydraulic radius (m) and = $A/P$
$P$ = wetted perimeter (m)
$S$ = slope of gradient of the stream bed
$n$ = roughness coefficient

More accurate values for discharge can be obtained when a permanent gauging station has been established on a stretch of a river where there is a stable relationship between stage (water level) and discharge, and this has been measured and recorded. Once this relationship is established, readings need only be taken of stage, because the discharge may then be read from a stage–discharge curve.

Water quality samples do not have to be taken exactly at a gauging station. They may be taken a short distance upstream or downstream, provided that no significant inflow or outflow occurs between the sampling and gauging stations. The recommended distance is such that the area of the river basin upstream of the sampling station is between 95 per cent and 105 per cent of the area of the river basin upstream of the gauging station.

The stage–discharge relationship, or rating curve, usually includes the extremes of discharge encountered in a normal year. The rating curve should be checked periodically, ideally once a year, since minor adjustments may be necessary to take account of changes in the cross-section of the stream or instability in the flow characteristics, or to eliminate errors in previous measurements. Systematic variation as a result of unstable flow may be apparent if the stage–discharge relationship for a single flood event is examined. The discharge during the rising phase of a flood event is usually greater than that during the falling phase of the same flood event. While unstable flow can produce a loop in the stage–discharge plot for an individual storm event, it is not usually apparent in the annual rating curve that is commonly used in hydrological survey programmes. Unstable cross-sections cause stage–discharge variability and can produce sudden and significant shifts in the rating curve as a result of erosion or deposition of material in the river bed.

### 12.1.1 Measuring stream flow

If samples are to be taken at a point where the stage–discharge relationship is either unknown or unstable, discharge should be measured at the time of sampling. Discharge measurement, or stream gauging, requires special equipment and, sometimes, special installations. The measurements should be made by an agency that has staff with expertise in the techniques of hydrological survey. The most accurate method is to measure the cross-sectional area of the stream and then, using a current meter, determine the average velocity in the cross-section. If a current meter is not available, a rough estimate of velocity can be made by measuring the time required for a weighted float to travel a fixed distance along the stream.

For best results, the cross-section of the stream at the point of measurement should have the following ideal characteristics:

- The velocities at all points are parallel to one another and at right angles to the cross-section of the stream.
- The curves of distribution of velocity in the section are regular in the horizontal and vertical planes.
- The cross-section should be located at a point where the stream is nominally straight for at least 50 m above and below the measuring station.
- The velocities are greater than 10–15 cm $s^{-1}$.
- The bed of the channel is regular and stable.
- The depth of flow is greater than 30 cm.
- The stream does not overflow its banks.
- There is no aquatic growth in the channel.

It is rare for all these characteristics to be present at any one measuring site and compromises usually have to be made.

Velocity varies approximately as a parabola from zero at the channel bottom to a maximum near the surface. A typical vertical velocity profile is shown in Figure 12.1. It has been determined empirically that for most channels the velocity at six-tenths of the total depth below the surface is a close approximation to the mean velocity at that vertical line. However, the

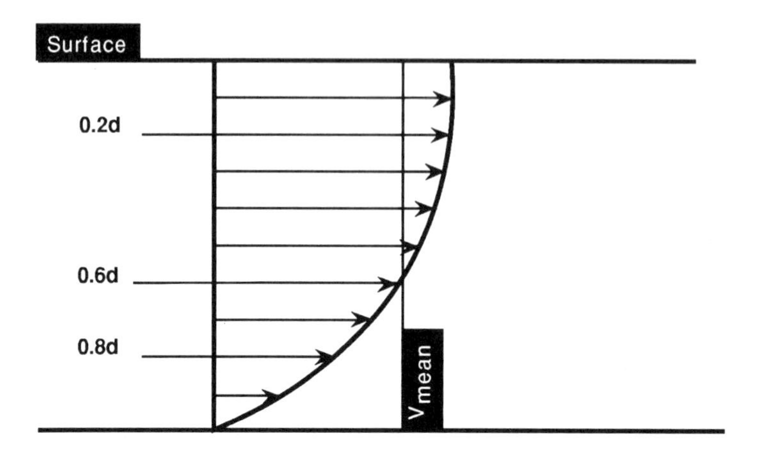


**Figure 12.1** Typical river velocity profile in the vertical plane

average of the velocities at two-tenths and eight-tenths depth below the surface on the same vertical line provides a more accurate value of mean velocity at that vertical line.

Velocity also varies across a channel, and measurements must, therefore, be made at several points across the channel. The depth of the river varies across its width, so the usual practice is to divide the cross-section of the stream into a number of vertical sections as shown in Figure 12.2 and measure velocity at each of these. No section should include more than 10–20 per cent of the total discharge. Thus, between 5 and 10 vertical sections are typical, depending on the width of the stream.

*Procedure for measuring discharge*

1. All measurements of distance should be made to the nearest centimetre.
2. Measure the horizontal distance $b_1$, from reference point 0 on shore to the point where the water meets the shore, point 1 in Figure 12.2.
3. Measure the horizontal distance $b_2$ from reference point 0 to vertical line 2.
4. Measure the channel depth $d_2$ at vertical line 2.
5. With the current meter make the measurements necessary to determine the mean velocity $v_2$ at vertical line 2.
6. Repeat steps 3, 4 and 5 at all the vertical lines across the width of the stream.

The computation for discharge is based on the assumption that the average velocity measured at a vertical line is valid for a rectangle that extends half of the distance to the verticals on each side of it, as well as throughout the depth at the vertical. Thus, in Figure 12.2, the mean velocity $\bar{v}_2$ would apply to a rectangle bounded by the dashed line p, r, s, t. The area of this rectangle is:

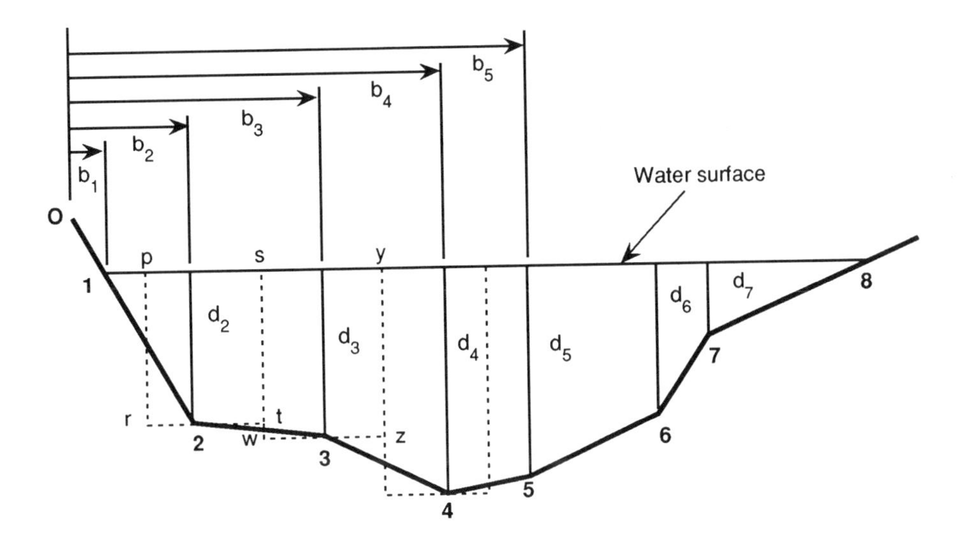


**Figure 12.2** Cross-section of a stream divided into vertical sections for measurement of discharge

$$a_2 = \frac{b_3 - b_1}{2} \times d_2$$

and the discharge through it will be:

$$Q_2 = a_2 \times \bar{v}_2$$

Similarly, the velocity $\bar{v}_3$ applies to the rectangle s, w, z, y and the discharge through it will be:

$$Q_3 = \frac{b_4 - b_2}{2} \times d_3 \times \bar{v}_3$$

The discharge across the whole cross-section will be:

$$Q_T = Q_1 + Q_2 + Q_3 \ldots Q_{(n-r)} + Q_n$$

In the example of Figure 12.2, $n = 8$. The discharges in the small triangles at each end of the cross-section, $Q_1$ and $Q_n$, will be zero since the depths at points 1 and 8 are zero.

If the water is shallow, the operator may wade into the stream holding the current meter in place while measurements are being made. Where the water is too deep for wading (more than 1 metre) the current meter must be lowered from a bridge, an overhead cableway or a boat. The section where flow measurement is made does not have to be at exactly the same place as either the monitoring station or the water level indicator provided that there is no significant inflow or outflow between these points along the stream.

Bridges are preferred as stream gauging stations because they usually allow easy access to the full width of the stream, and a water level indicator can be fastened to a bridge pier or abutment. Aerial cableways are often located at places where characteristics of the stream cross-section approach the ideal. However, they necessitate a special installation, and this is often impractical for a water quality monitoring team. Velocity measurements made from a boat are liable to yield inaccurate results because any horizontal or vertical movement of the boat will be identified as velocity by the current meter. In shallow streams the water velocity close to the boat will be affected and this may distort the meter readings.

Floats should be used for velocity measurement only when it is impossible to use a current meter. A surface float will travel with a velocity about 1.2 times the mean velocity of the water column beneath it. A partially submerged float made from a wooden stick with a weight at its lower end (so that it floats vertically) may be used. The velocity of a float of this type will be closer to the mean velocity and a correction factor of about 1.1 is appropriate if the submerged part of the float is at one-third to one-half the water depth. The velocity of any float, whether on the surface or submerged, is likely to be affected by wind.

Discharge measurement may also be made by ultrasonic or electromagnetic methods, by injecting tracers into a stream, or by the construction of a weir. None of these methods is recommended because they either interfere with the natural water quality or they are prohibitively expensive and complex to install and operate.

## 12.2 Lakes and reservoirs

In lakes and reservoirs, hydrological information (particularly water residence time) is needed for the interpretation of data and the management of water quality (for further discussion see also section 2.1). These hydrological measurements are required in two different situations:

- when samples are to be taken from tributaries and outflowing streams, and
- when samples are to be taken from the lake or reservoir itself.

Both types of sampling may aim at the estimation of the mass flow of some variable in the water body and, consequently, hydrological data are essential.

### 12.2.1 Sampling from tributaries and outflowing streams

In outflowing streams, hydrological measurements should be obtained in the same manner as described above for rivers. In tributaries, the location of sampling stations and flow measurement stations should be selected so that backwater effects (water backing up the river from the lake) will be avoided. If this is not possible, the water level at the mouth of the tributary should be measured and recorded to provide data on the magnitude of the backwater and its variation with time.

### 12.2.2 Sampling from a lake or reservoir

The water level at the time of sampling must be measured. If the water surface is calm and a water level gauge has been installed, a single reading may be sufficient. If there is no official gauge, the water level should be recorded in relation to a conveniently located, identifiable point on a rock outcrop, large boulder or other landmark that is reasonably permanent. If there is any reason to suspect that this water level marker might move or be moved, reference should be made to a second landmark. The use of landmarks as a water level reference is a temporary measure and a water level gauge should be installed as soon as possible.

Waves and the inclination of the water surface (seiches) may cause problems in observing water levels. High waves may make it difficult for the observer to see the gauge and the continual motion of the water makes it impossible to determine the exact water level. In such conditions the observer should try to record the highest and lowest positions of the changing water level and calculate the average of the two. The wind condition should also be noted, together with an estimate of the height of the waves.

In certain conditions, current measurements in lakes or reservoirs provide information that is helpful in the interpretation of the results of analyses of water and sediment samples. Currents may cause water quality to vary appreciably within short distances or time periods. The flow velocities that normally occur in lakes are measured with sensitive recording current metres anchored at given depths. Sometimes, however, a rough estimate of the flow field (the general pattern of flow within a lake) can be made by observing the motion of surface floats. In reservoirs, the operation of valves or sluices can create localised currents which can affect the water quality in their vicinity. The special characteristics of reservoirs and their operation, in relation to water quality monitoring, are described in *Water Quality Assessments*.

## 12.3 Mass flux computation

Water quality monitoring often seeks to obtain two main types of estimations related to the physical and chemical variables being measured:

- the instantaneous values of the concentrations of water quality variables, and
- the mass flux of water quality variables through river sections, lakes or reservoirs over specified periods of time.

For a river, if only one sample has been taken at a point, provided the concentration $c$ of the variable of interest has been determined and the instantaneous discharge $Q$ has been obtained, the instantaneous mass flux $Q_m$ of that variable can be calculated from the formula $Q_m = cQ$.

When samples have been taken and discharge has been measured in each of the vertical sections across a stream, the instantaneous mass flux is given by:

$$\sum_{i=1}^{n} C_i \overline{V}_i A_i \quad \mathrm{g\ s^{-1}}$$

where $n$ = number of vertical sections
$C_i$ = concentration of the variable in section $i$ ($\mathrm{mg\ l^{-1}}$)
$\overline{V}_i$ = mean velocity in section $i$ ($\mathrm{m\ s^{-1}}$)
$A_i$ = cross-sectional area of section $i$ ($\mathrm{m^2}$).

The average concentration at the cross-section can be obtained from:

$$\frac{Q_m}{Q} \quad \mathrm{mg\ l^{-1}}$$

where $Q$ = the instantaneous discharge ($\mathrm{m^3\ s^{-1}}$)
$Q_m$ = the instantaneous mass flux of the variable ($\mathrm{g\ s^{-1}}$).

Calculation of the mass flux over a period ($t_0$ ... $t_m$) may be determined quite accurately, but this requires many measurements of water quality and the use of a complex formula. Normally, water quality determinations are carried out at relatively long time intervals (weekly or monthly). By contrast, discharges are often determined daily, based on daily observation of the water level and the stage–discharge relationship. The simplest way to estimate daily concentrations of a variable is to assume that each measured value of concentration is valid for half of the preceding and following intervals between the collection of samples. However, this assumption is valid only when variations in the concentration of a variable are small. It is usually necessary to use complicated interpolation techniques.

Regression techniques are appropriate if a reliable relationship can be established between the concentrations of the variable of interest and some other chemical or physical variable that is measured at frequent intervals. The most suitable variable will often be discharge (determined easily from water level). Thus the consistency of the relationship between the concentration of a variable and the discharge should always be checked. If the relationship is reasonably consistent, discharge can serve as the basis on which to estimate the mass flux of a variable. If it is inconsistent, however, some other relationship should be sought.

In small streams and during flood peaks, the discharge may vary considerably over a 24-hour period. A variation in the concentration of a variable in excess of an order of magnitude is typical, and both sampling and discharge measurements need to be frequent. If the daily maximum flow is two to three times the average flow, a 4-hour interval between samples is recommended. If the maximum is greater than three times the average, discharge should be measured and samples taken every hour. This information can be obtained during a short-term pilot study. Examples of the measurement of discharge in river and particulate matter monitoring are given in *Water Quality Assessments*.

## 12.4 Groundwater

Groundwater is important as a significant water resource, in its own right, and also because of its interaction with surface water. Groundwater recharges streams and rivers in some areas, while in others it is itself recharged by surface water.

The hydrogeological conditions of water-bearing rocks or aquifers have a significant influence on the quality of groundwater in much the same way that hydrological conditions influence surface water quality. The rate of flow through the aquifer, residence time, inflows and outflows from the aquifer all influence the groundwater quality (a full discussion is available in *Water Quality Assessments*).

As the rate of flow of groundwater is much lower than that of surface water, there is a significant risk that contaminants can build up in aquifers to the point where the water becomes unusable. This could force the abandonment of boreholes and result in a permanent reduction in the quantity of usable groundwater.

There are certain influences that are unique to groundwater quality. These are the nature of the aquifer, the presence or absence of contaminants in unsaturated layers above the aquifer, the presence of naturally occurring contaminants in the aquifer and the interaction of groundwater, surface water and contaminant movement. Further detail can be found in section 2.2.

### 12.4.1 Groundwater flow

Detailed investigation of groundwater flows over a large area requires specialist knowledge and equipment and the manipulation of predictive formulae. Groundwater flow is three-dimensional and, therefore, more difficult to predict than surface water. In addition, whereas surface water flow direction can be easily predicted by topographical survey, groundwater flow direction depends on aquifer type and hydraulic conditions in the aquifer and is difficult to assess without carrying out pump tests and tracer studies.

Information on the direction of groundwater flow can be obtained by mapping out water levels in boreholes within the same aquifer. This gives an indication of the hydraulic gradient (or piezometric surface) and, thus, an idea of groundwater movement.

Groundwater flow information will assist in the prediction of contaminant movement in groundwater, in particular the spread and speed of movement of contaminants after a polluting event. However, this prediction is a complex procedure which is often inaccurate and is complicated further by the lack of knowledge of contaminant behaviour in groundwater.

Flows within aquifers on the medium scale may be assessed through tracer studies, which will indicate direction and rate of flow. The rate of movement of water into particular wells can be quite easily evaluated by pump tests. These tests will also provide information on the depression of groundwater level around a well during pumping.

The usual procedure for a field pump test is that water is pumped from a production borehole at a constant rate, which is controlled and measured. This may be done using a flow meter on the discharge pipe. The change in the level of the water table (or piezometric surface) around the borehole is monitored by measuring the change in depth to the water table at observation wells surrounding the borehole.

Pump tests provide valuable information for calculating aquifer properties which, in turn, provide information of importance to water quality, such as rate of movement of groundwater into boreholes, the area of the aquifer that will be exploited by the borehole and where land use restrictions may need to be applied.

Full inventories of boreholes should be prepared and should include information concerning the pumping depth, yield, aquifer transmissivity and storage coefficient. This will permit groundwater movement and water quality to be modelled. Pumping from adjacent boreholes will influence the yield and draw-down of a production borehole.

It is important to collect information on any changes of static water level. In some areas where initial intensive groundwater abstraction leads to a drop in the water table, there has been significant contamination of the unsaturated soil layers overlying the aquifer. When abstraction was later decreased, the water level rose and became contaminated through the dissolution of contaminants held in the soil profile. This may cause problems if water is withdrawn in the future and may also increase the risk to the rest of the aquifer, thus potentially reducing the quality of groundwater available.

## 12.5 Source literature and further reading

Chapman, D. [Ed.] 1996 *Water Quality Assessments. A Guide to the Use of Biota, Sediments and Water in Environmental Monitoring.* 2nd edition, Chapman & Hall, London.

Dickinson, W.T. 1967 *Accuracy of Discharge Measurements.* Hydrology Paper No. 20, Colorado State University, Fort Collins.

Herschy, R.W. [Ed.] 1978 *Hydrometry. Principles and Practices.* John Wiley and Sons, Chichester.

O'Connor, D.J. 1976 The concentration of dissolved solids and river flow. *Water Resources Research,* **12**(2), 1279–1294.

Solomon, S.I. 1966 Statistical association between hydrologic variables. In: *Proceedings of Hydrology Symposium No. 5, Statistical Methods in Hydrology.* National Research Council of Canada, Ottawa, 55–113.

Steele, T.E. 1971 The role of network design in the management and control of streamflow water quality. In: V. Yevjevich [Ed.] *Systems Approach to Hydrology.* Colorado State University, Fort Collins, CO, 395–423.

Task Committee on Preparation of Sedimentation Manual. Committee on Sedimentation of Hydraulics Division 1962 Sediment measurements techniques: a fluvial sediment. *Journal of Hydraulics Division, A.S.C.E,* **95**(5).

Todd, D.K. 1980 *Groundwater Hydrology.* 2nd edition, John Wiley, New York.

US EPA 1973 *Handbook for Monitoring Industrial Waste Water.* United States Environmental Protection Agency, Washington, D.C.

WMO 1974 *Guide to Hydrological Practices.* Publication No. 168, World Meteorological Organization, Geneva.

WMO 1980 *Manual on Stream Gauging. Vol I, Fieldwork. Vol II, Computation of Discharge.* Operational Hydrology Report No. 13; WMO — No. 519, World Meteorological Organization, Geneva.

# Chapter 13

# SEDIMENT MEASUREMENTS

Sediments play an important role in elemental cycling in the aquatic environment. As noted in Chapter 2 they are responsible for transporting a significant proportion of many nutrients and contaminants. They also mediate their uptake, storage, release and transfer between environmental compartments. Most sediment in surface waters derives from surface erosion and comprises a mineral component, arising from the erosion of bedrock, and an organic component arising during soil-forming processes (including biological and microbiological production and decomposition). An additional organic component may be added by biological activity within the water body.

For the purposes of aquatic monitoring, sediment can be classified as deposited or suspended. Deposited sediment is that found on the bed of a river or lake. Suspended sediment is that found in the water column where it is being transported by water movements. Suspended sediment is also referred to as suspended matter, particulate matter or suspended solids. Generally, the term suspended solids refers to mineral + organic solids, whereas suspended sediment should be restricted to the mineral fraction of the suspended solids load.

Sediment transport in rivers is associated with a wide variety of environmental and engineering issues which are outlined in Table 13.1. The study of river suspended sediments is becoming more important, nationally and internationally, as the need to assess fluxes of nutrients and contaminants to lakes and oceans, or across international boundaries, increases. One of the most serious environmental problems is erosion and the consequent loss of topsoil. Although erosion is a natural phenomenon, the rate of soil loss is greatly increased by poor agricultural practices which result, in turn, in increased suspended sediment loads in freshwaters. Loss of topsoil results in an economic loss to farmers, equivalent to hundreds of millions of US dollars annually, through a reduction in soil productivity. Good environmental practice in agriculture, which may include contour ploughing and terracing, helps to protect against soil loss and against contamination of surface waters.

*This chapter was prepared by E. Ongley*

**Table 13.1** Issues associated with sediment transport in rivers

| Sediment size | Environmental issues | Associated engineering issues |
|---|---|---|
| Silts and clays | Erosion, especially loss of topsoil in agricultural areas; gullying | |
| | High sediment loads to reservoirs | Reservoir siltation |
| | Chemical transport of nutrients, metals, and chlorinated organic compounds | Drinking-water supply |
| | Accumulation of contaminants in organisms at the bottom of the food chain (particulate feeders) | |
| | Silting of fish spawning beds and disturbance of habitats (by erosion or siltation) for benthic organisms | |
| Sand | River bed and bank erosion | River channel deposition: navigation problems<br>Instability of river cross-sections |
| | High sediment loads to reservoirs | Sedimentation in reservoirs |
| | Habitat disturbance | |
| Gravel | Channel instability when dredged for aggregate | Instability of river channel leads to problems of navigation and flood-control |
| | Habitat disturbance | |

Water users downstream of areas of heavy soil run-off may have to remove suspended sediment from their water supplies or may suffer a reduction in the quantity of water available because of reservoir siltation. The rapid reduction in the storage capacity of reservoirs due to siltation is a major sediment-related problem world-wide. Moreover, the availability of water for irrigation from the reservoir leads to more intensive land use and increased soil erosion. These effects may also be exacerbated by desertification (impoverishment of vegetative cover and loss of soil structure and fertility), whether anthropogenic or climatic in origin. In addition, gradual enrichment of reservoir waters with nutrients (some of which also arise from agricultural practices) leads to enhanced production and increased sedimentation of organic material originating from the water column (from decaying plankton) or littoral zones (from decaying macrophytes). Consequently, the rate of reservoir siltation often greatly exceeds that predicted during design. Monitoring data for sediment transport to, and productivity within, reservoirs are therefore required for accurate calculations of sediment transport and deposition and for the management of major reservoirs. Further information on monitoring and assessment approaches for reservoirs is given in the companion guidebook *Water Quality Assessments*.

In order to protect surface water resources and optimise their use, soil loss must be controlled and minimised. This requires changes in land use and land management, which may also have an impact on water quality. Control of the siltation rate in reservoirs requires that adequate data are available at the design stage. This, in turn, demands an understanding of sediment transport and appropriate methods for measuring sediment load and movement.

Recognition of the importance of sediments and their use in monitoring and assessment programmes is increasing and methods are constantly being refined. For the purposes of water quality monitoring a distinction can be made between measuring sediment quantity and sediment quality. Some of the techniques available for studying sediment quality (as a component of water quality studies) are not yet widely accepted or used and have not been standardised. Although they may be suitable for special surveys, some methods are too complex and costly for routine monitoring programmes. A full discussion of the role of sediments and particulate material in water quality monitoring and assessment is available in the companion guidebook *Water Quality Assessments*. This present chapter concentrates on some of the fundamental procedures required for the more common sediment measurements necessary for water quality monitoring programmes.

## 13.1 Types of sediment transport

Sediment transport is a direct function of water movement. During transport in a water body, sediment particles become separated into three categories: suspended material which includes silt + clay + sand; the coarser, relatively inactive bedload and the saltation load.

*Suspended load* comprises sand + silt + clay-sized particles that are held in suspension because of the turbulence of the water. The suspended load is further divided into the wash load which is generally considered to be the silt + clay-sized material (< 62 μm in particle diameter) and is often referred to as "fine-grained sediment". The wash load is mainly controlled by the supply of this material (usually by means of erosion) to the river. The amount of sand (> 62 μm in particle size) in the suspended load is directly proportional to the turbulence and mainly originates from erosion of the bed and banks of the river. In many rivers, suspended sediment (i.e. the mineral fraction) forms most of the transported load.

*Bedload* is stony material, such as gravel and cobbles, that moves by rolling along the bed of a river because it is too heavy to be lifted into suspension by the current of the river. Bedload is especially important during periods of extremely high discharge and in landscapes of large topographical relief, where the river gradient is steep (such as in mountains). It is rarely important in low-lying areas.

Measurement of bedload is extremely difficult. Most bedload movement occurs during periods of high discharge on steep gradients when the water level is high and the flow is extremely turbulent. Such conditions also cause problems when making field measurements. Despite many years of experimentation, sediment-monitoring agencies have so far been unable to devise a standard sampler that can be used without elaborate field calibration or that can be used under a wide range of bedload conditions. Even with calibration, the measurement error can be very large because of the inherent hydraulic characteristics of the samplers and the immense difficulty with representative sampling of the range of sizes of particles in transit as bedload in many rivers. Unless bedload is likely to be a major engineering concern (as in the filling of reservoirs), agencies should not attempt to measure it as part of a routine sediment-monitoring programme. Where engineering works demand knowledge of bedload, agencies must acquire the specialised expertise that is essential to develop realistic field programmes and to understand the errors associated with bedload measurement. Local universities or colleges may be able to assist in this regard.

*Saltation load* is a term used by sedimentologists to describe material that is transitional between bedload and suspended load. Saltation means "bouncing" and refers to particles that are light enough to be picked off the river bed by turbulence but too heavy to remain in suspension and, therefore, sink back to the river bed. Saltation load is never measured in operational hydrology.

## 13.2 Sediment measurement

While the underlying theory is well known, the measurement of sediment transport requires that many simplifying assumptions are made. This is largely because sediment transport is a dynamic phenomenon and measurement techniques cannot register the ever-changing conditions that exist in water bodies, particularly in river systems. Some of the sources of extreme variability in sediment transport are discussed below.

### 13.2.1 Particle size

Knowledge of the size gradient of particles that make up suspended load is a prerequisite for understanding the source, transportation and, in some cases,

**Table 13.2** Particle size classification by the Wentworth Grade Scale

| Particle description | Particle size (mm) | Cohesive properties |
|---|---|---|
| Cobble | 256–64 | Non-cohesive sediment |
| Gravel | 64–2 | |
| Very coarse sand | 2–1 | Non-cohesive sediment |
| Coarse sand | 1–0.5 | |
| Medium sand | 0.5–0.25 | |
| Fine sand | 0.25–0.125 | |
| Very fine sand | 0.125–0.063 | |
| Silt | 0.062–0.004 | Cohesive sediment |
| Clay | 0.004–0.00024 | |

environmental impact of sediment. Although particles of sizes ranging from fine clay to cobbles and boulders may exist in a river, suspended load will rarely contain anything larger than coarse sand, and in many rivers 50–100 per cent of the suspended load will be composed only of silt + clay-sized particles (< 62 μm). The size of particles is normally referred to as their diameter although, since few particles are spherical, the term is not strictly correct. Particle size is determined by passing a sample of sediment through a series of sieves, each successive sieve being finer than the preceding one. The fraction remaining on each sieve is weighed and its weight expressed as a percentage of the weight of the original sample. The cumulative percentage of material retained on the sieves is calculated and the results are plotted against the representative mesh sizes of the sieves. A series of eight sieves can be used for sediment analysis, with mesh sizes from 1.25 mm to 63 μm or less. Further details of these methods are available in the appropriate literature (see section 13.6).

Clay particles are plate-like in shape and have a maximum dimension of about 4 μm. Silt particles, like sand, have no characteristic shape; their size is between those of clay and sand with diameters ranging from 4 μm to 62 μm. Since the smallest mesh size of commercially available sieves is about 40 μm, the sizes of clay and small silt particles cannot be determined by sieving, and sedimentation techniques are used instead. The sedimentation rate of the particles is measured and their diameter calculated from the semi-empirical equation known as Stokes' Law.

There is no universally accepted scale for the classification of particles according to their sizes. In North America, the Wentworth Grade Scale (see Table 13.2) is commonly used; elsewhere, the International Grade Scale is preferred. There are minor differences between the two scales and it is, therefore, important to note which scale has been selected and to use it consistently.

The boundary between sand and silt (62 µm) separates coarse-grained sediments (sand and larger particles) from fine-grained sediments (silt and clay particles). Coarse-grained sediments are non-cohesive, whereas fine-grained sediments are cohesive, i.e. the particles will stick to one another as well as to other materials. Particle cohesiveness has important chemical and physical implications for sediment quality.

Sedimentology and water quality programmes have adopted a convention that considers particulate matter to be larger than 0.45 µm in diameter; anything smaller is considered to be dissolved. This boundary is not entirely valid because clay particles and silt can be much smaller than 0.45 µm. For practical purposes, however, the boundary is convenient, not least because standard membrane filters with 0.45 µm diameter pores can be used to separate suspended particles from dissolved solids. A general procedure for the measurement of total dissolved solids (TDS) is described in Chapter 7, section 7.24.

### 13.2.2 Composition of sediment

The amount and nature of suspended load in a water body is affected by the availability of sediment as well as by the turbulent forces in the water. The sand component of the suspended load in a river originates mainly from the river bed. As discharge increases, so do the turbulent forces that cause the sand to be taken into suspension. Sand particles tend to settle quite rapidly because of their shape, density and size. Therefore, the concentration of sand is highest near the bed of a river and lowest near the surface. The curves for medium and coarse sand in Figure 13.1 show this variation of concentration with depth. In lakes, coarser material is deposited rapidly at the point where the river enters the lake and is only resuspended and redeposited under highly turbulent conditions (such as generated by high winds).

The bed sediment of a river contributes only a small portion of the clay and silt-sized particles (< 62 µm) present in the suspended load. Most of this fine material, which may be 50–100 per cent of the suspended load in many rivers, is eroded and carried to the river by overland flow during rainstorms. This fraction does not easily sink in the water column, and slight turbulent forces keep it in suspension for long periods of time. As a consequence, the silt + clay fraction tends to be fairly evenly distributed throughout the depth of a river as illustrated by the vertical profile for silt + clay in Figure 13.1. In lakes and reservoirs, fine suspended material originates from river inputs, shoreline and lake bed erosion and organic and inorganic material generated within

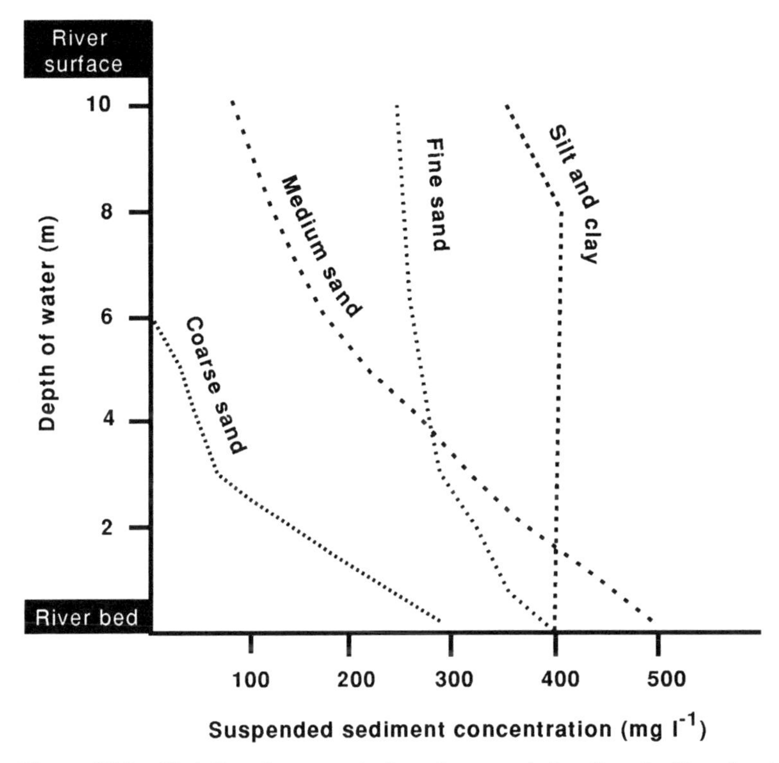


**Figure 13.1** Variations in concentration of suspended sediment with water depth for sand, silt and clay as measured at one field site

the lake by biological activity. In eutrophic waters the latter source can be quite significant. Fine material can be repeatedly resuspended by lake currents (generated by wind stress) until it is eventually deposited in an area where water movements are insufficient to resuspend or remobilise it. Such depositional basins in lakes or reservoirs are important for sediment quality studies because they can indicate the history of anthropogenic influences on the composition of the sediment.

### 13.2.3 Hysteresis effects

A rainstorm causes an increase in discharge and an associated increase in turbulence in a river. This turbulence takes bed sediments into suspension leading to relatively high concentrations of suspended material in the water. During prolonged rainstorms, discharge and turbulence may remain high but there is usually a progressive decline in the quantity of suspended material present in the water. This is because the quantity of sediment on a river bed,

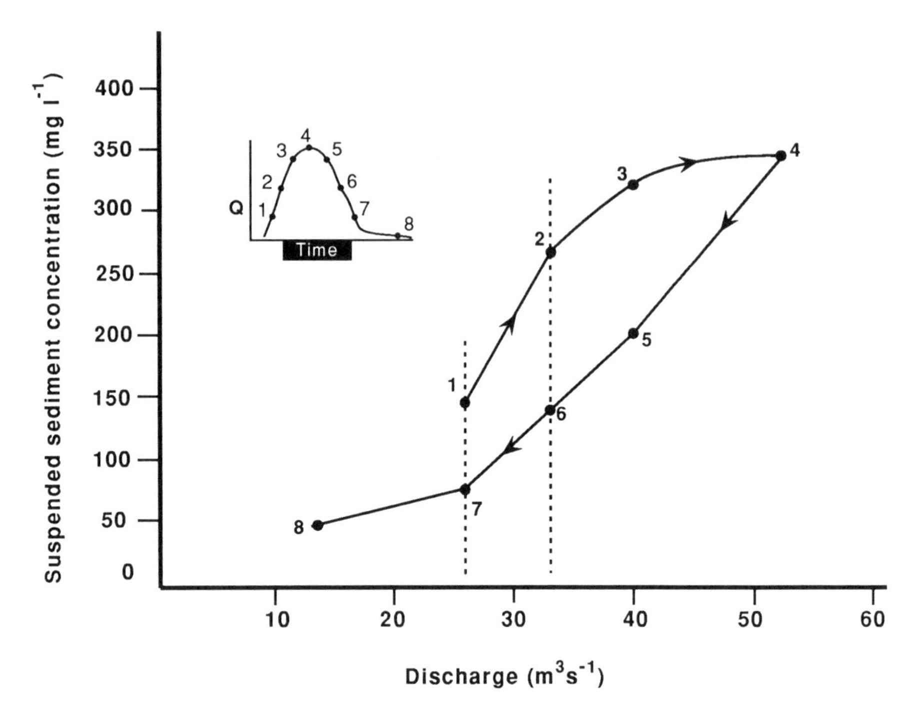


**Figure 13.2** Typical hysteresis effect observed in suspended sediment. Sample numbers are those noted in the storm hydrograph illustrated in the insert.

and which is introduced into the river by erosional processes, is limited and the amount of sediment available to be taken into suspension gradually diminishes during a storm event. When a series of discharge measurements and water samples are taken at intervals throughout a storm event (when flow increases, reaches a peak, and then decreases), the graphical plot of the concentration of suspended sediment against discharge will often take the form of a hysteresis loop. This is shown in Figure 13.2, where samples 1 and 7 were taken at the same discharge rate but sample 7 (taken late in the discharge event) has a lower concentration of suspended sediment than sample 1. Similar differences in concentration are evident for samples 6 and 2. In Figure 13.2, the inset graph shows the time sequence of sampling in relation to the discharge.

Hysteresis may also be observed in plots of seasonal data. This reflects periods of the year when sediment may be more readily available than at other times. Higher concentrations may occur, for example, after a long, dry period or in dry months when vegetation is not able to hold back soil particles that are being eroded.

## 13.3 Sampling for sediment

The methods and equipment used for sampling suspended sediment are different from those used for deposited sediments. Also sampling methods for measurements of the quantity of sediment in transport are different than for measurement of sediment quality. The reason for these differences reflects the fact that sediment quantity must include the sand-size fractions which are unequally distributed in depth, whereas sediment quality focuses on the silt + clay fraction which is not depth-dependent. These differences are more fully described in the appropriate literature (see section 13.6).

For bottom sediments it may be necessary to collect deposited sediments with minimum disturbance in order not to lose the fine material on the sediment surface, or because the vertical distribution of the sediment components is important (such as during establishment of historical records or depositional rates). In deep waters this necessitates the use of grabs or corers (see also section 11.2.2), but in shallow water a scoop or spatula may be used. Further discussion of the relative merits of different sampling techniques is available in *Water Quality Assessments* and other relevant publications.

There are four main types of samplers for suspended sediments:

- integrated samplers,
- instantaneous grab samplers,
- pump samplers, and
- sedimentation traps.

It was shown in Figure 13.1 that the concentration of the coarser fractions of suspended sediment increases towards the bottom of the river channel. This segregation of material by particle size requires that, for the purposes of measuring quantity of suspended sediment, a depth-integrating sampling technique is used to obtain a sample that accounts for different sediment concentrations throughout the vertical profile of a water body. Many types of sampler have been designed for depth-integrated sampling of suspended sediment. Some are available commercially but are rather expensive. All of them have a number of features in common:

- Each has a water inlet nozzle and an air outlet. As the water and suspended sediment enter, air is displaced through the air outlet.
- Each permits isokinetic sampling. That is, water velocity through the inlet nozzle is equal to the water velocity at the depth of the sampler. This is important for larger particles, such as sand, because the sampler would otherwise tend to over- or under-estimate the amount of suspended sediment. Errors caused by lack of isokinetic sampling are minimal for small particles (< 62 μm) and for practical purposes can be ignored.

- Each has a metal body (for weight) that encloses a glass or plastic bottle for retaining the sample. The bottle is changed after each sample is taken.
- The diameter of the water inlet can be selected (or changed) so that the sampler will fill more or less quickly, depending on the depth of the river.

In practice, depth-integrating samplers are lowered to the river bottom, then immediately raised to the surface; lowering and raising should be done at the same rate. The objective is to fill the sampler to about 90 per cent capacity; if the sampler is completely full when it emerges from the water the sample will be biased because the apparatus will have stopped sampling at the point at which it filled up.

Large, heavy samplers are usually only necessary when samples must be obtained from a bridge, boat or similar situation. In shallow streams, where all points can be reached by wading, a bucket (if nothing else is available) or a small sampler attached to a metal rod can be used. It is possible to make a simple depth-integrating sampler for use in shallow streams, using a wide-mouth, 1-litre bottle, a rubber stopper and short pieces of rigid tubing. The tubing forms the water inlet and air outlet. The lengths and diameters of the tubes may require experimentation but, in general, the air outlet tube should be of a smaller diameter than the water inlet. The bottle is secured to a metal rod or wooden pole, then lowered and raised as outlined above. An example of a home-made sampler is shown in Figure 13.3. The sample must be taken facing upstream so as to avoid sampling bottom sediment that was resuspended by the operator's feet.

For measurements of sediment quality, such as for phosphorus, metals, pesticides, etc., it is generally only the silt and clay-size material that is relevant. Also, the amount of sample required is often much larger (e.g. several grams in some instances) than is necessary for the physical measurement of suspended sediment concentration. Therefore, it is often necessary to use methods which concentrate the suspended material from a large volume of water. For some types of chemical analyses it is possible to obtain values from digestion and analysis of the sediment which is retained on a filter with pore diameters of 0.45 μm. However, the errors in the results obtained can be quite large. Alternatively, where the sampling site is close to a laboratory, large volumes of water can be left to settle for several days and the resultant supernatant water siphoned off, leaving the solids on the bottom of the container. This method can also lead to errors in results due to contamination and/or the chemical and microbiological modification of compounds during the settling period. More commonly, raw water samples are taken to the laboratory where they are centrifuged by standard or continuous-flow

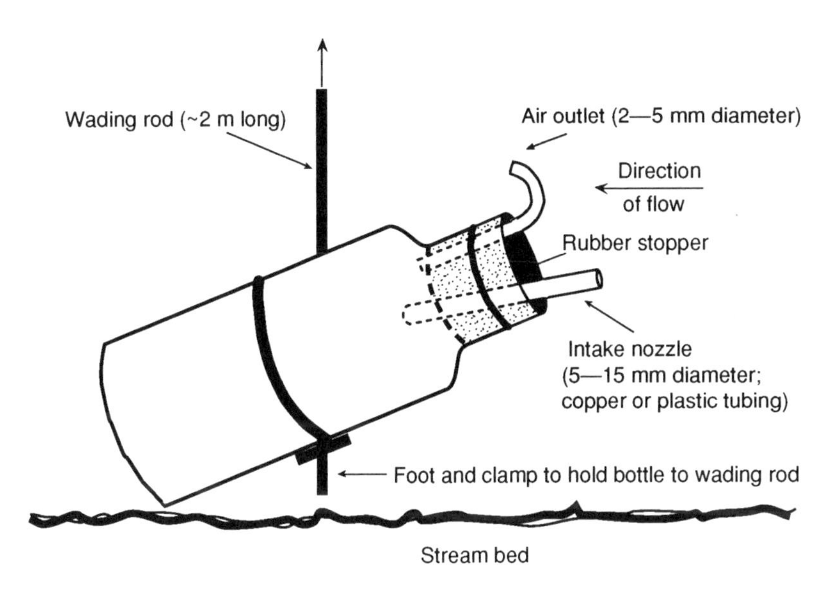


**Figure 13.3** Home-made suspended sediment sampler

centrifuges. In more remote locations portable continuous-flow centrifuges can be used. However, these are expensive and require mains or portable electrical power. It is most important to note that there is no single recommended method for collection and analysis of the chemistry of suspended sediment (see also section 13.5).

In lakes, the use of sediment traps provides a simple and cheap alternative for the collection of suspended material, although the samples may not be totally representative (due to the continual microbial processes within the trap and the efficiency of the particle capture). A simple sediment trap can be made from a plastic, glass or metal cylinder open at the top and closed at the bottom and with a height to width ratio greater than five. There should be nothing obscuring the opening and the trap should be suspended vertically with the aid of a weight and mooring system. A trap should not be left in place for longer than one month without removing the sample.

## 13.4 Measuring suspended sediment

Particle size distribution and concentration not only vary in the vertical section, but may also vary considerably across a river section. Therefore, measuring suspended sediment concentration must take into account these variations. This becomes especially important when suspended sediment concentration is being measured for the purpose of calculating sediment load in a river.

For determining suspended sediment load, it is necessary to consider all particle sizes (sand + silt + clay). Therefore, a depth-integrating sampler must be used as described in section 13.3 to ensure that the depth-dependent sand-sized fraction is correctly sampled. There are two generally accepted methods for measuring suspended sediment concentration for load determination as described below.

*Equal-discharge-increment method*

This method requires first that a complete flow measurement be carried out across the cross-section of the river. Using the results, the cross-section is divided into five (more on large or complex rivers) increments (i.e. vertical sections) having equal discharge. The number $n$ of increments is based on experience. Depth integrated suspended sediment sampling is carried out at one vertical within each of the equal-discharge-increments, usually at a location most closely representing the centroid of flow for that increment. The sediment concentration for each equal-discharge-increment is measured according to section 13.4.1. The mean discharge-weighted suspended sediment concentration ($SS_C$) is obtained by taking the average of the concentration values $C$ obtained for each interval $i$.

$$SS_c = \frac{\sum_{i=1}^{n} C_i}{n}$$

The discharge-weighted suspended sediment load ($SS_L$), in tonnes per day, for the river cross-section is obtained by multiplying the concentration, $C$, in ppm (mg $l^{-1}$) by the discharge, $Q$, in $m^3\ s^{-1}$ of each equal-discharge-increment, $i$, and summing for all increments. This method is very time-consuming, but is that most used by sediment agencies.

$$SS_L = \sum_{i=1}^{n} (C_i Q_i) \times 0.0864$$

*Equal-width-increment-method*

This method is used without making flow measurements and is usually used in small to medium rivers and especially rivers that are shallow enough for wading. The operator marks off 10–20 equal intervals across the river cross-section. At the deepest point, the operator takes a depth-integrated sample, noting the transit rate of the sampler (i.e. the uniform speed at which the sampler is lowered, then raised to the surface). Using that same transit rate, a suspended sediment sample is taken at each of the intervals. Because each vertical will have a different depth and velocity, the sample volume will vary with each vertical sampled. Note that the bottle must never be over-filled. All samples are composited into a single container which is then agitated and sub-sampled, usually two or three times, and analysed for suspended sediment concentration. The average of these analyses is the mean cross-sectional suspended sediment concentration. In this method, the results are corrected for differences in discharge at each section by virtue of using the same transit rate (and the same nozzle diameter) at all sections — i.e. a shallow section with less discharge will produce a proportionally smaller suspended sediment sample than a deep section having greater discharge.

For suspended sediment quality (section 13.5), where the primary interest is the chemistry associated with the silt + clay ($< 0.63$ μm) fraction, sampling can be greatly simplified because this fraction is not normally depth-dependent (Figure 13.1). While there are no universally accepted rules for sampling, many scientists will collect a grab sample from a depth of 0.5 m at the point of maximum flow in the cross-section. For larger rivers, or rivers where there is concern over cross-sectional variation, grab samples can be taken from several locations across the section and integrated. For more exacting work where accurate loads are required, especially for micro-pollutants, sampling should be carried out using either of the methods noted above. It is particularly important to avoid sampling near river banks (or lake shores) where elevated concentrations of suspended matter occur and which are often contaminated by garbage and other anthropogenic materials.

### 13.4.1 Laboratory procedures for measuring sediment concentration

The concentration of suspended sediment is usually determined in the laboratory using the method described in section 7.25 for the determination of total suspended solids (TSS). When this method is followed, the filters used should have a pore size of 0.45 μm. Note, however, that if a sample which has been filtered for TSS determination will also be required for sediment quality analysis, a filter appropriate to the further analyses should be used (see section 13.5.1 below). Where the concentration of sediment is high, it may be difficult and time-consuming to filter a large enough volume of sample. In this case it is possible to evaporate a measured portion of sample to dryness in a pre-weighed dish and to determine the weight of the residue, although it must be recognised that any dissolved salts in the sample will increase the value obtained (this bias may be quite great if the water is saline).

Another possibility is to make measurements with a field turbidity meter that has been calibrated against natural samples, preferably from the sampling site where it is being used (this relationship would not be valid for other sites). Provided that most of the suspended sediment is fine grained, there is usually a good relationship between turbidity and suspended sediment concentration.

The size boundary between sand and silt + clay, i.e. 62 µm, is important if the nature of infilling of a reservoir is to be determined or if sediment quality is of interest. Sand will settle to the bottom almost immediately when a river enters a reservoir and velocity is reduced. By contrast, silt and clay will stay in suspension much longer and move further within the reservoir. Furthermore, it is the $\leq 62$ µm fraction of suspended sediment that is mainly responsible for the transport of chemicals adsorbed on particles.

Measurement of the quantity of silt + clay (i.e. $\leq 62$ µm) in a sample requires that a known volume [$V_{sample}$ (ml)] of well mixed sample is first passed through a 62 µm mesh sieve. The sample is wet-sieved, i.e. distilled water is used to rinse the sample through the sieve. All of the water that passes through the sieve (original sample plus rinse water) is collected and filtered through a membrane filter of 0.45 µm pore size and of known weight. The sand collected on the sieve is dried and weighed [$W_{sand}$ (g)] and the silt and clay collected on the filter paper is dried and weighed [$W_{clay+silt}$ (g)]. The results can be expressed as follows:

Concentration of sand (mg $l^{-1}$) $= (W_{sand}/V_{sample}) \times 10^6$

Concentration of clay + silt (mg $l^{-1}$) $= (W_{clay+silt}/V_{sample}) \times 10^6$

Total suspended load (mg $l^{-1}$) $= [(W_{sand} + W_{clay+silt})/V_{sample}] \times 10^6$

If sand concentration is not required separately, then filter a known volume of raw water through a pre-weighed 0.45 µm pore diameter filter paper. The suspended sediment concentration is then the dry weight (in grams) of the filter paper + retained sediment, minus the original weight of the filter paper, all divided by the volume (ml) of the sample, as:

Total suspended sediment concentration (mg $l^{-1}$) $= [W_{sand+silt+clay}/V_{sample}] \times 10^6$

### 13.4.2 Estimating suspended sediment concentration

Water agencies often need to calculate suspended sediment load on an annual basis, but wish to reduce the amount of field sampling required to determine suspended sediment concentration. In some rivers there is a moderately good relationship between suspended sediment concentration and discharge, i.e. the higher the discharge the higher the suspended sediment concentration. It is possible, therefore, to develop a rating curve which is a regression of suspended sediment concentration $Y$ as a function of discharge $X$. This relationship is, however, subject to a high degree of variability and error, especially when the suspended sediment is comprised mainly of silt + clay. Nevertheless, when used carefully, and the rating curve is checked frequently for stability, agencies can use measurements of discharge to estimate suspended sediment concentration ($C_{estimated}$) for the purpose of calculating suspended sediment load ($SS_L$) in tonnes per day as:

$$SS_L = Q_{observed} \times C_{estimated} \times 0.0864$$

This procedure is carried out daily, and the daily loads are summed for the year.

The alternative to extrapolating from a rating curve is some form of statistical estimation in which concentration data are clustered so that they represent a specified flow interval (e.g. by dividing the annual flow record into 10 per cent intervals of discharge). Total load is then calculated by assessing the statistical probability of each interval of flow occurring, multiplying it by the average suspended sediment concentration and discharge for that range, and then summing for all flow intervals. A common application of this approach is the so-called Beales Ratio Estimator. However, there is no general consensus on the best estimator to use, since each may produce different results in different situations.

## 13.5 Sediment quality

Fine grained sediment (silt + clay) is responsible for a significant proportion of the annual transport of metals, phosphorus, chlorinated pesticides and many industrial compounds such as polynuclear aromatic hydrocarbons, polychlorinated biphenyls, dioxins and furans. Of the 128 priority pollutants listed by the United States Environmental Protection Agency, 65 per cent are found mainly, or exclusively, in association with sediment and biota. Consequently, water quality programmes that focus only on the water phase miss most of the more toxic contaminants (see also Figure 2.5). In North America it has been found that up to 95 per cent of the annual phosphorus load in rivers is transported in association with suspended sediment. Organic micropollutants are mainly bound to the organic component of the suspended matter, which is commonly measured as total organic carbon (see section 8.4).

Sediment measurements are not necessary for the assessment of certain water quality issues of major concern in some countries, such as faecal contamination. Nevertheless, other issues such as eutrophication of lakes and reservoirs and fluxes of nutrients and contaminants to coastal waters and oceans, which are becoming more important in all world regions, are increasing the need for sediment-quality measurements. Unfortunately, water quality agencies throughout the world still tend to pay little or no attention to suspended sediment quality. The following are the most frequently identified reasons for this:

- Lack of awareness of current techniques in sedimentology.
- Poorly defined objectives for monitoring programmes (see section 3.3).
- Programmes reflect more established concerns (e.g. major ions, faecal contamination).
- Lack of funding, expertise, equipment, etc.

Only the last of these is a valid reason because the initial investment in sampling and analysis of suspended sediment can be high. Nevertheless, when a programme is developed with well defined objectives, a suspended sediment programme, together with water chemistry, can greatly increase the amount of useful information available to water managers, and may ultimately result in agencies saving money. When combined with an inexpensive bioassay programme (see section 11.4), limited sediment and water sampling can significantly reduce the number of expensive chemical analyses now being undertaken by agencies in industrialised, and rapidly industrialising, countries.

### 13.5.1 Sample collection and preparation

Collection and storage of sediment samples for sediment quality analysis require special handling in order to avoid contamination, especially if the variables to be measured occur in very low concentrations (e.g. $\mu g\ g^{-1}$). The necessary precautions depend on the analyses to be performed. All materials which come into contact with samples intended for metal analyses should be plastic, plastic coated or glass (i.e. samplers, sample storage containers, filtration equipment). All equipment should be acid-cleaned and thoroughly rinsed with pure (double distilled) water. If it is necessary to use a metal sampler for sediments intended for metal analysis (e.g. a grab), the portion of sample in contact with the sides of the sampler should be discarded. Plastic equipment should be avoided for any sampling and handling procedures for sediments intended for analysis of organic micropollutants; appropriately cleaned metal equipment is most suitable (preferably stainless steel). Samplers, filtration equipment and storage containers for samples intended for phosphorus analysis can be made of metal, plastic or glass but should be cleaned with a phosphate-free detergent.

Samples are collected as described in section 13.3 paying particular attention to the requirements for reducing the risk of contamination according to the the analyses to be performed. Suspended sediment is normally concentrated by ultra centrifugation or by filtration in order to collect the fine particulate matter, with which most nutrients and contaminants are associated. Standard 0.45 μm pore size filters are commonly used. Inorganic filters (polycarbonate, cellulose acetate) with low trace element contents are recommended where inorganic substances will be determined, and glass-fibre filters are recommended for organics, particulate organic carbon and chlorophyll determinations.

It is preferable that samples requiring storage prior to analysis should be dried and weighed before placing in a refrigerator (if intended for metal or nutrient analysis) or a freezer at –20 °C (if intended for hydrocarbons or other organic analyses).

Chemical analyses should be performed on dry material of known weight. The precise preparation and analytical details depend on the elements or compounds to be analysed and, therefore, only the basic principles are described here. Samples for metal determinations (including the filters used to separate the suspended sediments) are digested in acid on a hot plate and diluted to a known volume with distilled water. The diluted digest is analysed using standard techniques such as atomic absorption spectrophotometry (AAS) (see section 8.1). The acid used for digestion depends on the proportion of organic matter in the sediment sample and whether it is necessary to determine total metal content or only the proportion which is thought to be most environmentally available (e.g. for bioaccumulation). It is important, when digesting filtered samples that blank filters are also included in each batch of digestions as a means of checking for contamination and allowing for any metals present in the filters themselves (see also Chapter 9 on Analytical Quality Assurance).

Samples for determination of organic micropollutants are extracted with organic solvents and analysed using standard techniques such as gas chromatography or high pressure liquid chromatography. Filters may need to be ground or pulverised in a blender prior to extraction. Many analyses can be performed on unfiltered samples.

Results of sediment quality analyses are usually expressed as mg $g^{-1}$ or μg $g^{-1}$ dry weight or mg $l^{-1}$ or μg $l^{-1}$ of sample volume. Since coarser particles tend to have low concentrations of metals, nutrients and organic micropollutants, a high proportion of coarse particles (e.g. sand) in a sample tends to dilute the contaminant concentration for the whole sample. This is known as the matrix effect. Provided that the fraction less than 63 μm (with which most metals tend to be associated) accounts for at least 30–40 per cent of the total sample, the following correction can be applied:

$$\text{Corrected value } (\mu\text{g g}^{-1}) = \frac{\text{Trace metal concentration } (\mu\text{g g}^{-1})}{\text{Percent of sample } < 63\ \mu\text{m}}$$

The percentage less than 63 µm can be determined by wet sieving a sample of known dry weight through a pre-weighed 63 µm screen and re-weighing the screen and content after drying. This correction is only useful when there is a need to know the likely concentration of the pollutant on that fraction of the sample (the < 0.63 µm component) which actually carries the pollutant, or when the chemistry of different sediment samples is being compared, in which case the correction normalises the sample chemistry for grain size effects (i.e. eliminates the matrix effect). For calculation of the load of a pollutant the correction is not necessary because the pollutant concentration in suspended sediment will be multiplied by the mean suspended sediment concentration (sand + silt + clay) that has been measured as noted in section 13.4.

## 13.6 Source literature and further reading

Allan, R.J. 1986 *The Role of Particulate Matter in the Fate of Contaminants in Aquatic Ecosystems.* Scientific Series No. 142, National Water Research Institute, Environment Canada, Burlington.

Bogen, J., Walling, D.E. and Day, T. [Eds] 1992 *Erosion and Sediment Transport Monitoring Programmes in River Basins.* Publication No. 210, International Association of Hydrological Sciences, Wallingford.

Chapman, D. [Ed.] 1996 *Water Quality Assessments. A Guide to the Use of Biota, Sediments and Water in Environmental Monitoring.* 2nd edition, Chapman & Hall, London.

Golterman, H.L., Sly, P.G. and Thomas, R.L. 1983 *Study of the Relationships Between Water Quality and Sediment Transport. A Guide for the Collection and Interpretation of Sediment Quality Data.* Technical Papers in Hydrology No. 26, United Nations Educational, Scientific and Cultural Organization, Paris.

Graf, W.H. 1984 *Hydraulics of Sediment Transport.* McGraw-Hill, New York.

Gregory, K.J. and Walling, D.E. 1973 *Drainage Basin Form and Process.* Edward Arnold, London.

ISO 1968 *Test Sieving — Part 1: Methods Using Test Sieves of Woven Wire Cloth and Perforated Metal Plates.* International Standard ISO 2591-1, International Organization for Standardization, Geneva.

ISO 1985 *Liquid Flow in Open Channels — Sediment in Streams and Canals — Determination of Concentration, Particle Size and Relative Density.* International Standard ISO 4365, International Organization for Standardization, Geneva.

ISO 1990 *Representation of Results of Particle Size Analysis — Part 1: Graphical Presentation.* International Standard ISO 9276-1, International Organization for Standardization, Geneva.

Linsley, R.K., Kohler, M.A. and Paulhus, I.L.H. 1988 *Hydrology for Engineers*. McGraw-Hill, New York.

Meybeck, M. 1992 Monitoring of particulate matter quality. In: M. Allard, [Ed.] *GEMS/WATER Operational Guide*, Third edition, Unpublished WHO document GEMS/W.92.1, World Health Organization, Geneva.

Ongley, E.D. 1992 Environmental quality: changing times for sediment programs. In: J. Bogen, D.E. Walling and T. Day [Eds] *Erosion and Sediment Transport Monitoring Programmes in River Basins*. Publication No. 210, International Association of Hydrological Sciences, Wallingford, 379–390.

Ongley, E.D. 1993 Global water pollution: challenges and opportunities. Paper presented at the Stockholm Water Symposium, August 1993.

Ongley, E.D., Yuzyk, T.R. and Krishnappan, B.G. 1990 Vertical and lateral distribution of fine grained particles in prairie and cordilleran rivers: sampling implications for water quality programs. *Water Research*, **24**(3), 303–312.

Ongley, E.D., Krishnappan, B.G., Droppo, I.G., Rao, S.S. and Maguire, R.J. 1992 Cohesive sediment transport: emerging issues for toxic chemical management. *Hydrobiologia*, **235/236,** 177–187.

U.S. Soil Conservation Service. 1983 *National Engineering Handbook. Section 3 — Sedimentation*. United States Department of Agriculture, Washington, DC.

WHO 1992 *GEMS/WATER Operational Guide*. Third edition. Unpublished WHO document GEMS/W.92.1, World Health Organization, Geneva.

WMO 1988 *Manual on Water-Quality Monitoring*. Operational Hydrology Report No. 27, WMO Publication No. 680, World Meteorological Organization, Geneva.

Yalin, M.S. [Ed.] 1977 *The Mechanics of Sediment Transport*. Pergamon, Toronto.

# Chapter 14

# USE AND REPORTING OF MONITORING DATA

The whole system of water quality monitoring is aimed at the generation of reliable data, i.e. data that accurately reflect the actual status of the variables which influence water quality. It is acknowledged that simply generating good data is not enough to meet objectives. The data must be processed and presented in a manner that aids understanding of the spatial and temporal patterns in water quality, taking into consideration the natural processes and characteristics of a water body, and that allows the impact of human activities to be understood and the consequences of management action to be predicted. This is not to say that water quality information must be presented in a way which requires the user to appreciate the full complexity of aquatic systems. The information should provide the user with the understanding necessary to meet the objectives behind the monitoring programme. The importance of setting objectives in the design and implementation of monitoring programmes is discussed in Chapter 3. Objectives imply that the activity has a purpose that is external to the monitoring system; e.g. a management, environmental policy, public health or research purpose. The intent is to use the information to explain water quality, to communicate the information more widely, or to control water quality. Consequently, there is little point in undertaking a monitoring programme unless the resultant data are to be used in fulfilling the objectives.

For some information-users, complex graphical or statistical analysis may be essential, such as for long-term management planning and pollution control in a lake system. For other users, a single variable statistic may be adequate for management purposes. For example, BOD measurement (see Chapter 7) or a water quality or biotic index (see Chapter 11) could be used to evaluate the effectiveness of treatment of domestic or agricultural wastes discharged into a river or stream.

This chapter provides a brief introduction to some of the principal approaches to use and reporting of monitoring data. Full treatment of data handling and presentation for assessments, including details and worked examples of some key statistical methods, is covered in the companion guidebook *Water Quality Assessments*.

*The preparation of this chapter was co-ordinated by A. Steel with contributions from M. Clarke, P. Whitfield and others*

## 14.1 Quality assurance of data

Quality assurance of data is an important precursor to data analysis and use. Effective quality control procedures during sampling and analyses help to eliminate sources of error in the data. However, a second series of data checks and precautions should be carried out to identify any problems that might lead to incorrect conclusions and costly mistakes in management or decision-making. The procedures involved commonly rely on the identification of outlying values (values that fall outside the usual distribution), and procedures such as ensuring that data fall within the limits of detection of a particular method of measurement and checking that normal ion ratios are present. Any anomalous values should be checked because there could be a problem with calculations or with analyses, in which case remedial action should be taken. Genuine outlying values do occur, however, and may be important indicators of changes in water quality.

### 14.1.1 Basic data checks

Basic data checks may be carried out by the laboratory that generates the data and/or by the person or organisation that uses or interprets them. The most common errors in reports of analytical data are faults in transcription, such as incorrect positioning of the decimal point or transcription of data relating to the wrong sample. The number of times that data must be copied before the final report is composed should, therefore, be kept to a minimum. Other errors may include:

- Omission of a major ion. Analysis for nitrate, in particular, is sometimes omitted, although nitrate may represent a significant proportion of anions.
- Incorrect identification of samples.
- Double reporting of ions. This can occur, for example, when alkalinity titrations record ions that are also analysed separately, such as silicate and phosphate.
- Faulty analytical technique.

Some examples of basic data checks that may be readily employed are described below. Table 14.1 illustrates a series of validity checks on a set of results for some commonly measured variables in rivers.

*Major ion balance error*

Percentage balance error = 100 × (cations – anions)/(cations + anions).

For groundwaters, the error should be 5 per cent or less unless the total dissolved solids (TDS) value is less than 5 mg $l^{-1}$, in which case a higher error is acceptable. For surface waters, an error of up to 10 per cent is acceptable; if it exceeds 10 per cent, the analysis should be checked for errors in transcription or technique.

**Table 14.1** Checking data validity and outliers in a river data set

| Sample number | Water discharge (Q) | Elec. Cond. ($\mu S\ cm^{-1}$) | $Ca^{2+}$ ($\mu eq\ l^{-1}$) | $Mg^{2+}$ ($\mu eq\ l^{-1}$) | $Na^+$ ($\mu eq\ l^{-1}$) | $K^+$ ($\mu eq\ l^{-1}$) | $Cl^-$ ($\mu eq\ l^{-1}$) | $SO_4^{2-}$ ($\mu eq\ l^{-1}$) | $HCO_3^-$ ($\mu eq\ l^{-1}$) | $\Sigma+$[1] ($\mu eq\ l^{-1}$) | $\Sigma-$[1] ($\mu eq\ l^{-1}$) | $NO_3$-N ($mg\ l^{-1}$) | $PO_4$-P ($mg\ l^{-1}$) | pH |
|---|---|---|---|---|---|---|---|---|---|---|---|---|---|---|
| 1 | 15 | 420 | 3,410 | 420 | 570 | 40 | 620 | 350 | 3,650 | 4,440 | 4,620 | 0.85 | 0.12 | 7.8 |
| 2 | 18 | 405 | **3,329.4** | 370 | 520 | 35 | 590 | 370 | 3,520 | 4,254 | 4,480 | **0.567** | **0.188** | **7.72** |
| 3 | 35 | **280** | 2,750 | 390 | **980** | 50 | **1,050** | 260 | 2,780 | 4,150 | 4,090 | 0.98 | 0.19 | 7.5 |
| 4 | 6 | 515 | 4,250 | **5,200** | 620 | 50 | 680 | 510 | 4,160 | 5,440 | 5,350 | **0.05** | **0.00** | 8.1 |
| 5 | 29 | 395 | 2,950 | 420 | 630 | **280** | 670 | 280 | 2,800 | **4,280** | **3,770** | 0.55 | 0.08 | **9.2** |
| 6 | 170 | 290 | 2,340 | 280 | 480 | 65 | **930** | 250 | 2,550 | **3,165** | **3,730** | 1.55 | 0.34 | 7.9 |
| 7 | **2.5** | 380 | 3,150 | 340 | 530 | 45 | 585 | **3,240** | **375** | 4,065 | 4,200 | 0.74 | **3.2** | 7.6 |

Questionable data are shown in **bold**

[1] $\Sigma+$ and $\Sigma-$ sum of cations and anions respectively

Sample number

1 Correct analysis: correct ionic balance within 5 per cent, ionic proportions similar to proportions of median values of this data set. Ratio $Na^+/Cl^-$ close to 0.9 eq/eq etc.

2 Excessive significant figures for calcium, nitrate, phosphate and pH, particularly when compared to other analyses.

3 High values of $Na^+$ and $Cl^-$, although the ratio $Na^+/Cl^-$ is correct — possible contamination of sample? Conductivity is not in the same proportion with regards to the ion sum as other values — most probably an analytical error or a switching of samples during the measurement and reporting.

4 Magnesium is ten times higher than usual — the correct value is probably 520 $\mu eq\ l^{-1}$ which fits the ionic balance well and gives the usual $Ca^{2+}/Mg^{2+}$ ratio. Nitrate and phosphate are very low and this may be due to phytoplankton uptake, either in the water body (correct data) or in the sample itself due to lack of proper preservation and storage. A chlorophyll value, if available, could help solve this problem.

5 Potassium is much too high, either due to analytical error or reporting error, which causes a marked ionic imbalance. pH value is too high compared to other values unless high primary production is occurring, which could be checked by a chlorophyll measurement.

6 The chloride value is too high as indicated by the $Na^+/Cl^-$ ratio of 0.51 — this results in an ionic imbalance.

7 Reporting of $SO_4^{2-}$ and $HCO_3^-$ has been transposed. The overall water mineralisation does not fit the general variation with water discharge and this should be questioned. Very high phosphate may result from contamination of storage bottle by detergent.

Source: Demayo and Steel, 1996

An accurate ion balance does not necessarily mean that the analysis is correct. There may be more than one error and these may cancel each other out. As a result additional checks are needed.

*Comparison of electrical conductivity and total dissolved solids*
The numerical value of electrical conductivity (μS) should not exceed that of TDS (mg $l^{-1}$). It is recommended that conductivity be plotted against TDS and values lying away from the main group of data checked for errors. The relationship between the two variables is often described by a constant (commonly between 1.2 and 1.8 for freshwaters) that varies according to chemical composition. For freshwater the normal range can be calculated from the following relationship:

$$\text{Conductivity} \approx \text{TDS} \times a \qquad \text{where } a \text{ is in the range 1.2–1.8}$$

Typically, the constant is high for chloride-rich waters and low for sulphate-rich waters.

*Total dissolved solids (calculated) v. total dissolved solids (dry residue)*
The calculated value of TDS and that of TDS determined by dry residue should be within 20 per cent of each other after correction for loss of $CO_2$ on drying (1 g of $HCO_3$ becomes approximately 0.5 g of $CO_2$). However, dry residue can be difficult to determine, especially for waters high in sulphate or chloride.

*It must be scientifically possible*
Totals of any variable must be greater than the component parts as in the following example:

- Total coliforms must be greater than thermotolerant coliforms.
- Total iron must be greater than dissolved iron.
- Total phosphate must be greater than dissolved phosphate.
- Total phosphate must be greater than phosphate alkalinity.

The species reported should be correct with regard to the original pH of the sample. Carbonate species will normally be almost all $HCO_3^-$, high $CO_3^{2-}$ levels cannot exist at a neutral pH and $CO_3^{2-}$ ions and $H_2CO_3$ cannot coexist.

*Departure from expected values*
Distinguishing between a valid result and an error in the data requires experience. Visually scanning a spreadsheet of data can help to identify analyses in which some variables are much higher or lower than in others, although the values may not necessarily represent errors and may be of interest in their own right. For analytical techniques it is recommended that control charts (see Chapter 9) be prepared to identify values than lie away

from the main group of data. Otherwise, specialised techniques for identifying outliers need to be used (see *Water Quality Assessments* for details).

*Anomalous results*

If nitrate is present in the absence of dissolved oxygen, the value for one or the other is likely to be incorrect, since nitrate is rapidly reduced in the absence of oxygen. There may have been a malfunction of the dissolved oxygen meter or loss of oxygen from the sample before analysis.

If nitrite and $Fe_2^+$ are present in the same sample, either the analysis is incorrect or the sample is a mixture of aerobic and anaerobic waters. For a lake, it could be a representative sample, but for a groundwater it would probably mean that the water was derived from two different aquifers.

If there are high levels of iron or manganese and of aluminium (of the order of 1 mg $l^{-1}$) at a neutral pH, they are probably the result of colloidal matter which can pass through a 0.45 µm filter.

*Analytical feasibility*

If the results do not fall within the analytical range of the method employed, errors of transcription or of analytical technique should be suspected.

## 14.2 Data handling and management

The computer software used in data handling and management falls into three principal classes:

- Statistical software which processes numerical data and performs statistical tests and analyses.
- Spreadsheets which handle both numerical data and text, and usually include powerful graphical and statistical capabilities (thus overlapping with the purely statistical software).
- Database software which is designed to manage the input, editing and retrieval of numerical data and text. No statistical or graphical capabilities are in-built, but the power of the programming language allows the skilled user enormous scope for data manipulation, sorting and display.

In an ideal situation all three classes of software can be used together in a complementary fashion. Nevertheless, it is essential that the eventual statistical methods to be applied are taken into account fully in the original monitoring programme design. This should help to ensure that the data produced are adequate for the statistical techniques which are to be applied.

A fourth class of data-handling software, known as GIS (geographical information systems), has recently been developed. This is specifically designed to relate data to geographical locations and output them in the form of maps. Data on different aspects of, in this case, a water quality monitoring programme can be superimposed. For example, data on groundwater quality such as chemical variables can be overlaid with data on land use, etc. This allows the relationships between the selected aspects to be studied and a map, for example of groundwater protection zones, to be generated. The overwhelming constraints on the wider use of GIS software are its cost and its need for sophisticated hardware and highly skilled operators.

The scope and nature of computerised data-handling processes will be dictated by the objectives of the water quality monitoring programme. Generally speaking, however, a database offers the best means of handling large quantities of data and it should be capable of exporting data in formats that are accepted by all good statistical, spreadsheet and GIS packages.

An experienced programmer can create a relational database management system (RDBMS). Such systems allow the creation of data tables that can be related to other, associated, data tables through unique keys or indexes. A data table is a collection of data records and a record comprises a series of data fields or information variables. A simple analogy for a database is provided by a telephone directory in which each record, or entry, consists of various information fields, i.e. family name, initials, address, etc., and telephone number. Entries are listed in alphabetical order of family name and then of initials (or forenames). However, since many people may share the same family name, the entry "Smith", for example, cannot be considered a unique index. The only unique index or key in the directory is the individual telephone number.

In creating a database, the first decision to be made concerns what fields of information are required and the second, how many data tables are needed for their efficient storage. It is a mistake to group all data in a single table unless, like the telephone directory, the database is to be very simple. A useful approach is to separate fields of information that relate to variable data and those that deal with unchanging details, such as the name, description and location of a sampling site.

All data tables relating to a single sampling site must be identifiable as such, regardless of the type of data they contain. To avoid the need for including extensive details of site description and location in every such table, each sampling site should be identified by a unique index, usually a code of no more than 10 characters, which appears in every data table relating to that particular site. Then, if it becomes necessary to amend any details of site description, only one data table, the one concerned with the site itself, will need to be changed. All other related tables will continue to carry the unique, and unaltered, code.

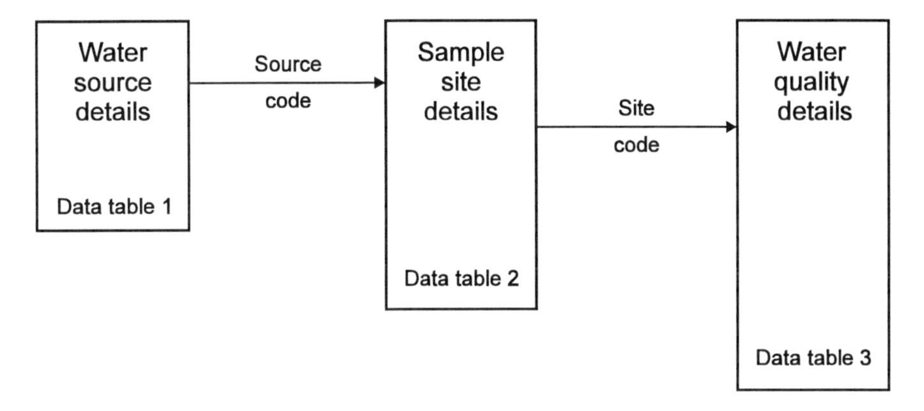


**Figure 14.1** Relationship between tables for a simple water quality database

Other unique indexes, or keys, can be created to relate data tables to each other. It may be valuable to create a key based on sampling dates, for example, which would allow rapid retrieval of all data recorded for a particular date or between two specified dates. Consequently, it becomes obvious that the desired format of the output from the database dictates the keys that should be created, such that selecting the appropriate key or index yields data in a particular order or for a particular variable.

Figure 14.1 illustrates the linking of tables described above by providing a suggested scheme for a simple water quality database. Three data tables have been used to store the required information. The first table is concerned with the water source to be tested, which might be a lake, river, borehole, water supply scheme or even an industrial treatment plant. The name, type and location of the water source are recorded in this table, and possibly a brief description. In addition, the table must contain a source code, the primary "look-up" key, that links it to the second table.

The second table contains more detailed information about sampling sites. A river, for example, may have two or more designated sampling sites, the locations for which must be accurately defined. This table carries the source code or look-up key from table one, plus unique codes for each sampling site that provide links to the third table. Typically, the second table would contain the following information for each site:

- Source code: the look-up key to table one.
- Site code: the link to the third table.
- Location: geographical location of the sampling site.
- Map number: a reference to the map covering the site location.

- Map reference: the latitude and longitude of the site or a national grid reference.
- Description: a brief textual description of the site.

Data on water quality (temperature, pH, conductivity, chemical and microbiological variables, etc.) are stored in the third table. Many details will be recorded for each sample from each site. Additional information should cover how and when each sample was collected and when the various analyses were performed. Ideally, each sample record should include a unique key for purposes of identification, cross-reference, and data retrieval; for example, this might be the analysing laboratory's reference number for the sample.

Certain precautions relating to the deletion and/or amendment of information must be rigorously observed. In the example of Figure 14.1 deletion of a record from table one might leave related data in table two "orphaned", that is, without linkage to "parent" data. Deletions should, therefore, be made only of material which is not shared by a related table. Moreover, amendment of a key field in one table must be reflected in parallel changes to related records in other tables. These simple rules are essential for preserving the referential integrity and validity of the database, both of which are crucial to efficient data management. Many modern databases track and maintain their referential integrity, thereby reducing the likelihood of creating orphaned data tables.

A final point concerns the designs of screens for the input of data to the database. Where standardised forms already exist for written records, these should be mimicked as far as possible for computerised data entry. The greatest advantages of the computerised database are ease of manipulation, ease of retrieval and, particularly, ease of dissemination of data to all interested parties. It is this last feature, which is one of the principal objectives of the water quality monitoring programme, that gives computerised databases their enormous advantage.

## 14.3 Basic statistical analysis

Producing meaningful information from raw data normally requires an initial statistical analysis of that data; for example, to determine the magnitudes of variables, their variability, any time trends, etc. It is also necessary to be able to give some indication of the "confidence" the user may have in the statistical outputs. For example, if a sample average is 50 per cent greater than a previous estimate, could that be expected as a reasonable chance occurrence, or is it indicative of real change? In order that such results are valid, it is essential that only statistical tests which are suited to the data (defined by its type, manner and frequency of collection, etc.) are used. With the ready availability of computerised statistical packages it can be too easy to apply inappropriate analysis to data, especially when they can be automatically collected, stored and presented for analysis. This handbook does not aim to provide detailed methods for statistical analysis, but only to show the potential of such analysis and of effective presentation of data. Worked examples of some of the statistical techniques discussed here are available in *Water Quality Assessments*.

### 14.3.1 General considerations

Many powerful, traditional statistical tests rely on the data conforming to an underlying pattern or frequency distribution. Most commonly, this is the "normal" distribution, or variants thereof. Such statistical tests are termed parametric to indicate the requirement that the data conform to some understood and describable (by its parameters) distribution. The mean and standard deviation, for example, are of this type. Another fundamental class of statistical tests are non-parametric (or distribution-free) which, as their description implies, make no assumptions about the data to which they may be applied. The median and the percentiles are examples of non-parametric tests.

When data conform to the appropriate distribution requirements, the parametric statistical tests are usually more powerful than their non-parametric counterparts (where they exist). However, any advantage can be swiftly eroded once the data distribution becomes distorted and cannot be corrected by transformations, for example by using logarithms of the data instead of the raw data. Water quality data sets are often not easily definable, therefore there has been much recent development of non-parametric statistics, both to provide complementary tests to their parametric counterparts and to improve their power. An important, early step in programme design is, therefore, to assess which statistical tests would best (and most simply) serve the information needs of the programme, to evaluate any data requirements this may imply and, thus, to ensure that the monitoring programme is designed to provide for those data needs.

### 14.3.2 Outline of statistical techniques

Statistical analysis of water quality data may be very broadly classified in two groups:

- Descriptive statistics.
- Inferential statistics.

These classifications suggest that, initially, the task is to provide an accurate and reliable statistical description of the data set. Following that it may be possible to give some account of how well the data set, and/or its interactions, conforms to some theory as to its origin and so, possibly, identify potential means of modification. Thus the "what?" comes before the"how?".

The basic statistical techniques which constitute these two classes can be generally identified as:

- Analysis of data distributions.
- Testing assumptions about data sets.
- Specifying data magnitudes and variability.
- Estimating reliability of data statistics.
- Comparisons of data sets.
- Associations between data sets.
- Identifying trends and seasonality within data sets.
- Testing theories relating to the water quality data.

The various tests could be further sub-divided by virtue of being parametric or non-parametric, single or multi-variable, single or multi-sample, discrete or continuous data, etc.

*Analysis of data distributions*

Analysis of data distributions is normally carried out to check that the data set, either in its raw state or following transformation, conforms to a definable data distribution which is appropriate to the parametric statistics to be applied to it. Knowing the distribution may help in accepting or rejecting assumptions about the sampling conditions which generated the data and may also allow some predictions to be made about the data. Formal analysis of distribution can be a difficult task, particularly with sparse and variable data. However, the first activity (as with almost all water quality data) is to plot the data in an appropriate manner (e.g. scatter plot, frequency histograms, probability plots, see section 14.5). This will often give a reasonable indication of the distribution type, even though statistical difficulties may be encountered. Usually, once a general idea of the distribution is gained, the assumption would be tested against the data available.

*Testing assumptions about data sets*

The tests in this category usually involve procedures which assess the statistics derived from some other test against values which are characteristic of some presumption about the data. For example, are the frequency histogram data consistent with a normal distribution? Traditionally the $\chi^2$-test has been used, but other tests (such as the *W*-test, see *Water Quality Assessments*) provide some advantages over this traditional approach, depending on the nature of the data.

*Specifying data magnitudes and variability*

This category would include all the analyses leading to the mean, variance, standard deviation, and coefficient of variation (parametric) statistics, together with the median and appropriate percentiles of non-parametric statistics. The 25th and 75th percentiles are the most commonly used. The 25th and 75th percentiles are the values below which 25 per cent and 75 per cent, respectively, of the data values occur.

*Estimating reliability of data statistics*

The purpose of these methods is to provide some measure of the possible errors which may be associated with the estimated statistics. Thus, for example, the standard error of a mean allows some judgement of how great a difference in mean may occur in a repeated sampling. The availability of such error estimates allows confidence limits to be constructed for a variety of statistics, and from which some conclusions about the reliability of the data may be drawn. Other procedures in this category would be to deal with outliers. These are data which do not conform to the general pattern of a data set. They may be anomalous, in which case they must be excluded because of the distortion (particularly for parametric tests) they may cause. However, they may also be highly informative, proper, extreme values of the data set and excluding them would lead to incorrect conclusions. Critical testing is, therefore, necessary for such data.

*Comparisons of data sets*

Monitoring programmes often attempt to compare water quality conditions at one site with that at another (perhaps acting as a control site). Various statistical methods are available to compare two or more such data sets. These range from such basic tests as the Student's *t*-test to the Mann-Whitney *U*-test and the Kruskal-Wallis test for multiple comparisons.

*Associations between data sets*

Once a programme begins to address the issues of forecasting, or establishing causes-and-effects in water quality variables, then it is essential to be able to quantify the relationships amongst the variables of interest. Two primary classes of such assessment are correlation analysis and regression. Correlation establishes the statistical linkage between variables, without having to ascribe causation amongst them whereas regression seeks the functional relationship between a supposed dependent water quality variable and one, or many, variables which are termed "independent".

*Identifying trends and seasonality within data sets*
Water quality variables frequently exhibit variability in time. This variability may be cyclical with the seasons, or some other established variation over time. Special statistical tests have been developed to deal with these possibilities. Many are complex and too advanced for a basic water quality monitoring programme. Nevertheless, the Mann-Kendall test for trends, and the Seasonal Kendall test for seasonality, are amongst the more general tests which may be applied.

*Testing theories relating to the quality data*
Once basic data have been analysed and tentative conclusions have been drawn, almost all water quality monitoring programmes require that those conclusions are validated as far as possible. Generally, the approach is to define null-hypotheses, broadly proposing the alternative to the conclusion of interest. The null-hypothesis is then subjected to testing using the statistics available. If refuted, then the hypothesis of interest is accepted within the probability defined for the testing procedure. The test procedures available include the *t*-test, *F*-test and the analysis of variance (ANOVA).

This brief review has allowed only very few, basic methods to be identified. Nonetheless, many can routinely form part of sensitive statistical analyses of water quality data. Almost all methods mentioned will be available as part of computer statistics packages, in addition to their detailed description in applied statistical texts (see section 14.8).

## 14.4 Use of data and the need for supporting information

The achievement of the common objectives of monitoring programmes cannot rely solely on water quality data and the derived statistics. Information on water quality must be placed in the context of the natural and man-made environment and may need to be combined with quite different types of data. This is particularly true if the ultimate intention is regulation or control of water quality. For example, a monitoring programme might be designed to assess the effectiveness of pollution control of domestic sewage in relation to the quality of well waters. It is relatively easy to visit a number of wells, take samples, produce laboratory data and report the degree of contamination of the different samples. However, this does not constitute assessment; it merely provides some of the information needed for assessment. Assessment of pollution control would also require information on the different strategies for sewage collection and treatment, and for source protection at the different sites. It would then be necessary to determine whether the patterns of waste generation were similar and whether the groundwater bodies were comparable in their hydrogeology. While it might seem relatively easy to identify deterioration or improvement at a particular site, it is far more difficult to explain exactly why those changes have occurred and to suggest actions for management. The changes that followed management action would also need to be monitored in an effort to identify the most effective action.

The above example is intended to show that water quality data must often be combined with other data and interpreted in a way that specifically addresses the objectives of the end-user of the information. Some water quality monitoring programmes generate excellent data but lack supporting information on the possible influences on water quality. This poses a problem for management because control of water quality frequently requires intervention in human activities, such as agriculture, industry or waste disposal. Point sources of pollution may be easy to identify, but effective control of diffuse sources requires more extensive data. If the information about the influences on water quality at a particular location is incomplete, it may not be obvious where management action would be most effectively deployed. Water quality assessment and management is, therefore, concerned not only with monitoring data but also with a wide range of supporting and interpretative data that may not always be available. Sometimes priorities for action must be selected on the basis of limited information.

*Water quality standards and indices*

Pure analytical data may not always convey to non-experts the significance of poor water quality or the possible impacts of elevated chemical concentrations. The typical administrator may have a limited technical background, and it may be necessary to use special "tools" to communicate relevant information in a manner that will prompt suitable interventions.

The simplest tool is a chemical standard or "trigger" value. When the trigger value is exceeded, the administrator (or other information user) is prompted to take action. That action, for example, may be designed to prevent people from drinking water that does not meet health standards or it may be a management intervention to regulate discharges in an operational situation. A variant of the trigger approach is based on a percentage of values exceeding the standard. When a certain number of values (e.g. 5 per cent) fall outside an acceptable standard, the water is judged to be of unacceptable quality for specific uses.

Standards are also important as objectives for quality. They may function as targets for quality control programmes, and the relative success of a management strategy may be judged by comparing the monitoring data withestablished standards. Standards are, therefore, convenient means of stating expert judgements on whether water is of an acceptable quality for a particular use.

Information may also be summarised and presented to the non-expert in the form of water quality indices. A water quality index combines data for certain variables that have an impact on water quality. An index may simply aggregate the values or it may score or weight data according to their relative importance for water quality. Examples include general water quality indices (combining a range of chemical and physical variables), use-oriented indices (including variables of importance for specific uses) and biological indices (combining scored information for various species sensitive to changes in water quality, see section 11.2).

## 14.5 Simple graphical presentation of results

Graphs can communicate complex ideas with clarity, precision and efficiency. They will reveal patterns in sets of data and are often more illustrative than statistical computations. The general objective in graph construction is to concentrate a large amount of quantitative information in a small space so that a comprehensive overview of that information is readily available to the viewer. This can be achieved if:

- the graph is uncluttered,
- lines, curves and symbols are clear and easy to see,
- appropriate scales are chosen so that comparison with other graphs is possible, and
- clearly different symbols are used to represent different variables.

There are a wide range of graphical methods which are suitable for reporting data from monitoring programmes. In deciding the most appropriate form of graphical presentation, four basic types of plot are usually involved: a scatter plot, a bar graph (histogram plot), a time series plot, and a spatial plot. All of these may be presented in several different ways. Time series plots are designed to illustrate trends with respect to time, together with any seasonality effects. Spatial plots can be used to illustrate vertical and longitudinal profiles and, by using maps and cross-sections, can demonstrate geographic and local quality distributions. Survey results are often very effectively displayed on maps. These basic graph types can be used with both raw data and derived statistics. There are also other plot types, such as box plots, pie diagrams and trilinear diagrams (rosette diagrams) which are most suited to displaying summary data. It may also be necessary, in some instances, to

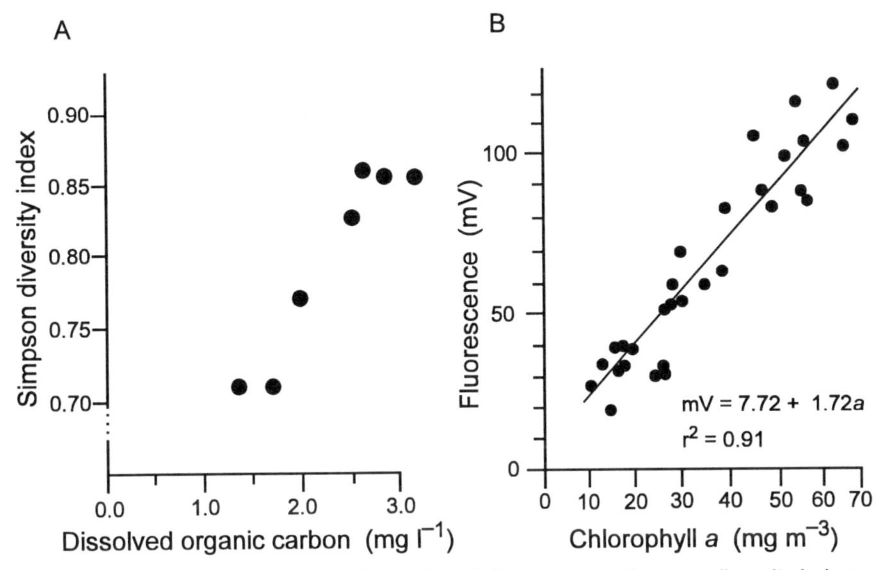


**Figure 14.2** Example of scatter plot. **A**. Correlation between Simpson diversity Index and dissolved organic carbon (After Pinder, 1989) **B**. Regression of chlorophyll *a* and fluorescence (After Friedrich and Viehweg, 1987)

display supporting, ancillary information such as river flows, sample numbers and frequency, etc.

Examples of scatter plots are shown in Figure 14.2. They each illustrate two quality variables sampled at the same time. There may, or may not, be implied cause-and-effect in such graphs. At its most basic, the scatter plot gives an indication of the co-variation (or lack-of) between two variables. The scatter plot matrix is a development of this basic concept that can be applied to a variety of different data types and to illustrate differing responses. It is a simple, but elegant, solution to the difficult problem of comparing a large collection of graphs. In the scatter plot matrix, the individual graphs are arranged in a matrix with common scales. An example is shown in Figure 14.3. A row or column can be visually scanned and the values for one variable can be seen relative to all others. Thus column A shows the variation of conductivity, TON, $SiO_2$ and $SO_4$ with flow, whereas row B shows the conductivity associated with flow, TON, $SiO_2$ and $SO_4$. Such a graph can often demonstrate the significance of the data much more dramatically than extensive statistical analyses.

A bar graph is shown in Figure 14.4. The example deals with summary values, and so confidence limits may also be applied to the top of the bars to

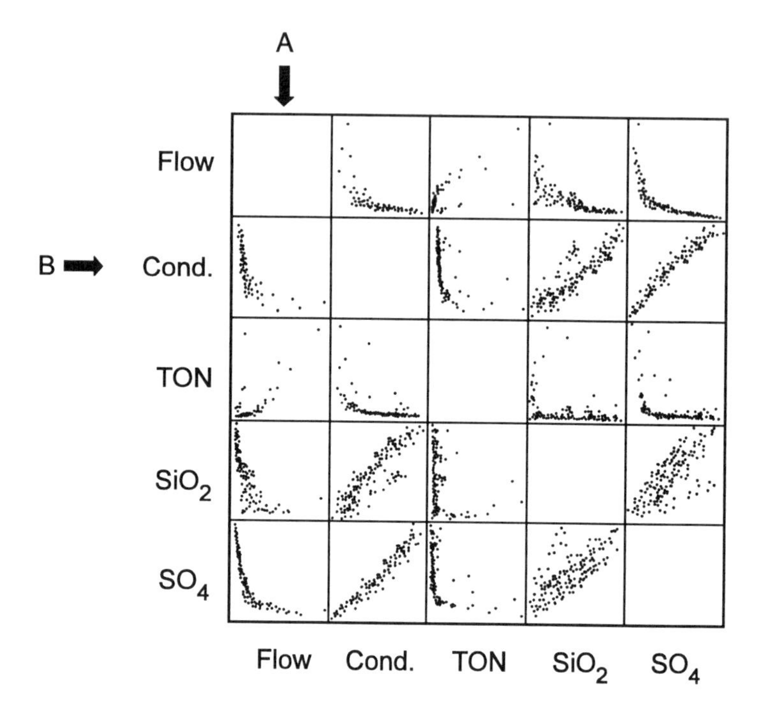


**Figure 14.3** A scatter plot matrix for five variables

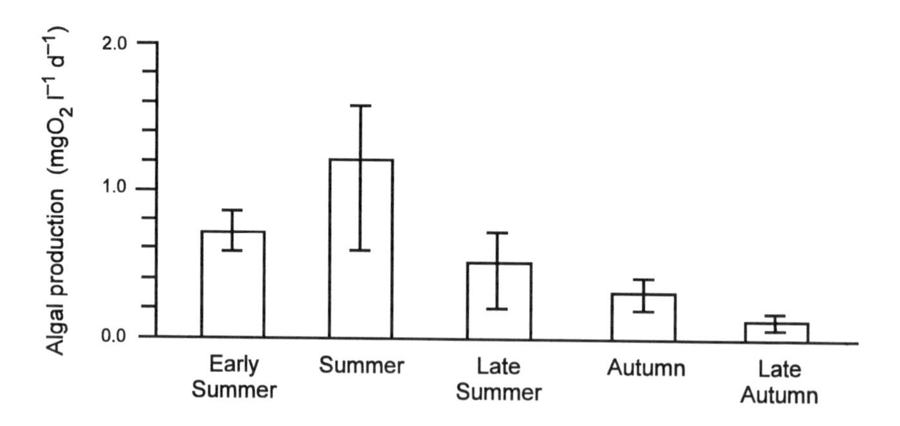


**Figure 14.4** Example of a bar graph with 95 per cent confidence limits indicated as vertical lines (After Spodniewska, 1974)

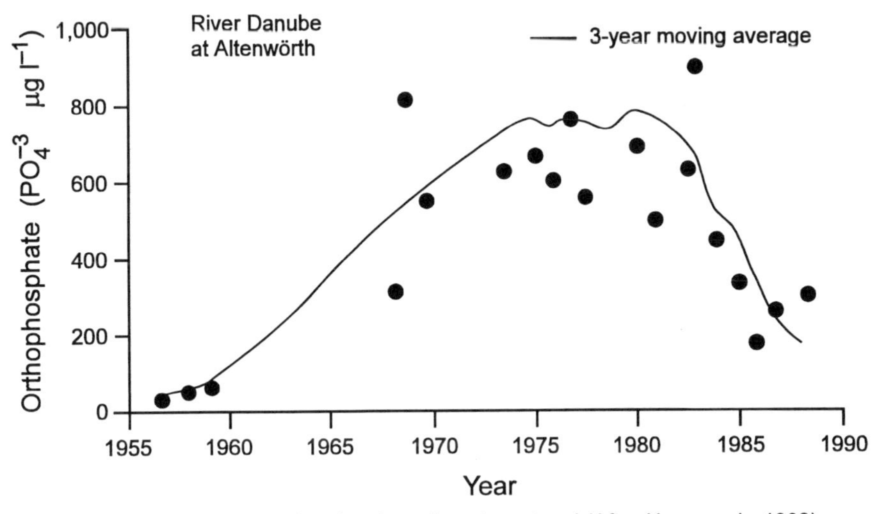


**Figure 14.5** A time series plot showing a long-term trend (After Humpesch, 1992)

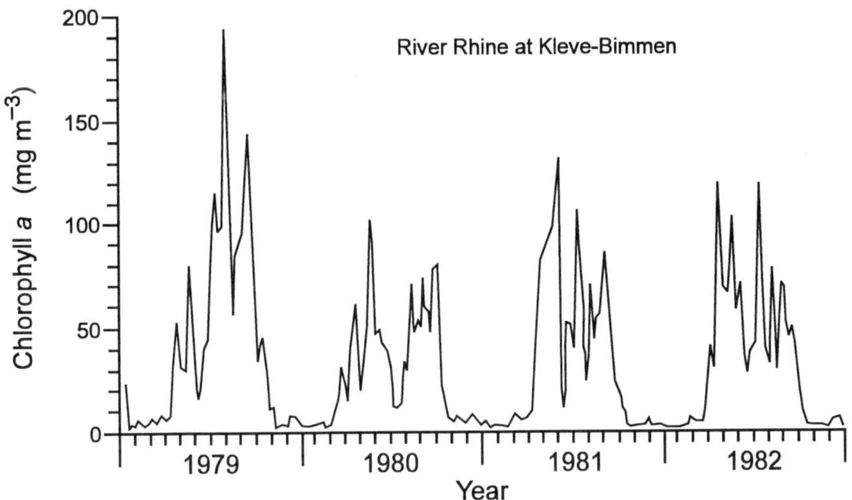


**Figure 14.6** A time series plot indicating seasonal variations (After Friedrich and Viehweg, 1984)

give an indication of the variability of the data. In another type of bar graph the bars may also be "stacked" in order to display relative proportions.

Figures 14.5 and 14.6 show different uses of time series plots. The concentrations of the variables in a series of samples collected over a time period are plotted against the times at which the individual samples were obtained. These show clearly how the concentrations of the variables change with time and can be used to indicate possible trends (Figure 14.5) and seasonal

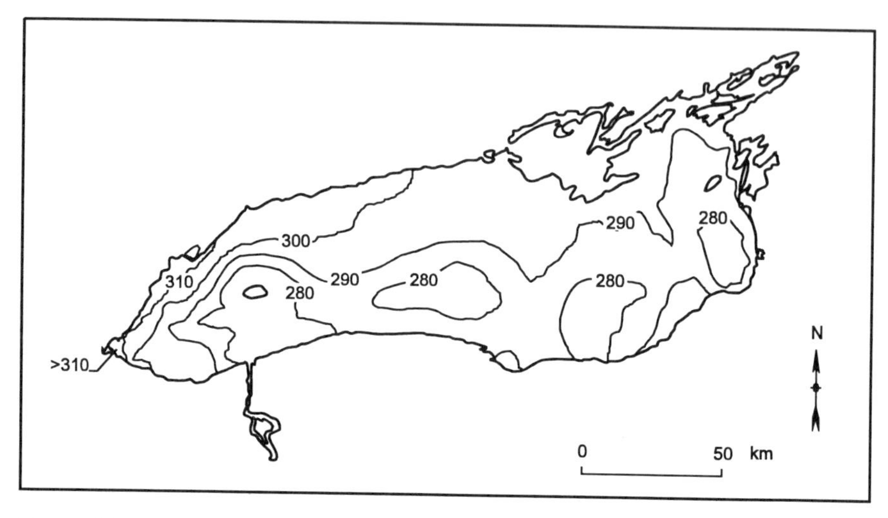


**Figure 14.7** A spatial plot showing isopleths of conductivity in Lake Ontario

variations (Figure 14.6). The information which may be gained is governed by the frequency of sampling in relation to the variability of the water quality characteristics being studied.

Spatial plots can be used to show horizontal or vertical variations in water quality variables. Figure 14.7 shows an example of the horizontal distribution of conductivity in a large lake. Profiles through space or time summarise large quantities of information in a single figure. Interpolation between data points to obtain isopleths (lines joining points where the value of a variable is the same) as in Figure 14.7 may be done either manually or by computer. This type of graph can also combine temporal and spatial data, and is particularly useful, for example, for illustrating seasonal changes in the vertical composition and characteristics of lake waters (Figure 14.8).

The findings of a survey can be effectively displayed by superimposing the monitoring results on a map of the area surveyed. The survey results in each sub-area may then be presented in a graphical form (framed rectangle, pie chart, etc.), in which the measured value of the variable is in proportion tosome identifiable value. In the example of framed rectangle graphs in Figure 4.9, the population dependent on groundwater is shown as a percentage of the total population in each sub-area. This example also shows the dependence on groundwater for the entire area, so that the relative dependence in each administrative subdivision is immediately obvious. If a water quality variable was the item of interest, its measured concentration could be shown in relation to the water quality standard or some other absolute value

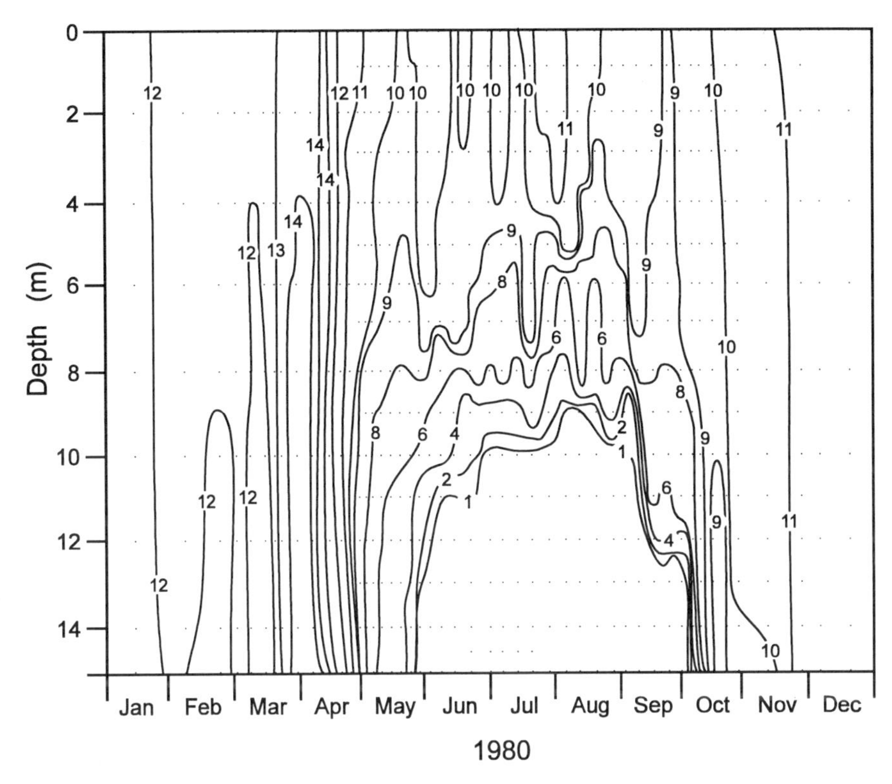


**Figure 14.8** Combination of spatial and temporal data to show vertical variations in a lake. Isopleths of oxygen in a lake which is seasonally stratified

for that variable. Some reference must always be provided so that the viewer can readily understand the significance of the reported result.

Box and whisker plots are a useful way to compare distributions because they permit the comparison of corresponding percentiles. They are used to indicate the basic statistical attributes (maximum, minimum, median, upper and lower quartiles) as shown in the inset to Figure 14.10, which illustrates the use of such plots to display time series information.

The discharge volume of a river is a major factor in understanding the river system. In many rivers, flow will change with time by as much as several orders of magnitude. If discharge is plotted on a time series graph, flood peaks will dominate the discharge scale but will be infrequent on the time scale. Similar features may exist for the concentration of one or more variables in a river, such as suspended sediment concentration. The relationship between concentration of a variable and river discharge is clearly demonstrated in a hysteresis plot in which concentration is plotted against flow as

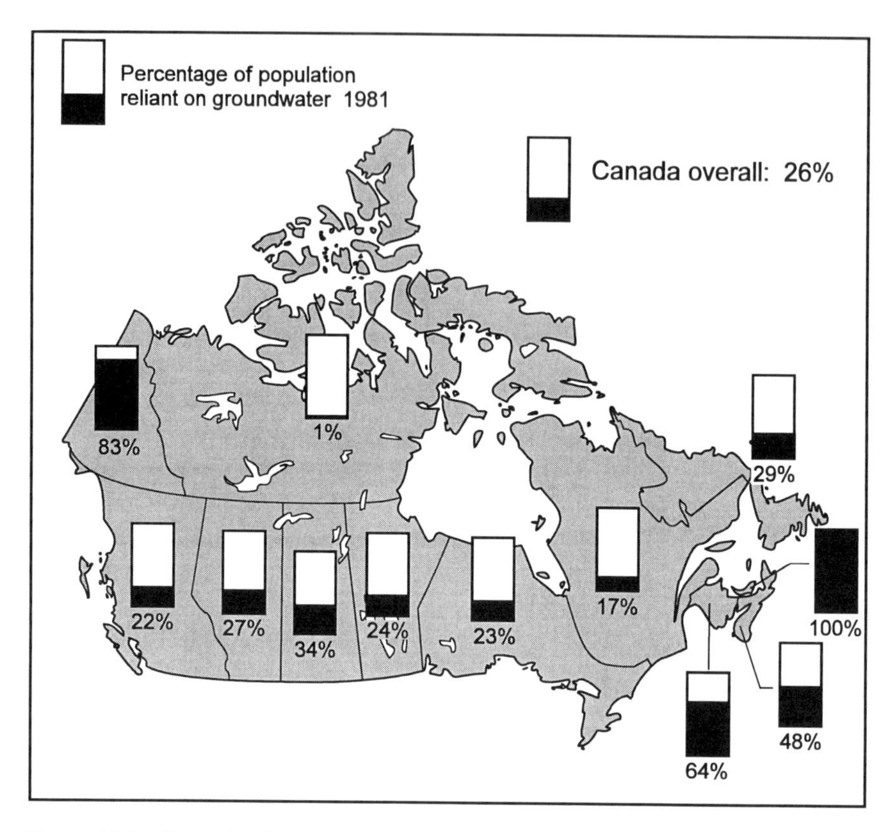


**Figure 14.9** Example of framed rectangle graphs on a map

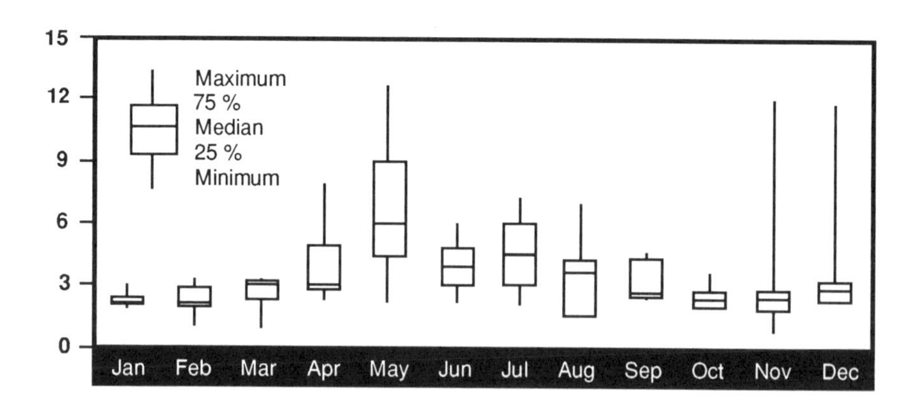


**Figure 14.10** The use of box and whisker plots to display time series information. Inset shows the construction of a box and whisker plot

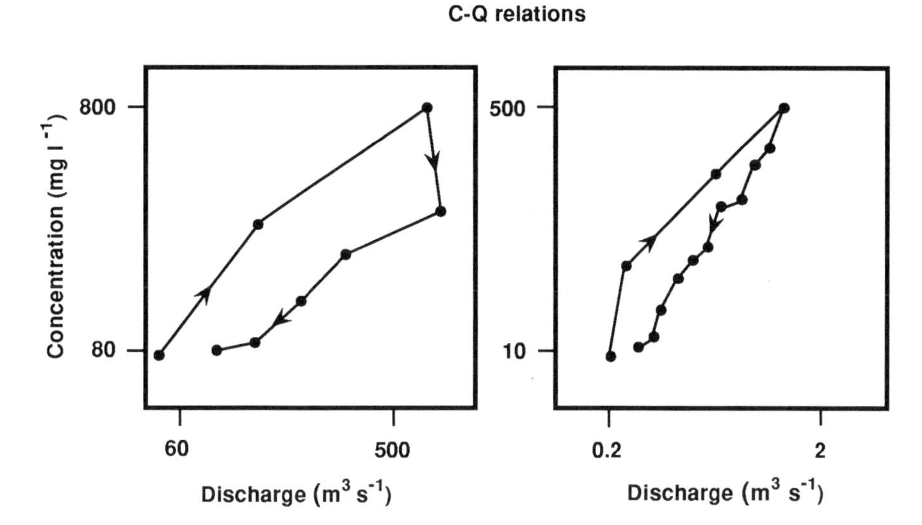


**Figure 14.11** Examples of hysteresis plots of concentration versus discharge

shown in the example of Figure 14.11. The concentration–flow pairs are plotted in the order in which they are observed and the points are connected in the same order. In many cases the graph is more illustrative when logarithmic scales are used for both flow and concentration. The hysteresis plot, showing each data point, indicates that the concentration of a variable can be different when stream discharge is decreasing from what it is at times of increasing discharge.

A very useful overview technique for survey data is the superimposition of other types of graphical analyses (such as those presented above) on a map of the study area. This permits a geographical perspective on differences that are observed at different monitoring sites. This and other types of analysis and presentation are easily done with computer software such as the RAISON/GEMS package used in the GEMS/WATER Programme (Figure 14.12).

## 14.6 Reporting

Reporting should focus on the synthesis of the data collected, rather than on the individual numbers that make up the data. It is important that the resultant interpretation gives the broad view developed from, and supported by, the fine details. In order to convey information effectively to managers, politicians and the public, reports must clearly describe the environmental situation in terms that can be readily understood. Most people who read

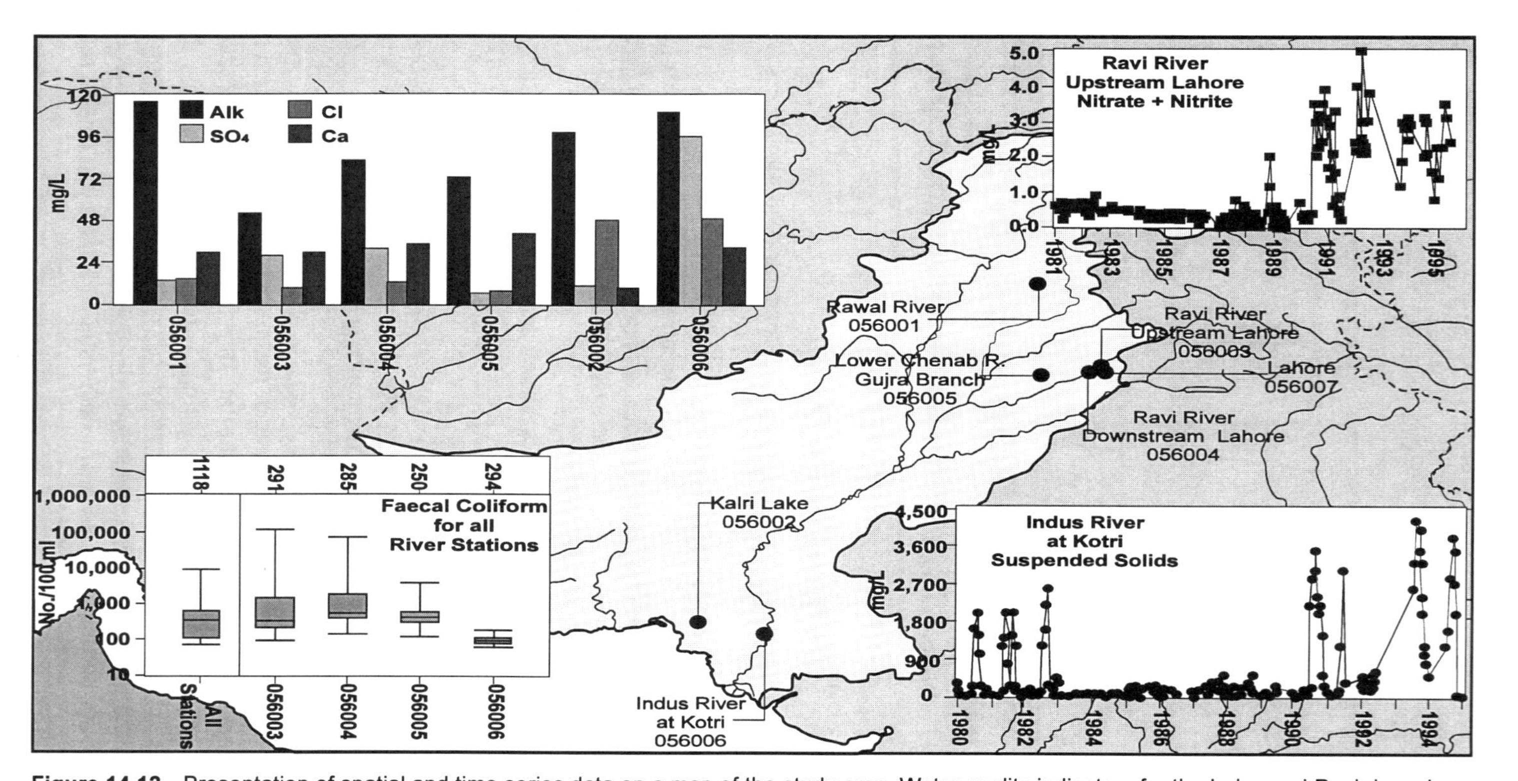


**Figure 14.12** Presentation of spatial and time series data on a map of the study area. Water quality indicators for the Indus and Ravi rivers in Pakistan. Alkalinity is reported as mg $l^{-1}$ $CaCO_3$. Nitrate is reported as mg $l^{-1}$ $NO_3^-$. Data assessment by the UNEP GEMS Collaborating Centre for Freshwater Monitoring and Assessment, Canada. Analysis was undertaken using the GEMS/RAISON system

reports on monitoring are not specialists and simple, clear explanations are essential for effective communication.

Monitoring results, reported in terms of the effects that processes, events and mechanisms have on water quality, should be readily understandable to their diverse audiences. Authors must recognise the broad range of readership and endeavour to present material in the simplest possible terms.

### 14.6.1 Types of report

The several different types of report that are possible have certain elements in common, including the statement of objectives and the description of the study area. To avoid any misunderstandings it is important that all elements of the monitoring programme are precisely and clearly described. The four principal types of report are described below.

*Study plan report*

The study plan report (or monitoring programme document, see Chapter 3) defines the objectives of the monitoring programme, including the questions to be addressed and the present understanding of the environment to be studied. It also defines the sampling and data review strategy that will be followed to meet the objectives. In this regard it must be noted that data collected for a specific purpose may not be applicable to other issues and data collected for no clearly stated purpose may not answer any questions at all. If scarce resources are to be used effectively, it is essential that a study plan is prepared. The study plan is analogous to the workplan for a building, i.e. it clearly documents the processes so that all parties understand their roles and adopt common goals for the overall study.

*Protocol and methods report*

The protocol and methods report, sometimes known as the Standard Operating Procedure (see Chapter 9), describes methods and equipment in sufficient detail for other scientists to be able to assess the scientific validity of the results reported. Without this documentation, work might be viewed as merely semi-quantitative and the conclusions regarded as suspect, if not completely invalid. Detailed descriptions of all methods must be included, i.e. sample collection, sample preservation in the field, field analyses, laboratory procedures, quality assurance procedures, data recording, and computer processing. All reporting procedures should be covered, including those for reporting quality assurance results and the results of various analytical measurements. The report should include provisions for the protection of all data against loss or damage by the regular filing of duplicate copies in a safe and secure location.

The report should be prepared at the beginning of a programme and must be regularly reviewed, revised and updated. Review helps to ensure the detection of problems that could invalidate the results. International protocols should be used to the greatest possible extent. All members of the study team should contribute to the preparation of the protocol and its subsequent revisions because collective knowledge and understanding of procedures enhance the quality of the data collected.

*Data report*

In some circumstances it may be necessary to report data without detailed interpretation, and a data report permits the early distribution of information before interpretation is complete. However, it may contain unverified or uncertain results and, as a result, will become outdated as the main body of data is corrected and amended. If data are stored on a computer, data reports can distributed in machine-readable form.

The primary purpose of a data report (or indeed of any report) is to transmit information to an audience. The information should be assembled in a well organised format so that the reader can easily review it. This is particularly true of a data report, which will usually consist of tabulated material with few supporting statements. Suppose, for example, that samples taken at a site once a month for a year were analysed for 30 variables and that the results formed the basis of a data report. If this report were arranged with all 30 results for each sampling date on a separate page, a reader looking for seasonal trends would have to turn pages 12 times per variable. If, however, the results were arranged chronologically in columns, patterns or extreme conditions could be readily observed. Graphical presentation and statistical summaries can also be used to improve understanding of data.

*Interpretative report*

Interpretative reports provide a synthesis of the data and recommend future actions. Different approaches may be used, for example process identification, problem-solving, and achievement of desired goals, are some alternatives. The synthesis of the data should be targeted to the specific objectives for which the data were collected. Interpretative reports should be produced regularly to ensure that programme objectives are being met and are currently valid. The time interval between successive reports should never be greater than two years.

### 14.6.2 Structuring a report

The structure described in the following paragraphs may be used as a general outline for the preparation of monitoring reports, although not all reports will necessarily contain all of the elements.

- *Summary*: The summary briefly outlines what was done, describes the significance of key findings, and lists the important recommendations. It should be written so that a non-scientist can understand the purpose of the work and the importance of the results. Its length should never exceed two or three pages (about 1,000 words). Often, the summary is the only part of the report which is read by senior managers.
- *Introduction*: The objectives and the terms of reference of the study should be presented in the introduction. The problems of issues being addressed should be clearly stated, previous studies should be described, and relevant scientific literature should be reviewed. Any major restrictions (personnel, access, finance, facilities, etc.) which limit the desirable scope of the programme should be identified.
- *Study area*: A summary of the geography, hydrology and other salient features (which, for example, may include land use, industry, and population distribution) of the area under study should be provided (see section 3.5). All locations, structures and features mentioned in the text should be identified on maps such as the example shown in Figure 14.13. In certain cases, profiles of river or lake systems may provide useful information to augment that provided on the maps.
- *Methods*: A separate section should provide details of the methods and procedures used for all aspects of the study. In most cases this can be covered by reference to the protocol and methods report (see section 14.6.1 and Chapter 9). A summary of the procedures to be followed for quality assurance should be presented (see Chapter 9).
- *Results*: Results should be presented in graphical form whenever possible, and graphs should clearly demonstrate the relationship of the data to the monitoring objectives. Some examples of suitable types of graph are shown in section 14.5. Graphs should summarise entire sets of data rather than individual observations. International units of measurement should be used wherever possible but if local units are used precise descriptions and conversions should be provided.
- *Analysis of results*: The statistical analysis of results should be described. The reliability of the statistics, and its implication, should also be given. Analytical procedures appropriate to the problem should be used and

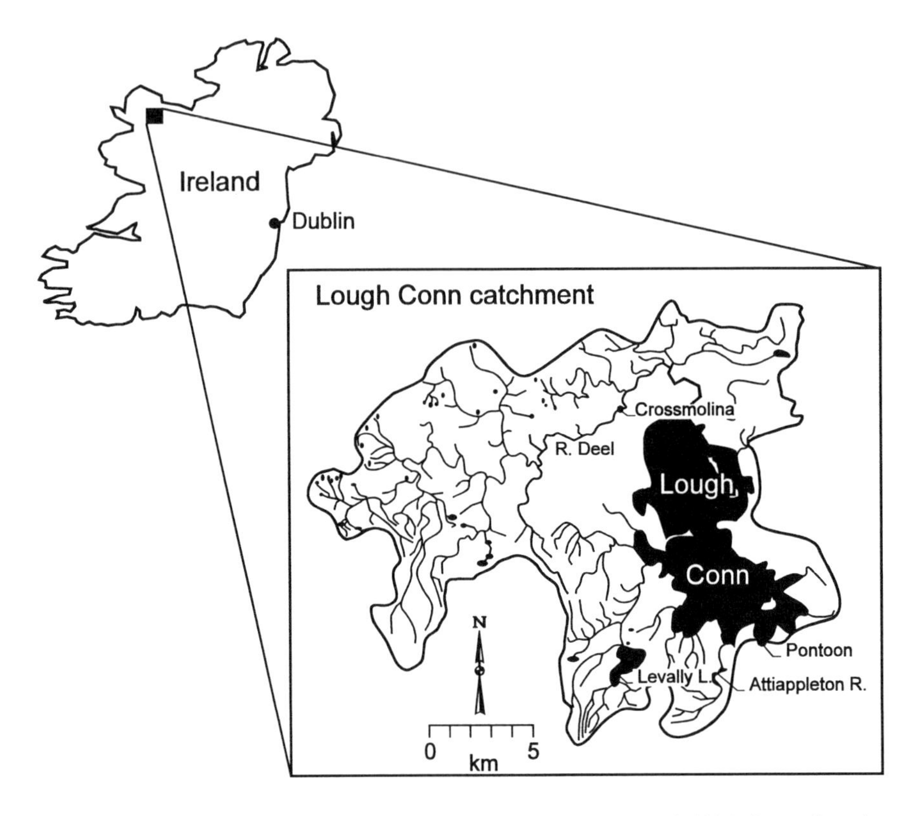


**Figure 14.13** Example of a study area map (After McGarrigle *et al.* 1994. Reproduced with permission from Mayo County Council, Ireland)

should focus on the goals of the study or the programme. Again, results should be presented in graphical, rather than in tabular, form.

- *Significance of results*: The report should include a section devoted to interpretation of the results in terms of the monitoring objectives. This will help ensure that the objectives are being met and that questions are being answered. This section should emphasise the need to provide information.
- *Recommendations*: Recommendations for proposed future activities should normally be listed in two categories: one concerned with scientific matters, the other with management issues. Ideally, these recommendations should be presented in order of priority.
- *Information sources*: All sources of information and all literature referred to should be correctly and fully cited so that a permanent record is available.

## 14.7 Recommendations

It will be clear from the foregoing that the appropriate use and reporting of monitoring data is a vital part of the overall monitoring and assessment programme. Without clear reporting of understandable and relevant results to programme controllers (managers and administrators), little will have been achieved. That requirement can only be met if it is fully taken into account at the very earliest stages of the overall programme objective definition and subsequent design. It is also essential to understand that at every stage of the monitoring process, the data needs of the analysis and reporting stages are recognised. Without good data, no useful information may be reportable, no matter how good the underlying analysis may have been.

## 14.8 Source literature and further reading

Chapman, D. [Ed.] 1996 *Water Quality Assessments. A Guide to the Use of Biota, Sediments and Water in Environmental Monitoring*. 2nd edition, Chapman & Hall, London.

Cleveland, W.S. 1985 *The Elements of Graphing Data*. Wadsworth Advanced Books and Software, Monterey, CA.

Daniel, W.W. 1990 *Applied Non-Parametric Statistics*. 2nd edition. PWS-Kent Publishing Company, Boston.

Demayo, A. and Steel, A. 1996 Data handling and presentation. In: D. Chapman [Ed.] *Water Quality Assessments. A Guide to the Use of Biota, Sediments and Water in Environmental Monitoring*. 2nd edition, Chapman & Hall, London.

Elliott, J.M. 1977 *Some Methods for the Statistical Analysis of Samples of Benthic Invertebrates*. 2nd edition. Scientific Publication No. 25, Freshwater Biological Association, Ambleside.

Friedrich, G. and Viehweg, M. 1984 Recent developments of the phytoplankton and its activity in the Lower Rhine. *Verhandlungen des Internationale Vereinigen für theoretische und angewandre Limnologie,* **22**, 2029–2035.

Friedrich, G. and Viehweg, M. 1987 Measurement of chlorophyll-fluorescence within Rhine-monitoring — results and problems. *Archiv für Hydrobiologie Beihefte Ergebnisse der Limnologie,* **29**, 117–122.

Gilbert, R.O. 1987 *Statistical Methods for Environmental Pollution Monitoring* Van Nostrand Reinhold Co., New York.

Humpesch, U.H. 1992 Ecosystem study Altenwörth: Impacts of a hydroelectric power-station on the River Danube in Austria. *Freshwater Forum*, **2**(1), 33–58.

McGarrigle, M.L., Champ, W.S.T., Norton, R., Larkin, P. and Moore, M. 1994 *The Trophic Status of Lough Conn. An Investigation into the Causes of Recent Accelerated Eutrophication*. Mayo County Council, Castlebar.

Pinder, L.C.V. 1989 Biological surveillance of chalk streams. *Freshwater Biological Association Fifty Seventh Annual Report*. Freshwater Biological Association, Ambleside, 81–92.

Snedecor, G.W. and Cochran, W.G. 1980 *Statistical Methods*. 7th edition. The Iowa State University Press, Iowa.

Sokal, R.R. and Rohlf, R.J. 1981 *Biometry*. 2nd edition. W.H. Freeman, San Francisco.

Spodniewska, I. 1974 The structure and production of phytoplankton in Mikolajskie Lake. *Ekologia Polska*, **22**(1), 65–106.

Williams, G.P. 1989 Sediment concentrations versus water discharge during single hydrological events in rivers. *Journal of Hydrology*, **III**, 89–106.

# Appendix 1

## PORTABLE FIELD KITS

Portable field kits are useful for the field testing of samples in a water quality monitoring programme. This appendix contains brief descriptions of portable kits available from some of the major suppliers. The reference to items of equipment by trade name and the inclusion of equipment in the following lists does not imply an endorsement of that equipment.

Telephone and telefax numbers of the various suppliers are provided where this has been possible. The number in square brackets, e.g. [33] is the country code, and the number in round brackets, e.g. (453) is an area code within the country.

**BDH Laboratory Supplies**..........................................................................................
Seldown Road
Poole
Dorset
England

Telephone: [44] (1202) 660444
Telex: 41186 and 418123 TETRA G
Fax Group III: [44] (1202) 666856
Cables: Tetradome Poole

This company distributes the Merck line of test strips and kits for water analysis. Mercko-quant, Aquamerck and Aquaquant are the trade names of the three available systems.

- *Merckoquant* test strips consist of a strip of paper, impregnated with reagents and bonded to a plastic backing. Colour develops on the strip after it is dipped in water and colour intensity is compared with a colour scale. This system can be used for the semi-quantitative determination of Aluminium, Ammonium, Calcium, Iron, Manganese, Nitrate, Nitrite, pH, Potassium, Sulphate and Total hardness.
- *Aquamerck* kits use either titrimetric or colorimetric methods to obtain semi-quantitative results in the analysis of Aluminium, Ammonium, Calcium, Carbonate hardness, Chloride, Iron, Nitrite, Dissolved oxygen, pH, Phosphate and Total hardness.
- The *Aquaquant* system involves colorimetric methods to obtain quantitative results in the analysis of Ammonium, Chloride, Iron, Manganese, Nitrite and Silicon.

*This appendix was prepared by R. Ballance*

**ELE International Ltd**..........................................................................................

Eastman Way
Hemel Hempstead
Herts
HP2 7HB
England

Telex: 825239 ELE LTD G
Fax: [44] (1442) 252474 and 219045
Tel: [44] (1442) 218355

This company assembles the following three standard systems under the trade name "Paqualab": systems 50, 25 and 25L. System 50 has two incubators powered by an internal rechargeable battery permitting simultaneous total and faecal coliform analyses, meters to measure pH, temperature, conductivity and turbidity, and a photometer for measuring chemical variables. System 25 is a single compartment incubator (37 °C or 44 °C) with internal rechargeable battery, meters and photometer the same as system 50. System 25L is a single compartment incubator powered from an external source of either 12 V or 24 V DC. All systems include a membrane filtration unit and a supply of aluminium Petri dishes. Only the system 25 is described below. Kits can be assembled to permit analyses for more than 30 physical and chemical variables. The company also sells several types of water samplers and two models of current meter. Most of the physical and chemical methods of analysis use either a photometer or an electrode and meter. Operators' manuals are available in English, French and Spanish. The analyses done by the different methods are:

- Photometric methods are used for Alkalinity, Aluminium, Ammonia, Boron, Chloride, Fluoride, Iron, Magnesium, Manganese, Nitrate, Nitrite, pH, Phosphate, Potassium, Silica and Sulphate.
- Electrode methods are applied to the measurement of Dissolved oxygen concentration, pH, Conductivity, Temperature, Total dissolved solids and Turbidity.

Special purpose units can be provided for:

- Determination of Biochemical oxygen demand by the Warburg and Sierp method. The unit has six stations, thus permitting simultaneous determinations on six samples.
- Determination of Chemical oxygen demand with digestion carried out by the bichromatic method. The thermal reactor unit can simultaneously test 18 samples plus one blank.

**Potapak Limited**..........................................................................................

Toomer's Wharf
Newbury
Berks
RG13 1DY
England

Telex: 846114 INTACL G
Fax: [44] (1635) 30844
Tel: [44] (1635) 30552

The Potolab kit produced and sold by Potapak Limited is designed to serve the needs of physical, chemical and bacteriological analyses in the field and can be purchased with equipment necessary to measure or analyse for Alkalinity, Aluminium, Ammonia nitrogen,

Boron, Calcium (hardness), Conductivity, Fluoride, Iron, Magnesium, Manganese, Nitrates, Nitrites, pH, Phosphate, Potassium, Silica, Sulphate and Temperature. The Potalab consists of twin incubators so that some samples can be incubated at 37 °C (total coliforms) and others at 44.5 °C (faecal coliforms) simultaneously. The physico-chemical component of the Potalab contains a pH/Temp/mV meter, a conductivity meter, a turbidity tube and a universal photometer, together with reagents necessary to analyse for more than 25 variables.

*Weight:* 25 kg (incubator, photometer and miscellaneous equipment and supplies for chemical, physical and bacteriological analyses).
*Dimensions* (cm): 56 x 26 x 34.
*Portability:* Carrying handles.
*Power:* 220 V AC, internal 12 V battery with built-in charger. One charge will serve up to two incubation periods.
*Variables:* Bacteriology — Total and faecal coliforms; Physico-chemical — Alkalinity, Aluminium, Ammonia, Boron, Fluoride, Iron, Manganese, Nitrate, Nitrite, pH, Phosphate, Potassium, Silica, Sulphate.
*Consumables:* Sufficient for 200 bacteriological analyses supplied with kit. Additional consumables approximately UK £3.00 per test.
*Manual language:* English, but the company can supply copies in French, Spanish or Portuguese on request.
*Manual contents:* A complete description of all tests and how to get the best performance from the kit. Individual components in the kit can be returned to the supplier for major repair.
*Cost:* complete portable lab: UK £3,950
bacteriological unit: UK £2,550 (1993 prices).

**Hach Company**.............................................................
International Marketing Dept.
P O Box 389
Loveland, Colorado 80539
USA

Telex: 160840
Fax: [1] (303) 669 2932
Tel: [1] (303) 669 3050

**Hach Europe, S.A./N.V.**.............
B. P. 229
B 5000 Namur 1
Belgium

Telex: 846-59027
Fax: [32] (81) 44 13 00
Tel: [32] (81) 44 53 81

The company makes field kits and equipment that can be used to analyse for most of the chemical and physical variables listed in Chapters 6 and 7. The manual describing procedures for use of the kits is available in English, French and Spanish. Colorimetric, photometric, titrimetric, gravimetric or electrode methods are provided for the various analyses as follows:

- Colorimetric method for analysis of Calcium and Magnesium (hardness).
- Photometric methods for analysis of Aluminium, Boron, Chloride, Fluoride, Iron, Manganese, Nitrate, Nitrite, Ammonia nitrogen, Kjeldahl nitrogen, Dissolved oxygen, Phosphorus, Potassium, Selenium, Silica and Sulphate.
- Titrimetric methods for analysis of Alkalinity, Chloride, Iron and Dissolved oxygen.
- Gravimetric method for analysis of Total suspended solids.
- Electrode method for analysis of Calcium, Fluoride, Nitrate, Dissolved oxygen, Potassium and Sodium and for the measurement of Temperature, Conductivity and pH.

**Macherey-Nagel GmbH & Co KG**.................................................................................
P O Box 10 13 52
D-52313 Düren
Germany

Telex: 8 33 893 mana d
Fax: [49] (2421) 62054
Tel: [49] (2421) 6980

The company makes a broad range of kits and equipment that can be used to analyse for many of the chemical and physical variables listed in Chapters 6 and 7. Customised kits can be assembled on request. Colorimetric, titrimetric, photometric or electrode procedures are available for the various determinations and analyses as follows:

- Colorimetric methods for the analysis of Ammonia, Iron, Manganese, Nitrate, Nitrite, pH, Phosphate, Potassium, Silica and Sulphate.
- Titrimetric methods for the analysis of Alkalinity, Calcium, Chloride, Hardness and Dissolved oxygen.
- Photometric methods for the analysis of Aluminium, Ammonia, Chemical oxygen demand, Chloride, Fluoride, Iron, Manganese, Nitrate, Nitrite, Orthophosphate, Total phosphate, Potassium, Silica and Sulphate.
- Electrode methods for the determination of pH, Conductivity and Dissolved oxygen.

**The Tintometer Ltd**.................................................................................................
Waterloo Road
Salisbury
Wilts
SP1 2JY
England

Telex: 47372 TINTCO G
Fax: [44] (1722) 412322
Tel: [44] (1722) 327242

Although this company's speciality is equipment for colorimetric analyses, some of its equipment uses photometric or electrode methods. The colorimetric analyses are of two types; one is colour comparison with standard colour discs while the other depends on counting the numbers of reagent tablets required to produce a colour change in a measured volume of water. Users' manuals are available in English, French, Spanish and German. Application of the analytical systems is as follows:

- Colorimetric methods for analysis of Alkalinity, Aluminium, Boron, Chloride, Fluoride, Iron, Magnesium, Manganese, Ammonia nitrogen, Nitrate, Nitrite, Phosphorus, Silica, Sulphate, Dissolved oxygen and for the measurement of pH. Tablet count method for Alkalinity, Calcium, Chloride, Nitrite and Sulphate.
- Photometric methods for Aluminium, Fluoride, Iron, Ammonia nitrogen, Nitrate, Nitrite, Phosphate and pH.
- Electrode methods for Dissolved oxygen, pH, Conductivity, Temperature and Total dissolved solids.

**Robens Institute**..........................................................................................................
University of Surrey
Guildford
Surrey
GU2 5XH
England

Telex: 859 331 UNIVSY G
Fax: [44] (1483) 503 517
Tel: [44] (1483) 259 209

The Oxfam DelAgua field kit for bacteriological analysis was developed by the Robens Institute with the main objective of providing a quality product at low cost.

*Weight:* 10 kg.
*Dimensions* (cm): 37 x 14 x 26.
*Portability:* Carrying handles.
*Power:* 220 V and 110 V AC, internal 12 V battery with built-in charger, 72 hours between charges.
*Variables:* Faecal coliforms, pH, Free and total chlorine, Turbidity. Some optional extras can be supplied on request.
*Consumables:* Sufficient for 200 bacteriological analyses supplied with kit. Additional consumables approximately UK £0.24 per test.
*Manual language:* English, French, Chinese and Spanish.
*Contents:* Analytical methods, kit maintenance, troubleshooting, basic kit repair, circuit diagram. Repair kit provided. Return to supplier for major repair.
*Cost:* UK £1,050 (1996 price).

**Water and Environmental Engineering Group**...............................................................
Department of Civil Engineering
University of Leeds
Leeds
Yorks
LS2 9JT
England

Fax: [44] (1132) 33 23 08
Tel: [44] (1132) 33 22 76

The Leeds portable incubator was developed to fill the need for sturdy, low-cost equipment for carrying out bacteriological analyses in the field.

*Weight:* 4.5 kg.
*Dimensions* (cm): 35 x 21 x 26.
*Portability:* Carrying handles.
*Power:* 220 V AC, 12 V DC.
*Variables:* Aerobic bacteria, Total and faecal coliforms.
*Consumables:* Sufficient for 200 bacteriological analyses supplied with kit. Additional consumables approximately UK £0.30 per test.
*Manual language:* English.

*Manual contents:* Kit maintenance, troubleshooting, basic kit repair, circuit diagram.
*Repair:* Return to supplier for major repair.
*Cost:* UK £600 (1993 prices).

**CQ-3 Portable Kit**..........
7 Pan Jia Yuan Nan Li
Chao Yang District
Beijing 100021
China

*Weight:* 7 kg.
*Dimensions* (cm): 41 x 18 x 38.
*Portability:* Carrying handles and shoulder strap.
*Power supply:* 220 V AC, 12 V DC, built-in charger, 10 hours between charges.
*Variables:* Total bacteria, Total and faecal coliforms. Optional extras — *Shigella*, *Salmonella*, *Vibrio cholera*.
*Consumables:* Sufficient for 100 bacteriological analyses supplied with kit. Additional consumables approximately US $0.50 per test.
*Manual language:* Chinese and English.
*Manual contents:* Analytical methods, kit maintenance, troubleshooting, circuit diagram.
*Repair:* Repair kit provided. Return to supplier for major repair.
*Cost:* US$500 (1991 prices).

**Millipore Intertech**..........
B. P. 307
78054 St Quentin Yvelines
CEDEX
France

Tel: [33] (1) 30 12 70 00
Fax: [33] (1) 30 12 71 81
Telex: 698371

**Millipore Intertech**..........
P.O. Box 255
Bedford, MA 01730
USA

Tel: [1] (617) (275-9200)
Fax: [1] (617) (275-3726)
Telex: 4430066 MILIPR UI

This company manufactures literally thousands of articles that are associated with membrane filtration. Two of the kits and two optional incubators that may be used are described below.

*Portable Water Laboratory* includes incubator, filter holder assembly, sampling cup, vacuum syringe, adapter tube, forceps, alcohol bottle and a supply of consumables. This kit is used in accordance with the directions contained in the sections of Chapters 6 and 10 that refer to membrane filtration.

*Weight:* 7.9 kg (incubator only, 5.4 kg).
*Dimensions* (cm): 39.4 x 26.7 x 29.2.
*Portability:* One carrying handle — no shoulder strap.
*Capacity:* 18 plastic Petri dishes (47 mm diameter).

*Power supply:* Operates on 12 or 24 V DC and 115 or 220 V AC. Power cord with adapters for battery terminals, car cigarette lighter or mains outlet.
*Test variables:* Total coliforms is standard. Media for detection of other organisms can be provided. Temperature or incubation chamber adjustable between 27 °C and 60 °C, thermostatically controlled. Incubating temperature can vary ± 1 °C in stable ambients from–40 °C.
*Consumables:* Kit includes the following sterile consumables: 100 plastic Petri dishes, 25 plastic pipettes, 100 ampoules MF-Endo medium, 100 filters of 0.45 μm pore diameter with absorbent pads (47 mm diameter). Cost of additional consumables approximately US $3.00 per coliform test.
*Manual language:* English, French, Spanish.
*Manual contents:* Instructions for use of equipment. Gives no information concerning maintenance or repair of kit. Return equipment to supplier for major repairs.
*Warranty:* All products are warranted against defects in materials and workmanship when used in accordance with instructions, for one year from the date of shipment.
*Cost:* US $4,377 for complete kit. The cost of the incubator without the filter assembly and consumables is US $1,573 (1994 prices).

*Portable Water Analysis Kit*
This kit is designed for the field determination of total coliform bacteria in water by filtering samples through sterile 0.45 μm Millipore, 37 mm diameter, bacteriological analysis "monitors". The test sample is collected in the stainless steel cup and is drawn through a monitor by a valved syringe used as a vacuum pump. After sample filtration, ampouled culture medium is introduced into the monitor which is then incubated in the portable incubator (the incubator is the same as the one described above under Portable Water Laboratory).
*Weight:* 20.9 kg (kit complete with supplies), 5.4 kg (incubator only).
*Dimensions* (cm): 39.4 x 26.7 x 29.2.
*Portability:* Carrying handle — no shoulder strap.
*Capacity:* 30 monitors.
*Power supply:* 12 or 24 V DC, or 115 or 220 V AC.
*Consumables:* Sufficient for 100 total coliform tests supplied with kit. Replacement consumables cost US $4.81 per test.
*Manual:* Same as for Portable Water Laboratory (above).
*Warranty:* Same as for Portable Water Laboratory (above).
*Cost:* US$2,339 (1994 prices).

*Battery Powered Incubator (Rechargeable)*
*Weight:* 7.4 kg (with battery).
*Dimensions* (cm): 32.7 x 37.1 x 29.6.
*Portability:* Carrying handle and bag with shoulder strap.
*Capacity:* 48 Petri dishes or 75 monitors.
*Power supply:* 12 V DC supplied by rechargeable Ni/Cd battery, 12 V car battery (cigarette lighter or direct connection) or AC (115 or 220 V) power adapter. Battery charger and AC adapter included. Battery duration about 24 hours in 25 °C ambient when operating at 44.5 °C.
*Temperatures:* Fixed temperature settings: 30 °C, 35 °C, 37 °C, 41 °C (± 0.5 °C) and 44.5 °C (± 0.2 °C).
*Cost:* US $2,446, Charger and AC adapter included (1994 prices).

*Dual Chamber Battery Powered Incubator (Rechargeable)*
*Weight:* Approx. 12 kg with battery.
*Dimensions* (cm): 54 x 37.1 x 29.6.
*Portability:* Carrying handle and bag with shoulder strap.
*Capacity:* 48 Petri dishes in each chamber.
*Power supply:* As above.
*Temperatures:* Lower chamber: 30 °C, 35 °C, 37 °C, 41 °C.
Upper chamber: 41.5 °C (± 0.5 °C), 44.5 °C (± 0.2 °C).
The two chambers may be operated simultaneously at different temperatures.
*Cost:* US$4,184, Charger and AC adapter included (1994 prices).

# Index